Brazil in the Anthropocene

Brazil is considered one of the world's most important environmental powers. With a continental territory containing almost 70 per cent of the Amazon rainforest, along with a rich biodiversity and huge amount of natural resources, its geopolitical role in environmental decisions is crucial to ongoing global negotiations surrounding climate change.

Development policies based on extraction and exportation of raw materials by the mining and agribusiness sectors threaten the global environmental balance and the long-term sustainability of Brazil's economy. *Brazil in the Anthropocene* examines Brazil's role within the global ecological crisis and considers how national and international policy is influenced by the interdependence of social, political, ethical, scientific and economic factors in the modern age.

With chapters from a diverse range of international scholars this interdisciplinary volume will be of great interest to students and scholars of environmental politics, environmental sociology and the environmental humanities.

Liz-Rejane Issberner, economist, is Senior Researcher at the Brazilian Institute on Information in Science and Technology (IBICT) and Professor at the Post-Graduate Programme in Information Science (IBICT with Federal University of Rio de Janeiro, Brazil). Fellowship of research productivity (PQ) granted by the national council for scientific and technological development (CNPq). Her interdisciplinary research involves information and political ecology as well as eco-innovations and science, technology, and social change.

Philippe Léna, geographer and sociologist, is Emeritus Researcher at the Research Institute for Development (IRD-France) and National Museum of Natural History (MNHN, Paris, France). He has collaborated since 1980 with Brazilian research institutions like INPA (National Institute for Amazonian Research), MPEG (Museum Emílio Goeldi of Pará) and the Federal University of Rio de Janeiro (UFRJ), conducting research on social-ecological issues in Amazonia.

"This essential book was elaborated by a high level team of researchers with diversified formation and experience, who, beyond publishing and teaching, participated in the elaboration of public policies, emission inventories, mitigation and adaptation proposals, climatic and inclusive strategies. The texts are deep, controversial and complementary, and allow a wide historical and demographic understanding of socio-environmental conflicts, the limits and the contradictions of the regulatory action of the State and of measures to reduce impacts, in face of the accumulation and globalization dynamics. The chapters also address Amazonia, the dynamics of the legislative power, dominated by economic groups, the multiple forms of social, territorial and technological resistance. As a professor at the Federal University of Rio de Janeiro (UFRJ), congressman and former Minister of the Environment I think that this book will be of great utility for the academy (graduation and post-graduation) and for civil servants working in public agencies as well as for advisers of social movements and environmental and territorial actions of resistance."

Carlos Minc, Professor of Geography of the UFRJ,
state deputy, former Brazilian Minister of the Environment and
Secretary of the Environment of the State of Rio De Janeiro, Brazil

Routledge Environmental Humanities
Series editors: Iain McCalman and Libby Robin

The *Routledge Environmental Humanities* series is an original and inspiring venture recognising that today's world agricultural and water crises, ocean pollution and resource depletion, global warming from greenhouse gases, urban sprawl, overpopulation, food insecurity and environmental justice are all *crises of culture*.

The reality of understanding and finding adaptive solutions to our present and future environmental challenges has shifted the epicenter of environmental studies away from an exclusively scientific and technological framework to one that depends on the human-focused disciplines and ideas of the humanities and allied social sciences.

We thus welcome book proposals from all humanities and social sciences disciplines for an inclusive and interdisciplinary series. We favour manuscripts aimed at an international readership and written in a lively and accessible style. The readership comprises scholars and students from the humanities and social sciences and thoughtful readers concerned about the human dimensions of environmental change.

Brazil in the Anthropocene

Conflicts between predatory development and environmental policies

Edited by
Liz-Rejane Issberner
and Philippe Léna

LONDON AND NEW YORK

from Routledge

First published 2017 by Routledge

2 Park Square, Milton Park, Abingdon, Oxfordshire OX14 4RN

711 Third Avenue, New York, NY 10017

Routledge is an imprint of the Taylor & Francis Group, an informa business

First issued in paperback 2018

British Library Cataloguing-in-Publication Data
A catalogue record for this book is available from the British Library

Library of Congress Cataloging-in-Publication Data
Names: Issberner, Liz-Rejane, editor. | Léna, Philippe, editor.
Title: Brazil in the anthropocene : conflicts between predatory development and environmental policies / edited by Liz-Rejane Issberner and Philippe Léna.
Description: New York, NY : Routledge, 2016. | Series: Environmental humanities series | Includes bibliographical references and index.
Identifiers: LCCN 2016022164 | ISBN 9781138684201 (hbk : alk. paper) | ISBN 9781315544069 (ebk)
Subjects: LCSH: Environmental policy--Brazil. | Economic development--Environmental aspects--Brazil. | Sustainable development--Government policy--Brazil. | Climatic changes--Government policy--Brazil.
Classification: LCC GE190.B6 B733 2016 | DDC 338.981/07--dc23
LC record available at https://lccn.loc.gov/2016022164

ISBN: 978-1-138-68420-1 (hbk)
ISBN: 978-1-138-31590-7 (pbk)

Typeset in Bembo
by Saxon Graphics Ltd, Derby

Contents

Figures and tables

Contributors

Henri Acselrad is Professor at the Institute of Urban and Regional Research and Planning of the Federal University of Rio de Janeiro. Email: hacsel@uol.com.br

José Eustáquio Diniz Alves is Professor and Researcher at the National School of Statistical Sciences (ENCE) of the Brazilian Institute of Geography and Statistics (IBGE). Email: jed_alves@yahoo.com.br

Larissa Basso is PhD candidate at the Institute of International Relations and member of the Climate Change and International System in the Anthropocene Research Program, University of Brasília. Email: larissabasso@gmail.com

Gustavo Neves Bezerra is Professor at the Institute of Urban and Regional Research and Planning of the Federal University of Rio de Janeiro. Email: gustavonb@yahoo.com.br

Marcel Bursztyn is Full Professor at the Sustainable Development Centre of the University of Brasilia. Email: marcel.cds@gmail.com

Maria Augusta Bursztyn is Full Professor at the Sustainable Development Centre of the University of Brasilia. Email: dute.cds@gmail.com

Robert Davenport is PhD candidate at University of California, Santa Cruz, USA. Email: tropnevadr@gmail.com

Ladislau Dowbor is Economist, Professor at the Catholic University of São Paulo (PUC), and consultant to various UN agencies. Email: ldowbor@gmail.com

Philip M. Fearnside is Research Professor at the National Institute for Research in Amazonia (INPA-Manaus). Email: pmfearn@inpa.gov.br

Claudio Gesteira is Researcher at the Center for Integrated Studies on Climate Change and the Environment (Centro Clima), PPE/COPPE, Federal University of Rio de Janeiro (UFRJ). Email: claudio.gesteira@gmail.com

Carolina Grotera is Researcher at the Center for Integrated Studies on Climate Change and the Environment (Centro Clima), PPE/COPPE, Federal University of Rio de Janeiro (UFRJ). Email: carolinagrottera@gmail.com

Liz-Rejane Issberner is Senior Researcher of the Instituto Brasileiro de Informação em Ciência e Tecnologia (Brazilian Institute of Information in Science and Technology), IBICT/MCTI and Professor at PPGCI-IBICT/UFRJ. Email: lirismail@gmail.com

Emílio Lèbre La Rovere is Full Professor, Energy Planning Program (PPE), Institute of Graduate Studies and Research in Engineering (COPPE), Federal University of Rio de Janeiro (UFRJ), Head of the Center for Integrated Studies on Climate Change and the Environment (CentroClima). Email: emilio@ppe.ufrj.br

Philippe Léna is Emeritus Researcher of the Research Institute for Development (IRD-France) – National Museum of Natural History (MNHN, Paris). Email: philippe-lena@orange.fr

Jean-Pierre Leroy is Advisor to FASE – Federation of Organizations for Social and Educational Assistance and member of the Brazilian Network for Environmental Justice. Email: jpierre@fase.org.br

Violeta Refkalefsky Loureiro is Professor of sociology at the Federal University of Pará – UFPa. Email: violeta.loureiro@ig.com.br

Carlos Saldanha Machado is Full professor at the Oswaldo Cruz Foundation (Ministry of Health of Brazil) and a member of the advisory committee of Environmental Sciences from the Brazilian National Council for Scientific and Technological Development – CNPq. Email: saldanhamachado@gmail.com

Sérgio Margulis is an Economist, retired from the Brazilian Institute on Applied Economics Research, former lead economist at the World Bank and Secretary of Sustainable Development at the Secretariat of Strategic Affairs of the Presidency of Brazil. Email: margulis.sergio@gmail.com

Geovana de Oliveira Patrício Marques is a Lawyer at Fortaleza City. Email: geovanadambiental@gmail.com

George Martine is a Sociologist and demographer, senior fellow at the Harvard Center for Population and Development and director of UNFPA's Technical Team for Latin America and the Caribbean. Email: georgermartine@yahoo.com

Peter May is Professor at the Federal Rural University of Rio de Janeiro. Email: peterhmay@gmail.com

Leonardo Melgarejo is Agronomist, coordinator of the Working Group on Toxic Agricultural Chemicals and GMOs of the Brazilian Association of Agroecology (ABA). Email: leonardo.melgarejo@gmail.com

João Alfredo Telles Melo is a Lawyer, Professor of Environmental Law and Fortaleza City Councillor. Email: joaoalfredotellesmelo@gmail.com

Adriana Maria Magalhães de Moura is Senior Researcher at the Coordination of Studies in Environmental Sustainability at the Applied Economic Research Institute – IPEA. Email: adriana.moura@ipea.gov.br

Pedro Nogueira is Associate researcher at the Federal Rural University of Rio de Janeiro. Email: pedroecopantanal@gmail.com

Paulo César Nunes is Agronomist, specialist in forest carbon and in charge of the COOPAVAM, Juruena, Mato Grosso, Brazil. Email: paulojuruena@hotmail.com

José Augusto Pádua is Professor of Brazilian Environmental History at the Institute of History of the Federal University of Rio de Janeiro. Email: jpadua@terra.com.br

Natalie Unterstell is Public policy specialist; director of the Brazil 2040: scenarios and alternatives to climate change adaptation at the Secretariat of Strategic Affairs of the Brazilian Presidency and former negotiator on behalf of Brazil at the United Nations Framework Convention on Climate Change. Email: natalieunterstell@gmail.com

Rodrigo Machado Vilani is Associate professor at the Federal University of Rio de Janeiro State. Email: r_vilani@yahoo.com.br

Eduardo Viola is Full Professor at the Institute of International Relations and coordinator of the Climate Change and International System in the Anthropocene Research Program, University of Brasília. Email: eduviola@gmail.com

William Wills is Researcher at the Center for Integrated Studies on Climate Change and the Environment (Centro Clima), PPE/COPPE, Federal University of Rio de Janeiro (UFRJ). Email: wwills@lima.coppe.ufrj.br

Anthropocene in Brazil

An inquiry into development obsession and policy limits

Philippe Léna and Liz-Rejane Issberner

Of which Anthropocene are we talking?

Whatever might be the fate of the proposal of the Anthropocene Working Group to create a new geological epoch (likely to be answered in 2016) at the International Commission of Stratigraphy, the term has already been adopted by many scientific communities, in particular by the specialists of Global Change, and embraced by a wide audience. Despite the current debate surrounding its dating and causes, it has proven its usefulness to synthetically designate a global phenomenon, which until then had only been addressed in a fragmented way and which did not allow to account for its severity and complexity.

However, the issues are not the same for every field of research. For the stratigraphists, the question is whether the signals left by human activities in geological strata currently forming are global, clear and strong enough to justify the definition of a 'human epoch'. For the specialists of Global Change, what is in question is when it can be considered that one enters a human-driven planetary change, an anthropogenic shift of the Earth System. For the historical and social sciences, on the other hand, it is the choice of the term 'anthropocene' which is itself questionable, as it would place the responsibility of the current state of the planet on the human species in general and not on an economic, social and political system, i.e. capitalism (Malm 2015). Unlike the attempts made by some historians (Chakrabarty 2009, 2015) or naturalists to integrate natural history and human history, recent history and deep history, many researchers suggested a terminology that would point more clearly at the culprit. Some of the terms proposed were 'Occidentalocene' (Bonneuil 2015), 'Growthocene' (Chertkovskaya and Paulsson 2016), 'Technocene' (Hornborg 2015), 'Capitalocene' (Moore 2015) or 'new climate regime'[1] (Latour 2015). For the most part, these approaches are compatible with the original proposal by Crutzen and Stoermer (2000) to symbolically begin the Anthropocene with the invention and development of the steam engine by Newcomen and Watt (1784), which led to massive use of fossil fuels to power the industrial revolution. Beyond the technical aspect, it is also the moment when, despite the primitive accumulation of capital beginning well before that time, capitalism begins to drain an increasing share of the global resources and when emissions of GHGs are becoming significant.

Other disciplines, with at their frontline archaeologists, argued for an early Anthropocene, showing precocious impacts of the human species on the biosphere and possibly the lithosphere and atmosphere. Thus the human contribution to the extinction of large mammals during the Pleistocene (the Pleistocene/Holocene junction in the case of Brazil), or the erosion and deforestation processes and the first emissions of GHGs in the Neolithic with the spread of agriculture, domestication and irrigation; or the colonization of the Americas by Europeans (Lewis and Maslin 2015). Yet it seems that one should distinguish between, on the one hand, impacts of importance on the landscape and other species, but not necessarily synchronous or global, and which spread over centuries or millennia, and on the other hand the global and rapid changes currently affecting the Earth System. As Clive Hamilton said (2015: 103), 'The Earth System is not "the landscape", it is not "ecosystems", and it is not "the environment".' The approach in terms of the Anthropocene is therefore very different from that of 'environmental sciences' or studies of 'the impact of human activities on the environment' (op. cit.). The scale, the object and the temporality all considerably differ.

Scientists who are looking for the moment where the Earth System has globally deflected from its course because of human activities have proposed a date (1950), which corresponds to what is now called the Great Acceleration, when all forms of consumption and degradation started to increase exponentially (Steffen et al. 2007, 2011, 2015). From the beginnings of the industrial revolution to the immediate post-war period, the atmospheric CO_2 concentration had increased relatively little (from 280 ppm to 310 ppm) while it has since exceeded 400 ppm. Consumption of fossil energy exploded during this period. The Great Acceleration marks the entry into the era of development and growth policies, international programmes aiming at the inclusion of 'poor' countries into the world market, and thus the application to what used to be called the third world of economic policies from industrial countries, through institutions such as the World Bank and the IMF. The post-war period is the period during which two of the three main components of the ecological footprint, population growth and per capita consumption, increased considerably. Population growth mainly increased in the South and consumption in the North. This is also the time when technologies with high impact (third component) were developed on a large scale: nuclear, petroleum, synthetic chemistry and biology, new materials such as plastics, green revolution (new varieties, pesticides, chemical fertilizers and genetically modified seeds). Alongside globalization, the considerable increase in trade becomes the principal factor of surge of the ecological footprint.[2] Even though we shouldn't forget that the foundations that made this possible were historically built during the previous two or three centuries, the Great Acceleration has led in a very short time to exceed or threaten to reach in the short term the limits described by Rockström and Steffen et. al. (2009). The choice of this recent date as the beginning of the Anthropocene has the merit to underline the unsustainability of the current trends and the urgent need for a radical shift.

The latest proposal of the Anthropocene Working Group (Zalasiewicz J. et al. 2015) is compatible with the Great Acceleration. The Group has chosen as markers of entry in the Anthropocene the radionuclides from Atomic explosions (with two possible dates, 1945 or 1950). They are in fact present in all parts of the world and will be traceable in polar ice and sediments for millions of years.

From the point of view of social scientists and historians, it is nevertheless important to retain the two dates: a first Anthropocene (or first phase) corresponding to the industrial revolution (from 1784 or 1800 to 1945), the establishment of capitalist institutions and of the material and technical conditions for continuous acceleration; then a second Anthropocene (or second phase) corresponding to the spread of industrialization and the Great Acceleration. This second phase is underway, and if nothing is done, future impacts are likely to be even more serious than those that can be observed today. This division of Anthropocene in two phases – even if the stratigraphists, in obedience to the requirements of their discipline, only accept the second – also makes sense for most emerging or developing countries. This is the case for Brazil, which although having endured substantial environmental impacts in the seventeenth and eighteenth centuries, took little part in the first phase of the Anthropocene, at least on its territory (see Chapter 1).

'Development' and growth of the ecological footprint: a need for limits

Each country contributes to the overall human footprint based on various factors such as the size of its population, its consumption per capita, its level of industrialization and its place in world trade. First country in Latin America by its population (more than 205 million) and GDP (2,346 billion USD in 2014), Brazil was ranked seventh on the economic ranking of countries; it is however at risk of dropping by one or two places in 2016 due to a recession more severe than expected. From the point of view of international economic institutions, it is nevertheless one of the most promising emerging countries, in particular thanks to its important natural resources. Brazil also represents a major ecological issue. It hosts the most important biodiversity in the world, the biggest tropical forest and the major freshwater and arable soils reserves.

Despite the inadequacies of such indicators as the ecological footprint and the biocapacity, it is useful to recall that Brazil is part of a small number of countries with a net positive 'credit' in ecological terms (its biocapacity exceeding its ecological footprint), in a world where the number of countries whose footprint exceeds their biocapacity is growing[3] steadily. South America has the largest number of 'creditor' countries. As a result of its growth in population and consumption, but also of the export of primary products, the biocapacity of Brazil has dropped considerably since 1961, from 24 global hectares per capita to 9.2 in 2011 (see Chapter 2), leaving a positive balance of

6.3 gha (Global Footprint Network 2015). However, the average ecological footprint per capita of Brazil (2.9 gha) is higher than the global average (2.6 gha) and also higher than the global average which should not be exceeded to avoid overshoot (1.74 gha).

By comparison, the European Union accrued a deficit of 2.3 gha,[4] the USA and Japan 3.1 gha, and China 1.6 gha. These countries must drain resources from other countries to ensure their level of consumption[5]; they also 'use' the atmosphere (global warming) and the oceans (acidification) to absorb their releases of greenhouse gases, superior to the absorption capacity of the Earth System. Taking into account the material dimension of unequal exchange (land, water, mineral and agricultural resources, work, energy), and not simply the monetary one, shed a new light on the industrial revolution and the perpetuation of this asymmetry[6] (Hornborg 2011). However, this seems insufficient to describe certain aspects of the global contemporary ecological and economic impasse: regarding climate warming alone, if it were possible to cancel immediately and completely the CO_2 emissions of high-income countries (which is indispensable and urgent), this would not sufficiently reduce the overall carbon footprint to fit within the limits imposed by the biosphere by 2050. Global warming is itself only one aspect, certainly an important one and maybe the easiest one to resolve, of the global ecological crisis. There is therefore a challenge, a reorientation of the forms of production and consumption, which must be tackled by all countries at the same time, even if the distribution of efforts must be differentiated. This reality imposes a necessary international solidarity whereas the situation of every country, and therefore their interests in the short term, are divergent, as illustrated by the difficulties encountered by the various attempts to implement a global agreement on biodiversity conservation or reducing the emissions of greenhouse gases.

The fault line is not always, or not only, between 'developing' and 'developed' countries. Whatever the categories used to classify countries, they reveal themselves heterogeneous: the BRICS (Brazil, Russia, India, China, South Africa) comprise two countries with a net surplus of biocapacity (Brazil and Russia), and three with a net deficit; the group of high-income countries contains several with net surplus (Canada, Scandinavia, Australia, New Zealand) and most of them with great deficit. Among the high-income group, some of them (i.e. Australia, Canada) are also important suppliers of raw materials. But if we consider net physical exports of natural resources by the BRICS, Russia ranks second (at global level), Brazil third and South Africa tenth (Bednik 2016), while China and India have radically different profiles, as China exports mainly manufactured products (but occupies the first place as an importer of materials) and so does India to a lesser extent. Latin America in general has asserted itself as a great purveyor of materials from the 'commodity frontier' (Martínez-Alier et al. 2016). If we look at the list of the 24 countries regarded as 'emerging' by the IMF, their heterogeneity is even stronger and their interests even more divergent. But the common point between them all is the adoption of the same model of development, which leads to a considerable increase in the absolute demand for resources; meanwhile the human ecological footprint already exceeds by 50 per cent the capacity of regeneration and absorption of the planet and 80 per cent of its population lives in countries where the biocapacity is already below their ecological footprint, a situation only worsening.

Formulating the challenge of the Anthropocene seems easy: we have to dramatically reduce the human footprint. The difficulty is to get countries, businesses, institutions, political parties, social movements and citizens to undertake commitments to reduce the consumption of energy and matter. But this immediately raises the question of inequality between countries and within each country. This can be seen in all international negotiations and recently at the COP21. Who should pay the bill? What about responsibilities? What criteria to adopt? These issues hinder negotiations, each country seeking to assume the lower cost for itself. Inside countries, the consensus around the need for growth contributes to the social acceptance of inequality. How to restructure the different economic sectors around other conceptions of wealth without causing the secession of powerful interests and risking social explosion?

To overcome this challenge proves to be extremely complex and cannot be reduced to technical matters; it is rather an eminently political issue, mainly affected by economic pressures and social demands. If the 'right to development' is not questionable, it goes against what is the nature of so-called 'development'. Paths to achieve this have been the subject of much criticism since the 1960s. The focus used to be on the issues of distribution and inequalities (criticism of unequal exchange and imperialism). The report of MIT for the Club of Rome in 1972 (The limits to growth) was the first to raise the question of the physical volumes mobilized by 'development' (resources, waste, pollution). From this date onwards, the critics have further incorporated, each time, ecological issues and cultural diversity (territorial rights of indigenous peoples, for example), but failed to significantly influence the general dynamic. Mobilizations and protests have been in net decline following the two oil shocks (1973 and 1979), which paved the way to the dominance of neoliberal ideology and globalization. 'Development' always takes the form, and today more than ever, of an increase in the social metabolism, i.e. the consumption of more energy and raw materials and the increasing artificialization of the use of soil (deforestation, urbanization, infrastructures and so on). Thus, the curve of augmentation of the HDI per country seems to inevitably match the one of their ecological footprint (WWF 2014). This parallelism of the two curves is of course partly due to the weight of the GDP in the calculation of the HDI, but not limited to it. On the other hand, the decrease in the amount of matter and energy incorporated into each unit of product, GDP, or monetary value (real but limited to certain countries and certain sectors) is offset by the increase in overall consumption.

If all countries are involved to varying degrees in this dynamic (even the leaders of the industrialized countries talk about the needs of their 'development'), 'emerging' and some 'developing' countries also put forward an imperative of catching-up, a crucial (yet questionable) concept which exerts considerable pressure on the definition of public policies and officialises their allegiance to the dominant model (regardless of the posture of their political rulers). These countries therefore attempt to follow the footsteps of industrial countries in a shorter time rather than choosing a different path.[7]

Territorial expansion, 'development' and environmental degradation in Brazil

Brazil has relied on its hinterland, its vast spaces and resources to support its development policies. The booming of developmental policies remains associated with president Juscelino Kubitschek (1956–1961), although similar achievements must be attributed to previous Vargas Governments: the creation of the BNDES (*Banco Nacional de Desenvolvimento Econômico e Social*) and of large national companies such as Vale (mining) and Petrobras (oil), for example. President Kubitschek is responsible for the foundation of the new capital Brasilia, away from the coast, thus moving the centre of gravity of the country towards its hinterland and committing to ongoing territorial expansion. He is also the first president to give the country explicit objectives of growth. Since then, all following development plans have aimed at accelerating this growth. Most recently, these were the *Programa Brasil em Ação* (1996–1999), the *Programa Avança Brasil* (2000–2003) and the *Programa de Aceleração do Crescimento* (PAC 1 from 2007 to 2010, PAC 2 from 2010 to 2014 and currently extended). This last programme has seen the completion of important infrastructures such as the controversial Belo Monte dam (see Chapter 6). Future dams on the Tapajós River will have at least as much impact on Amerindian populations (Indians Territories) and the local environment (protected areas, National Park) and are already being challenged by social movements.

Territorial expansion as a solution to the impasses of 'development'

As an important supplier of raw materials, Brazil has made a contribution to the global ecological footprint almost since the beginning of European colonization (see Chapter 1). This has resulted in the almost total destruction of the Atlantic forest, the first biome to suffer the impact of the expansion of cattle ranching, sugar cane plantations and later on coffee plantations. It contributed directly to the industrial development of European countries and the United States by its exports of Amazonian rubber (end of the nineteenth century and beginning of the twentieth century), as well as to the Allied war effort from 1941. If the operating system of the rubber tappers was particularly unfair, it had the benefit of preserving much of the Amazon rainforest. But a great second phase of destruction began with the occupation and exploitation of the Amazon and the Cerrado Savanna project.[8] From its beginning, the military dictatorship (1964–1985) attached great importance to the integration of this immense region, which until then was marginal but rich in resources. In 1970, the National Integration Plan (NIP) devised the opening of roads (including the famed Transamazonian), the migration and settling of 500,000 farmers mainly from the Nordeste (a figure that would also never be reached, at least by the planned colonization), the granting of large tracts of land to private companies, the creation of a free zone in Manaus (1967), the construction of large dams such as the Tucuruí hydroelectric power plant opened in 1984, as well as grants and tax reliefs to render investments attractive in the region. This was done, almost always, disregarding the rights of indigenous and traditional peoples. These massive physical and energetic transfers have sometimes been referred to as internal colonialism (see Chapter 3). These policies would lead to significant deforestation and an increasing number of land conflicts. From the end of the 1970s, thanks to progress in satellite observation techniques, Brazil and the world discovered the extent of the damage. It is estimated that between 1977 and 1988, around 21,000 km² have been cleared annually. In the late 1980s, measures began to be taken to curb the deforestation, which proved largely ineffective. The absolute record in clearing was reached in 1995[9] with 29,059 km². This dynamic seemed uncontrollable and was the consequence of different factors, such as the price of land and commodities, the migration driven by the opening of roads or the completion of major infrastructure and mining projects. After stabilizing at a high level for a few years (around 18,000 km² per year), and despite the implementation of an extensive pilot programme funded by the G7 and the World Bank for the sustainable development of the region (PP – G7, coordinated by the Ministry of the Environment), these rates went upward from 2002, coinciding with the arrival in power of the Workers' Party (Partido dos Trabalhadores, PT) of Luiz Inácio Lula da Silva: 2002: 21,651 km²; 2003: 25,396 km²; 2004: 27,772 km². The government then took a series of measures to block the illegal appropriation of public land dynamics and to ensure better coordination between the Ministry of Environment (MMA), the Brazilian Institute of Environment and Renewable Natural Resources (IBAMA) and the Federal Police. It also created a series of protected areas in the most threatened regions, where the territorial expansion of the ranchers/foresters/grileiros[10] triptych was the most dynamic. Deforestation rates began to fall, without it being possible to assess with certainty the role of the economic conditions in this regression. These rates even reached 4,571 km² in 2012 before fluctuating: 5,891 km² (+ 29%) in 2013; 5,012 km² in 2014 and 5,831 km² (+ 16%) in 2015. In total, around 20 per cent of the Amazon has been cleared, and a roughly equivalent amount has been degraded. These figures make it clear that the bulk of Brazilian CO_2 emissions were for a long time the result of deforestation (see Chapter 12). Furthermore, in 2012 the congress approved the new Brazilian Forest Code after much controversy between environmentalists (who were opposing the substitution of the previous code) and the agribusiness defenders (who were in favour of the new code). More lenient with the substitution of natural vegetation by agricultural activities than the previous one, the new forest code fails to provide sufficient means and technical options for small farmers to maintain biodiversity in Legal Reserves – a fixed proportion of land under natural vegetation (see Chapter 16).

If the situation has, however, improved on the clearing front, it remains very fragile and sensitive to economic reversals. By the end of the 1960s to the mid-1990s, most of the deforestation was due to large landowners who came to the Amazon to develop cattle breeding. Directed and spontaneous colonization has also played an important role during this period. According to Schneider and Peres (2015), settlement projects converted into 'agrarian reform facilities' represent only 5.3 per cent of the region's area but 13.5 per cent of all deforestation carried out so far.[11] These dynamics, by rich landowners and family farmers both, are closely linked to the exploitation of wood (opening of access routes, financing of the first cropping). Mechanized farming of soybeans and corn also gave a new impulse to deforestation, pushing the livestock further in at the expense of the forest. Paradoxically, industrial agriculture is experiencing an intensification process that led to the decrease of this pressure of transfer. On the other hand, it is now located in the heart of the Amazon, taking advantage of the new port infrastructure built by the company Cargill in Santarém (State of Pará) and its storage capacity, despite soybeans being currently essentially produced in Mato Grosso (State mainly composed of *cerrado*).

Meat, soy, corn and timber exports have become more important due to continued growth in revenue from the exports and therefore gained political influence. The big landowners, whose influence heavily marked the history of Brazil, and major agricultural commodities producers, have an important and very influential parliamentary group in the Congress and are able to appoint ministers and to direct public policies according to their interests. The mining sector is also important and influential, given the volume of exports that it controls and also possesses a powerful lobby. The territorial expansion of these economic agents clashes often violently with the territories of indigenous peoples and protected areas (natural or dedicated to categories of traditional peoples) or even, in some cases, with family farmers from colonization or agrarian reform programmes. These conflicts, which had somewhat decreased, worsened over the past years, in parallel to the increase in the price of raw materials. For their part, the populations whose territories are threatened or invaded built resistance strategies (see Chapter 5) and, even if their struggle is unequal, can in many cases rely on the support of the Federal Public Ministry and a part of the national and international public opinion.

Acceleration of growth, extractivism[12] and neodevelopmentism

Driven by global growth and mainly that of emerging countries, among which are China, prices for raw materials increased sharply between 2000 and 2013. This increase allowed 'progressive governments' of Latin America to rely on their export earnings to finance social policies. Favourable prices encouraged farmers, especially in soy extensive monocultures, to bring new land under cultivation, and mine operators to increase their investments. Brazil is not the only country in the region where such a configuration could be observed.

Political leaders saw there an opportunity to increase their efforts to catch up and have strongly supported these sectors. They thus placed themselves in a position of double bind, well described by Melo and Rocha (2013). They have created a situation of increasing dependence on economic sectors whose interests are contrary to a part of their electorate. This generates the risk of losing popular support in the event of a drop in export earnings due to the cyclical nature of the fluctuation of the price of raw materials. This is what has been happening in Brazil since October 2014. The decreases in the prices of soybeans and iron, two major exports, and the drop of a few others, have contributed to the economic situation becoming worse. On top of this, the large offshore oil deposit (pre-salt layer) on which Brazil relied to shorten its catching-up period is no longer profitable given the current oil prices. Certainly, the global situation of excess oil and especially mining might not last, and demand could recover as well as prices (which is less likely for oil, due to the need to reduce CO_2 emissions and the increasing competition of renewable energies), but the model known as 'neodevelopmentist', characterized by the search for an acceleration of growth at all costs, could find itself considerably slowed down and delegitimized.

Extractivism revived the territorial expansion and increased the number of conflicts. Protected areas are perceived by the actors of extractivism as an obstacle for 'development', i.e. an obstacle to the valorization of capital. The hostility and violence against indigenous peoples has increased, to the point of worrying the rapporteur of the UN during its recent human rights visit[13] to Brazil. The murders of resistants to expulsions[14] caused by the increase in mining activities, dams, infrastructure, wood operators and landowners have also increased among traditional and peasant populations.[15] Proposals of law allowing the expansion of infrastructure and mining activities at the expense of protected areas and these categories of population are under consideration.

Yet once again, this situation is not unique to Brazil, as shown in the recent homicides of two resistants to the dam of Agua Zarca in Honduras and many others. The increasing pressure on resources and territories can be felt everywhere, including in old industrial countries. In Europe itself, the commodity boom has led to rehabilitating old mines or to the development of prospecting, in order to reduce external dependence (Bednik 2016), causing movements of protest and resistance. This resistance opposes the major works on infrastructure and controversial projects, which leads to the denaturalization of the territories and to expulsions.

On the other hand, since Brazil has taken advantage of this windfall to reduce inequalities, which allowed part of the poorest population to increase its consumption, the balance sheet therefore remains mixed. Abramovay (2016), criticizing the concept of 'green growth', highlights the gap between this increase in income and the quality of goods and public services that these peoples can get. He also shows that in the sector of energy production and transport, which thanks to hydroelectricity and ethanol[16] were regarded as the pillars of green growth, the term 'green' is largely inappropriate and that public policies did not promote innovation in these areas.

Extractivism, sometimes called 'neoextractivism' when it relates to the progressive governments of Latin America (Gudynas 2014a), also has other consequences, such as the constant pressure by its actors in favour of the relaxation of controls and of deregulation. Thus, public authorities have not been able (or have not desired) to oppose the rapid dissemination of genetically modified organisms (and particularly soybeans) (Chapter 7). They were also complicit in the strong growth in the use of pesticides in agriculture, especially through direct and indirect subsidies, which continues to seriously affect the health of citizens (Chapter 8). The absence or lack of control by public authorities is partly the cause of the most serious environmental disaster of the twentieth century in Brazil: the rupture, on 5 November 2015, of the dam of sludge extraction of the mining company Samarco (owned at 50 per cent by Vale) in Mariana, in the Minas Gerais state, costing the lives of 17 people and 'killing' the Rio Doce over 663 km, all the way to the sea. Other dams of this kind are threatening to break. International competition leads to lower production costs to maximize profits (in particular for shareholders). The occurrence of accidents (or slow and insidious contamination) is therefore not surprising. What is even more preoccupying is that the public authorization systems and control procedures do not work, or work very poorly, since the interests of governments are in many ways, and especially through the growth of exports, subdued to private interests. Moreover, despite disasters, accidents and denunciations, the legislative power (itself partly reflecting the interests of powerful lobbies) seeks to relax even further the environmental standards.

In the current international configuration, we are witnessing a global acceleration in the exploitation of natural resources, which does not seem likely to diminish due to causes other than exhaustion, a high cost of extraction or a severe recession. This acceleration is accompanied by the search for places where negative externalities will be more easily accepted or negotiated, allowing optimization of investments (see Chapter 4) with huge environmental and social consequences.

The limited and conflicting scope of environmental policies

No country can exclude itself from the current conditions of the competitive, productivist and growth-oriented economy, either as a main actor, an auxiliary or a 'victim'. Environmental policies can therefore only be, in the best of cases, a 'greening' of marginal dynamics and trends within this economy. Some countries have, however, aggravating circumstances, 'growth needs' and catching-up needs, which justify, in the eyes of politicians and of the companies concerned, the more predatory practices. Thus they agree to become the place of expansion of the commodities frontier on the pretext that it would only be a temporary situation to lay the foundations for their future development. Neodevelopmentism and neoextractivism thrive on the sustainable development and the ecological modernization rhetoric. It is therefore not surprising that the only visible ideology in the entrepreneurial sphere and within the governmental sphere is that of the green economy or green growth. Both key arguments of the green economy are the theory of decoupling between the consumption of energy and matter (or the emission of GHGs[17]) and GDP, as well as the dematerialization of the economy by the use of the NBIC (Nanotechnology, Biotechnology, Information Technology and Cognitive Science). Many previously published rebuttals have failed to damage these arguments insofar as they are at the centre of the stakes for the upcoming capitalism.

Just like in many other countries, Brazil is captive of a perverse financial dynamic that is not interested in 'sustainable development' (see Chapter 13). The alignment of the financial corporate world with the goals of sustainable development adopted in 2015 at the Third UN Conference on Financing for Development in Addis Ababa appear illogical given the current hegemonic system and, unsurprisingly, led to weak commitments by the signatory countries.

Despite a strong tradition of state regulation and regardless of the position of certain actors within governments, Brazilian environmental public policies cannot produce anything else than attempts to somewhat control what can be called a market environmentalism. According to the classification of Gudynas (2014b), it should be distinguished between forms of 'alternative development', which include some measures or principles within eco or sustainable development and 'alternatives to development', probably the only policies that could break with the idea of linear material progress that underlies the neodevelopmentism and its predatory dynamic. As in all other countries, governments can consider no other paths than that of light reformism, in line with the green economy. It is therefore not surprising that it is only in respect of climate policies that a matter of consensus can be reached: it is an area more easily quantifiable and which can be integrated easily into a market environmentalism (compensation, market of CO_2, REDD etc.).

Since the 1970s, Brazil has created an institutional and legislative apparatus in the environmental field. Global warming is considered to be a priority and gave rise to an institutional arsenal that has not ceased to be strengthened. However, as in most countries, policies put in place are often in contradiction with the objectives of other policies, in particular those of economic importance (see Chapter 14). Their effectiveness is often reduced or voided. Within the Government as well as at the National Environmental Council (CONAMA), an organization that brings together representatives of government institutions, the productive sector and the civil society, the interests are not only contradictory but sometimes opposing and irreconcilable (see Chapter 15).

The contradictions shaping domestic policies and favouring, in a nutshell, economic aspects at the detriment of environmental aspects, put at risk the long-term investments required to cope with climate change and the many risks it involves (see Chapter 11). Those contradictions in domestic policy are reflected in the Brazilian oscillating positions at international forums. The current Brazilian position has been weakened in the international community, and the country probably will not become the leader on climate regime it ought to be in the near future (see Chapter 9).

Making every use of the conflicts within the government and possibly at their source, lobbies reinforce their influence through the corruption that also affects the legislature. The most threatening lobbies for the environment are strongly represented in the National Congress. According to Carlos Machado e Rodrigo Vilani (see Chapter 10), 119 deputies intercede in favour of the interests of agribusiness, which represents 23 per cent of the Chamber of Deputies. The banks, meat and mining industries total respectively 197 (38 per cent), 162 (31 per cent) and 85 (16 per cent) members. Companies producing GMOs, pesticides or hydropower exert heavy pressure to gain influence on the Brazilian economy. The lack of independent information on GMOs does not counterbalance the dissemination of information in support of their use by the media. We are witnessing a trivialization of their use, achieved through the absence of debate as to their actual effects and their impact on health and the environment (see Chapter 7).

Questioning the inevitability of development, as suggested by Morin (2001), could well be the only way towards a sustainable future. According to the author, this would allow us to restore our freedom of choice and, depending on the circumstances, to elect whether we should globalize or not, grow or decrease, develop or regress, keep or transform.

Mumford ends his book, *Technics and Civilization* (1934), with a provocative and critical sentence on the way of life of the contemporary man: 'nothing is impossible'. The sentence conveys mankind's belief in the absence of technical limits. Nothing is an impossible problem for technique and that which is impossible today will certainly be solved tomorrow. The sentence also expresses the absence of ethical boundaries; there is no limit that cannot be transgressed in the name of progress.

The concept of Anthropocene allows us to establish a *nexus* between subjects as diverse as the assassination of Sister Dorothy Stang in the Amazon, the export of Brazilian pork to India, the drying of the southern rivers, the changes for the worse of the Brazilian Forest Code, the movements of resistance to pesticides and GMOs as well as many other topics in Brazil and worldwide. The path of development undertaken by mankind and the way we produce, organize our work, create value and distribute wealth confirms that we are still living as if nothing were impossible.

Notes

1 This term has the disadvantage of strengthening a current trend towards reducing the ecological crisis to its climatic dimension, consigning to the background, among others, the degradation of ecosystems and habitats, the loss of species, the widespread pollution and their complex interactions.

2 The search for high profits (in a context of decline since the 70s) led to an exacerbation and a misuse of the Ricardian mechanism of comparative advantages which became a source of social and environmental dumping, favoured by innovations in transportation such as the invention of the container.

3 Reviewing 50 years of theories and 'development' policies, the former Ambassador of Peru to the United Nations proposed to change the term 'developing countries'

for that of 'unsustainable national economies' (Rivero 2001), at least in the current economic system and its prospects for 'development'.

4 This deficit would be higher without taking into account the positive balance of Sweden and Finland.

5 The role of transfers of water resources in the production/consumption of agricultural and industrial products is not appropriately valued. Thus Brazil appears as the fourth exporter of water resources while it imports only very little; in comparison, Europe imports 40 per cent of its water footprint (Hoekstra and Mekonnen 2012).

6 This historical asymmetry is what founded the concept of ecological debt and its perpetuation is often referred to as ecological neocolonialism.

7 It should, however, be noted that certain conceptual initiatives introduced in the Constitutions of Ecuador and Bolivia ('Buen Vivir' and Pachamamma or Mother Earth) could be a draft for change, although this is not really the case today.

8 These vast savannas, occupying 22 per cent of the territory and more than 2 million km² (see Chapter 2) were mainly occupied by the agro-industry (in particular for the production of soybeans and corn). It is a biome whose biodiversity is generally underestimated (is there a relationship between the political weight of the actors in the region and this underestimation?) and which has already been destroyed at 50 per cent.

9 The aforementioned numbers are from the National institute for spatial studies (INPE-PRODES 2015) and are calculated on the basis of a timeframe between August to July. The 1995 clearing is thus referred to as the one from August 1994 to July 1995.

10 The *grileiros* are individuals who illegally appropriate public lands, either by forging false documents or by force and fait accompli.

11 For the year 2014–2015, data from the Ministry of Environment (MMA) and the National Institute of Colonization and Agrarian Reform (INCRA) differ, but the figures remain nevertheless high: 26.55 per cent of the total clearing for the MMA and 21 per cent for INCRA, which admits that they are increasing. These high numbers are the result of non-compliance with legislation requiring that 80 per cent of properties constitute a Legal Reserve forest. This overshoot is attributable to the move to cattle farming by about 80 per cent of family farmers. The latifundium and agribusiness remain of course the major contributors to Amazonian deforestation, but these figures question the agrarian reform policies carried out at the expense of the forest.

12 This term is widespread and we use it here while noting that it is used with a very different meaning in Brazil. It engrosses the collection by traditional populations of natural products for the market (rubber, Brazil nuts, vegetable oils, fibres, etc.), activities on which many projects rely on today to improve the conditions of life of these peoples. Used in the neodevelopmentist context, it signifies the industrial exploitation of resources mainly intended for export, whether agricultural commodities, mining or petroleum products, or hydroelectric dams where a significant part of the energy is intended for processing minerals.

13 The special rapporteur of the United Nations for the rights of indigenous peoples, Victoria Tauli-Corpuz, spoke of the "regression" and threats (including legislation) that affect the rights of these populations in Brazil. http://Amazonia.org. br/2016/03/ONU-alerta-Brasil-sobre-retrocessos-na-Protecao-dos-Direitos-Indigenas/?utm_source=akna&utm_medium=email&utm_campaign=not%EDcias +da+Amaz%F4nia+-+21+de+Mar%E7o+de+2016

14 Term coined by Sassen (2014) to designate a systemic dynamic of the deepening of the ongoing capitalist relations and which she characterized as a 'predatory formation' (in a somewhat different sense of 'dispossession' of D. Harvey). The

expulsion concerns both territories (land grabbing) and resources, social achievements, environmental destruction or austerity policies.

15 In an open letter, the Pastoral Land Commission (CPT) reported 50 killings of 'resistants' in 2015 and 13 between January and early April 2016, mainly in the Amazon. www.ecodebate.com.br/2016/04/12/Nota-publica-da-CPT-quem-Vai-deter-a-Violencia-contra-as-Comunidades-camponesas/

16 A recent Americano-Brasilian study (De Faria et. al. 2015) seems to confirm the high level of emissions by tanks dam in the Amazon, a thesis supported by Fearnside in this volume (Chapter 6)

17 Relative or absolute decoupling is easiest to obtain for the GHGs. It can be reached by renunciation to fossil fuels (at least in theory, as it is necessary to recognise the fossil energy required for the creation of a wind and solar farm). This is not the case in other areas (mining, agriculture, biodiversity etc.)

References

Abramovay, R. (2016) 'The green growth trap in Brazil', in Dale, G., Mathai, M. V. and Puppim de Oliveira, J. A. (eds) *Green Growth. Ideology, Political Economy and the Alternatives*, Zed Books, London.

Bednik, A. (2016) *Extractivisme. Exploitation industrielle de la nature: logiques, conséquences, résistances*, Le passager clandestin, Neuvy-en-Champagne.

Bonneuil, C. (2015) 'Tous responsables?' *Le Monde Diplomatique*, November 2015, 16–17. www.monde-diplomatique.fr/2015/11/BONNEUIL/54139

Chakrabarty, D. (2009) 'The Climate of History: Four Theses', *Critical Inquiry*, 197–222.

Chakrabarty, D. (2015) 'The Anthropocene and the convergence of histories', in *The Anthropocene and the Global Environmental Crisis: Rethinking Modernity in a New Epoch*, Routledge UK, 44–56.

Chertkovskaya, E. and Paulsson, A. (2016) 'The growthocene thinking through what degrowth is criticizing', http://entitleblog.org/2016/02/19/the-growthocene-thinking-through-what-degrowth-is-criticising/, accessed 10 February 2016.

Crutzen, P. J. and Stoermer, E. F. (2000) 'The Anthropocene', *Global Change Newsletter*, 41, 17–18.

De Faria, F. A. M. et al. (2015) 'Estimating greenhouse gas emissions from future Amazonian hydroelectric reservoirs', *Environ. Res. Lett.*, 10(2015)124019.

Global Footprint Network, National Footprint Accounts 2015 (data year 2011), www.footprintnetwork.org/fr/index.php/GFN/page/public_data_package, accessed 10 February 2016.

Gudynas, E. (2014a) 'Débat sur le développement et ses alternatives. Un guide bref et hétérodoxe', in Lang, M. and Mokrani, D. (eds) *Au-delà du développement: Critiques et alternatives latino-américaines*, Éditions Amsterdam, Paris.

Gudynas, E. (2014b) '"Churcar" Alternatives to Development, Alternautas', 1(1), 48–55, www.alternautas.net/blog/2014/7/28/churcar-alternatives-to-development, accessed 30 March 2016.

Hamilton, C. (2015) 'Getting the Anthropocene so wrong', *The Anthropocene Review*, 2(2), 102–107.

Hoekstra, A. Y. and Mekonnen, M. M. (2012) 'The water footprint of humanity', *Proceedings of the National Academy of Sciences*, 109(9), 3232–3237.

Hornborg, A. (2011) *Global Ecology and Unequal Exchange. Fetishism in a Zero-sum World*, Routledge UK.

Hornborg, A. (2015) 'The political ecology of the technocene: uncovering ecologically unequal exchange in the world-system', in Hamilton, C., Bonneuil, C. and Gemenne, F. (eds) *The Anthropocene and the Global Environmental Crisis: Rethinking Modernity in a New Epoch*, Routledge UK, 57–69.

INPE-PRODES (2015) Taxas de desmatamento 1988-2015.

Latour, B. (2015) *Face à Gaïa*, La Découverte, Paris.

Lewis, S. and Maslin, M. (2015) 'Defining the Anthropocene', *Nature*, 519, 171–180.

Malm, A. (2015) 'The Anthropocene myth', *Jacobin*, 30 March 2015, www.jacobinmag. com/2015/03/anthropocene-capitalism-climate-change/

Martinez-Alier, J., Demaria, F., Temper, L. and Walter, M. (2016) 'Trends of social metabolism and environmental conflict: a comparison between India and Latin America', in Dale, G., Mathai, M. V. and Puppim de Oliveira, J. A. (eds) *Green Growth: Ideology, Political Economy and the Alternatives*, Zed Books, London.

Melo, J. A. T. and Rocha, D. A. (2013) 'Une bougie pour Dieu, une autre pour le Diable. L'insoutenable politique de développement du gouvernement brésilien et la réponse de la société civile', *Ecologie et Politique*, 46, 67–82.

Moore, J. W. (2015) 'Putting Nature to Work', in Wee, C., Schönenbach, J. and Arndt, O. (eds), *Supramarket: A micro-toolkit for disobedient consumers, or how to frack the fatal forces of the Capitalocene*, Irene Books, Gothenburg, 69–117 and, forthcoming, Moore, J. W. (ed.), *Anthropocene or Capitalocene?: Nature, History, and the Crisis of Capitalism*, PM Press, Oakland.

Morin, E. (2001) *La Voie: pour L'avenir de l'Humanité*, Arthème Fayard, Paris.

Mumford, L. (1934) *Technics and Civilization*, Harcourt, Brace & Company, Inc., New York, p.149.

Rivero, O. (2001) *El mito del desarrollo, los países inviables en el siglo XXI*, Fondo de Cultura Económica, 2nd edn. Lima.

Rockström, J., Steffen, W. et al. (2009) 'A safe operating space for humanity', *Nature*, 461, 472–475.

Sassen, S. (2014) *Expulsions: Brutality and Complexity in the Global Economy*, Harvard University Press, Cambridge, Massachusetts – London, England.

Schneider, M. and Peres, C. A. (2015) 'Environmental Costs of Government-Sponsored Agrarian Settlements in Brazilian Amazonia', *PLoS ONE*, 10(8): e0134016. doi:10.1371/journal. pone.0134016.

Steffen, W., Crutzen, P. J. and McNeill, J. R. (2007) 'The Anthropocene: Are Humans Now Overwhelming the Great Forces of Nature?', *Ambio*, 36(8), 614–621.

Steffen, W., Grinwald, J., Crutzen, P. J. and McNeill, J. R. (2011) 'The Anthropocene: Conceptual and Historical Perspectives', *Philosophical Transactions of the Royal Society A*, 369, 842–867.

Steffen, W., Broadgate, W. et al. (2015) 'The trajectory of the Anthropocene: The Great Acceleration', *The Anthropocene Review*, 2(1), 81–98.

WWF (2014) *The Living Planet Report 2014*.

Zalasiewicz, J. et al. (2015) 'When did the Anthropocene begin? A mid-twentieth century boundary level is stratigraphically optimal', *Quaternary International*, 383, 196–203.

Part I

Development dynamics and social-environmental contradictions

1 Brazil in the history of the Anthropocene

José Augusto Pádua

From the late 1960s onwards, grievances and conflicts related to so-called "environmental problems" started growing swiftly in different regions of the planet. In the beginning, such problems were treated in a fragmented and isolated manner. They were externalities, dysfunctions, or accidents in the context of advancing urban–industrial modernization in different countries, and in international trade.

As the decades passed, the necessity to adopt a more integrative focus, which would allow dealing with the structural and persistent condition of such problems, becomes clear. It is not just a passing crisis. The environmental issue, in its different manifestations and intensity degrees, has been emerging in the wake of a "great transformation" in human history (to use the term coined by Karl Polanyi in 1944 to express the deeper dimension of the industrialization process and the commodification of work and nature from the eighteenth century on).

In search of a more integrative focus, it is possible to use many concepts already adopted by historians to think about the "great transformation," such as modernity, capitalism, or urban–industrial civilization. The concept of Anthropocene, however, introduced by Paul Crutzen (Chemistry Nobel Prize, 1995) in the year 2000, has become the most influential conceptual tool to think about the environmental intensity of that historical transformation. Its starting point is the radicalization of an integrative focus. Numerous environmental problems from the last decades come to be understood as a set of symptoms, signals, and indicators of a new historical period, a real shift in the material presence of human beings on Earth. Such vision, radically integrative, constitutes a great strength and also the main weakness of the concept of Anthropocene, as we will see.

Each of the concepts mentioned above, including the Anthropocene, has its limits and potentialities. The idea of Anthropocene is basically material and quantitative (although it is not difficult to reflect on its deep social and cultural implications). Its strength resides on the visceral absorption of the planet into human history, and of human history into the dynamics of the planet. In the trajectory of human societies, before the great transformation discussed here, the presence of the planet, whenever noted, was a mere abstraction. Human societies related to a number of spaces and specific sets of existing beings in the context of the ecological diversity of the planet. Such societies reproduced through appropriation and management of a relatively small portion of the matter and energy flows that exist in the planet's nature. But they did not touch the macrostructures of the Earth – and that is the Anthropocene's greatest news. The planet becomes the locus that measures the scale of human presence on Earth. Human action, perceived in a very aggregated way, acquires the weight of a geological agent.

The statement becomes more concrete in the context of a periodization of the Anthropocene. The empirical starting point has been the construction of graphs that compare, in the long term, aggregated indicators of human action (such as population growth, and energy consumption) and indicators of changes in the so-called "Earth system" (such as the loss of biodiversity, and the concentration of CO_2 in the atmosphere). These graphs have been revealing a strong upward turn in the curves of each variable from the nineteenth century on, and an extraordinary growth, a true uprighting of the curves, from 1945 on (Steffen et al. 2004). By 2005, the renowned environmental historian John McNeill, working with other researchers, started using the term "Great Acceleration" to identify that excessive growth from mid-twentieth century on. Incidentally, the initial use of the term was inspired on Polanyi's "Great Transformation" (Steffen et al. 2015, 2).

Based on that set of indicators, a very comprehensive preliminary periodization has been suggested (Steffen et al. 2011, 849). The first stage of the Anthropocene would be the industrial era formation, between 1800 and 1945. The dissemination of fossil fuel in the economic production, especially coal and oil, stands for the great ecological differential to be considered. Fossil fuels allowed for a large expansion of productive forces, fomenting a significant expansion of urban-industrial structures and of the consumption of natural resources. Of course, this energy foundation would not be able to define the period. The use of fossil fuels cannot be divorced from the technological, economic, institutional, and cultural changes pointed out by so many analysts of modernity (Kumar 1986). But it is also true that such remarkable increment in the global population and economy would be unthinkable without the existence of such an abundant energy source, of relatively easy extraction and transport in the context of available technical means from the eighteenth century on.

The second stage of the Anthropocene, which starts around 1945 and is still going on, refers to the Great Acceleration. In spite of the rise of new energy sources – such as large hydroelectric dams and nuclear plants – fossil fuels remain the bedrock of the system. In fact, what happened was an enormous quantitative expansion. It is as if the winds generated by the industrial revolution – which were already a rupture with pre-industrial rhythms and volumes of production and consumption – had become a hurricane capable of radically multiplying the environmental consequences of human action. The Great Acceleration was historically generated in the context of the post-Second World War, when availability of cheap and abundant oil – associated to the ascension of Middle East producers – interacted synergically with the dissemination of innovative technologies that catalyzed an explosion in mass consumption (telephones, cars, TVs etc.). Afterwards, new technological waves continued to contribute to further expand large-scale consumption, as in the case of computers and cell phones. In order to have a more realistic vision of the political challenges involved in the Anthropocene issue, one must consider that such growth increased in an unprecedented way the expectations and consumption patterns of the working class, especially at the vanguard of the industrialization process. The Great Acceleration is mostly the world of social democracy, where the distribution (rather than redistribution) of wealth and opportunities allowed a notable increase in the base level of consumption, without strongly reducing the concentration of wealth and the super-consumption of the rich (Przeworski 1986). The counterpart comes mostly through a rise in the destructive pressure on the ecological resources of the planet.

However, some analysts argue in favor of a third stage in the history of the Anthropocene, possibly called "Self-Conscious Anthropocene." It would be the moment when international public opinion, recognizing the risks inherent to its new planetary insertion, would promote a conscious debate toward finding feasible ways for sustainability. Dissemination of new ethical assumptions, new institutions, new technologies, and new socioeconomic configurations would allow a conscious transition to that goal (Steffen et al. 2007). It is clear that a third stage of that kind represents basically a will or a possibility. In concrete terms, we are still living in full the Great Acceleration. The total volume of goods moved through the oceans, for instance, including oil, minerals, and grain, grew from 2.6 billion tons in 1970 to 9.8 billion tons in 2014 (UNCTAD 2015, 6). As much as it might be real, though, the third stage is being generated in numerous meetings, studies, and debates around the world, all seeking a sustainable future. It is also present in countless social conflicts and experiments against the increase of environmental destruction and for sustainable ways of life and work. The volume and quality of such social mobilization cannot be overlooked, but it is still too soon to assess its possible consequences in the future.

Common but differentiated Anthropocene: the case of Brazil

In any case, the terms of the political debate about the Anthropocene are far from a definition. One of the main problems is precisely the character of the concept – excessively integrative. Social scientists may present numerous questions, such as "who is the 'anthropos' in the Anthropocene?" Although the database on global environmental changes is quite robust, the issue cannot be approached in a homogeneous and merely quantitative fashion. The idea of defining the "limits" of humanity on the planet, for instance, is not easily feasible, as it needs to consider important differences in the cultural and perception patterns (Palsson et al. 2013). Differences must also be noted concerning the social forces that promoted the historical construction of the Anthropocene, and the social and environmental consequences of its advancement.

Jason Moore, for instance, deems it historically superficial to think that the transformation occurred due to the action of a generic "anthropos," of "humanity as an undifferentiated whole." Thus, his proposition of calling the new moment "Capitalocene," clearly indicating that the intense biophysical transformations observed in the last few centuries are directly related to "a historical era shaped by relations privileging the endless accumulation of capital" (Moore 2014, 2–5). About this kind of analyses, Depesh Chakrabarty argued, in a thought-provoking way, that changes in the planet's environment, such as climate change, go beyond the history of capitalism. There is a deeper issue, "a question of human collectivity, an us, pointing to a figure of the universal that escapes our capacity to experience the world. It is more like a universal that arises from a shared sense of a catastrophe." That "us" has an unprecedented materiality, as nobody can escape this new relationship to the Earth. There are no lifeboats, not even for the rich and privileged (Chakrabarty 2009, 221).

These two perspectives are not completely antagonistic. One can say, using a formula often adopted in the current climate of diplomacy, that the responsibility for the Anthropocene is common but differentiated. In the Anthropocene, one needs to simultaneously recognize the integrative dimension of the problem – the aggregated impact of human action became a geological force – and the differentiated dimension of real human life. No one can escape or be alien to the new period in the interaction between humanity and the planet. But not all contributed equitably to its historical construction, and not all equally experience its consequences, not in the same shape or degree. In a sense, the new environmental reality unifies the whole of humanity. In another, there is a visceral inequality.

The difference could be thought about on the basis of numerous categories, such as class, culture, production and consumption patterns, geographical realities etc. Another possibility is to work the Anthropocene theme in the context of existing different countries. In my opinion, the importance of the latter is adamant in the face of the contemporary political reality. It is true that the current international system is far from being a model of rationality and balance. When the focus is on the environment, for instance, one realizes that the historical events of the last few centuries generated countries with different sizes, and very different availability of natural wealth. Nonetheless, in spite of many sociological prophecies about the weakening of national states in the context of globalization's dynamics, most political and economic decisions about the future of such wealth are still taken on the basis of each country's political reality, or stem from diplomatic maneuvers that generated various associations and treaties at the macro-regional or international levels. None of these associations or treaties, though, has the capacity to neutralize the relevance of political struggles at the national level. It is not about ignoring numerous social actors that move in a transnational way, and their dynamics. But the fact is, the logic of national states, including in the economic sphere, has been showing a considerable resilience. Thus, the history and the future of the Anthropocene as a historical period also needs to be examined in the context of each country.

Brazil is very suitable to develop that kind of analysis. It clearly reveals that the entry of different countries in the Anthropocene cannot be seen homogenously. There are remarkable differences in terms of historical timing and mode of insertion. The analysis needs to encompass at least three types of links: (1) the level of national societies' participation in the distinctive production and consumption patterns of the Anthropocene as a historical period; (2) the contribution of each national society, especially its intellectuals and scientists, in the formulation of the kinds of knowledge and ideologies that constitute what could be called the "Anthropocene culture"; and, (3) the role of each national economy as supplier of human and natural resources to enable the insertion of other countries and regions into the Anthropocene's patterns of production and consumption.

It is true that the Brazilian case has a few singularities that distinguish itself from the median historical situation of modern countries. Particularly its territorial and ecological dimension put it in a preeminent place within the debate about the future of the planet's environment.

Among the five largest national territories, Brazil is the only one completely within the tropical and subtropical world, including about 60 percent of the Amazonian gargantuan water-forest complex. It is a continental territory, not only large (around 8.5 million square kilometers) but also very rich in renewable natural resources, a feature that acquires new significance within the imperative to decarbonize the global economy in the next decades. Other features are the great concentration of tropical forests (around 30 percent of the remaining tropical forest cover on the planet), biodiversity (between 10 percent and 20 percent of the global stock), and a huge network of rivers grouped in eight large hydrographic basins that, added to at least two large aquifers, concentrate around 12 percent of the global fresh water stock (Santos and Câmara 2002, 32; Dabene and Louaut 2013, 38). Furthermore, the territory has a strong incidence of solar rays, and a great capacity to reproduce biomass and store carbon. Concerning minerals for industrial use, Brazil is the largest world producer of niobium, the second largest of iron, manganese, and tantalum, and the third largest of bauxite and graphite. Even in the case of oil, offshore findings indicate the existence of very expressive reserves (Mérenne-Shoumaker 2015, 76). It is not surprising, thus, that Brazil's international image be marked by its territory, either for its ecological wealth or for the destruction of such wealth.

It is important to note the ecological design of that territory, also to better understand some aspects of its history. The kaleidoscope of ecosystems existing within it has been aggregated, to facilitate a synthetic view, in six large biomes. It is clear that such division may not be taken rigidly or absolutely. Each biome is a set of different ecosystems, even if they are considerably similar. There are also many transition areas, with mosaics of different kinds of vegetation. To proceed to regional analyses, it is necessary to focus in more detail on the ecosystems' features and combinations. In a historical analysis of the country as a whole, though, the biome classification is very revealing. When the Europeans arrived in the sixteenth century, the present Brazilian territory, in the North and in the Northeast-South coastal axis, had two magnificent and continuous tropical forest coverings: the Amazon Forest (originally 4 million square kilometers, considering just the portion that is now part of Brazil), and the Atlantic Forest (originally about 1.3 million square kilometers). In between these two forest areas, there were large extensions of different kinds of savanna, especially the Cerrado, with its retorted trees and acid soils (about 2 million square kilometers), the semiarid Caatinga, subject to periodic droughts (about 840,000 square kilometers), and the Pantanal, abundant in humid areas and wildlife (about 150,000 square kilometers). In the extreme South, finally, there is the great plains of the Pampa (about 176,000 square kilometers). Also notable is the continuous Atlantic coast of more than 8,000 kilometers of sandbanks, mangroves, and other coastal formations (IBGE 2004; Scarano 2012).

How can a country with such territorial expression be thought of in the context of the Anthropocene's three stages mentioned above? The environmental dimension, of course, cannot be exclusive. The cultural and socioeconomic dimensions are also essential, because a country's territory does not exist by itself – it is always the result of complex and diversified interactions of social and natural movements throughout time. The sheer velocity of socioeconomic and ecological transformations experienced by Brazil in the last decades, in the context of the Great Acceleration, is also notable. An encompassing, though synthetic history of the country, based on the three stages of the Anthropocene, might demonstrate the relevance of thinking national histories from a planetary perspective.

Brazil in the first stage of the Anthropocene

The territory's size might be very deceptive to discuss the history of Brazil's economy and society. A point to be noted is that, considering the territory's formal magnitude, the population was relatively small, at least until the mid-twentieth century. It was around 4 million inhabitants by 1822, year of the independence from Portugal, and reached 17 million in 1900. By then, as a comparison, the United States already had more than 76 million inhabitants. It is true that the numbers in Brazil did not include most of the indigenous populations, who, even as inhabitants of the territory that the modern world recognized as "Brazil," could not be considered a part of that political entity. It is clear that many indigenous societies had been subjugated throughout the centuries, and forced to insert themselves in a subordinate role within the areas of Euro-descendant dominance. But other groups inhabited vast territories with virtually no Euro-descendant presence. It is impressive to think that, even today, in the early twenty-first century, there are tribes in deep Amazon that were not subjugated, and who do not even know of the existence of the country (Funai 2015).

To understand the phenomenon it is necessary to note a few peculiarities of the territorial formation in Brazil, in the context of new countries born from the demise of European colonial empires in America: (1) differently from Hispano-American countries, Brazil managed to insert in only one national political entity all regions composing Portuguese America; (2) differently from the US, Brazil did not need a great horizontal expansion, through negotiation or military conquest, to obtain a large territory. It inherited politically, at least from a formal point of view, the totality of Portuguese America, which already was an area almost the size of the current country. There it was a case of precocious territorial gigantism, even if that gigantism was only virtual or pretended, existing basically in imperfect maps, and diplomatic treaties.

In nineteenth-century Brazil, as well as in the three previous centuries of Portuguese colonialism, the areas under Euro-descendant dominance formed an "archipelago" of spots of territorial occupation, controlled by local elites and based on the exploration of various natural resources. These spots were concentrated on the Atlantic Forest, close to the coast, with much less dense occupation in the Caatinga, the Cerrado, the Pantanal, the Pampa, or along the Amazon River, and were surrounded by vast extensions of land that came to be called *sertões* (from the augmentative of *deserto*, or desert: *desertões* became just *sertões*, often translated into English as the backlands). These areas had great density of flora and fauna, where indigenous populations lived a very autonomous way of life, sometimes interacting with *quilombos* (the villages founded by African slaves who escaped forced labor in search of free lands), or even with some Euro-descendants who would choose to live in the *sertões*, far from oligarchical dominance.

Although the spots under Euro-descendant control were distributed throughout the continental territory, they must not be seen as isolated islands. There were, to start with, all the common cultural features, such as Catholicism and the Portuguese language. On the other hand, depending on the reality of each region, exchange flows existed in different intensity levels, either of products, people, or cultural practices. The social and cultural standards inherited from the Portuguese ancient regime, as well as various aspects of its material civilization, were reproduced and adapted in those new ecological contexts. More often than not, such practices were mixed to other techniques and cultural practices of indigenous or African origin. Thus, the country that we call Brazil today was molded by a set of culturally hybrid societies, dominated by regional oligarchies and by the logic of slavery. The construction and consolidation of a national state in the nineteenth century, under a monarchic regime, was able to include in one political frame all that diversity of occupied lands and *sertões*. Such political engineering was very difficult to undertake, as the central government had to be accepted by the many regional elites that kept the order in their dominions. The central political obsession was to keep the whole territory together, and to promote its gradual "self-colonization," incorporating the great backlands and their populations to the market and to "progress."

How to assess the contribution of such a social and geographic picture to the European "great transformation" that started the transition to the Anthropocene? It is true that the search for natural riches was one of the central elements of the European colonization of America. A list presented by the Spanish Crown to the navigator Vicente Yáñez Pinzón in 1501, pointing out what he should search for in his second journey through the Amazon River, is very suggestive: "Gold as well as silver, copper, or any other metal, pearls and precious stones, drugs, spices, and any other things from animals, fishes, birds, trees and herbs, and other things of any nature or quality" (Ribeiro and Moreira Neto 1993, 75). In that historic context, it must be noted, there was a clear hierarchy of economic wishes: precious minerals, organic materials that could generate profit, and, last but not least, anything surprising and valuable that could exist in the continent still almost unknown. In the case of Brazil, gold and diamond extraction only became a reality in the third century after the conquest. The first stage of the occupation process, thus, had to rely on the use of organic materials.

In the 1970s, Eduardo Galeano disseminated the powerful image of "open veins" to describe the colonial formation of Latin America. The predatory exploitation, and the transfer abroad of riches extracted from the natural world, would be the main cause of a history of poverty and political indigence in the region (Galeano 1997). There is some truth in that perception. Undoubtedly, there was a transfer of riches, and the colonial control of Latin America was relevant for Europe's socioeconomic future. Such relevance may be thought of in different levels. The region's colonial experience, for instance, was relevant for the construction of the Western scientific model (Safier 2008). The flow of material riches was also real, although it is not easy to point out its historical connections to the later industrialization. In the case of Brazil, the establishment of plantations and sugarcane mills, from the sixteenth century on, was fundamental to the modern invention of the so-called agricultural commodities (production directed to the international market). The territory's tropical ecology allowed the offering of "exotic" products, which generated considerable profits in European markets. Even though, many authors see the importance of colonial activities to the future of the European economy as limited because the "great transformation" would be more connected to endogenous factors, such as internal savings and technological innovation. One of the best-established connections, from an empirical point of view, would be the extraction of Brazilian gold from the eighteenth century on. The intense commercial relationship between England and Portugal transferred a large share of that gold to English bankers, leveraging local credit for industrial activities (Villela 2011, 9–13).

In any case, Brazil's colonial formation cannot be thought of exclusively on the basis of riches exports. The spots of occupation mentioned above had their own social logic, particularly the reproduction of local elites' hierarchic power. Moreover, many of the local economies were not aiming at foreign markets. From the perspective of environmental history, it is worth recalling the difference suggested by Wallerstein (1989) between "preciosities" and "bulk commodities." In the world of pre-fossil fuels, where transportation by sailing oceans imposed severe limits to quantity and weight of materials, the transfer of natural riches from America centered in products that had high exchange values in relatively small quantities (such as sugar, gold, fine woods etc.). Only from the nineteenth century, with the inception of steam ships and railways (and later on, in the twentieth century, huge cargo ships fueled by heavy oil), international trade started to promote an intense flow of materials, including bulk commodities such as oil and iron, which penetrate the very metabolism of economies.

Just a few formative areas in Brazil, often not too far from the coast, were involved in the production of preciosities for export (such as sugar, tobacco, gold, and diamonds). The riches produced in such areas soon materialized in sumptuous architecture in rural properties as well as urban spaces. In other areas, though, the occupation process was mostly based on the production for the domestic market, especially meats, dairy products, salt, manioc, beans, spirits, artisanal goods, and construction timbers.

The mode of occupation of the land generated by that model – established on a set of geographical, technological, and cultural factors that I discussed in detail in another paper (Pádua 2010) – was essentially devastating from an environmental point of view. Production methods were generally careless and extensive, based on a parasitic relationship with the natural world. Tropical forests, and other local ecosystems, were perceived as vast green oceans, always open to an ongoing occupation. The ubiquity of slash and burn clearings was the most obvious symbol of this mentality. Instead of mulching the soil in order to preserve its fertility, the choice was made to progressively burn new forest areas, since the richness of the resulting ashes would guarantee a few years of good harvests, after which the soil would decline in productivity, overtaken by weeds and ants. The forest, therefore, was not just a hindrance, but also a source of biomass to be combusted. The relative ease with which land was obtained encouraged careless use, and the subsequent near abandonment of the degraded areas. As a consequence, the push continued in the direction of unexplored forests. However, and in spite of the predatory treatment of conquered areas, the modest size of the economy as a whole allowed a limited aggregated environmental impact. Vast areas of the territory remained covered by its native vegetation.

Two expansion processes were especially important in the connection of Brazil to the first stage of the Anthropocene. The first process happened from the 1820s on, when the dense Atlantic Forest covering the valley of the Paraíba do Sul River, midway between the cities of Rio de Janeiro and São Paulo, was cleared to make room for coffee plantations. The small mountain slopes surrounding the valley were bared with axes and fire, and the plantations were designed in vertical lines uphill to facilitate overseeing the strenuous slave work. The result was soil erosion and short-lived plantations, which were soon abandoned in favor of new clearings. By the end of the nineteenth century, the general clearing of the forests contributed to the economic collapse of the region's farms, also impacting the abolition of slavery, the proclamation of the Republic, and the migration of coffee plantations to new frontiers, mainly to the West of the state of São Paulo (Dean 1995, Ch. 8; Brannstrom 2000).

Coffee may be understood as a preciosity. But its relevance to the urban-industrial way of life is considerable. It is one of modernity's "soft drugs," an energy-booster to survive its intense rhythm. In the words of Topik and Wells (2012, 222–224), the Brazilian production "not only largely satisfied growing world demand. Brazilians stimulated and transformed the place of coffee in overseas cafés and homes." Indeed, the country responded for about 80 percent of the coffee production expansion in the nineteenth century. In spite of rudimentary production technologies, the unbridled burning of copious forests was a great differential, along with the brutal exploitation of slaves. Even after the abolition of slavery in 1888, the availability of cheap and abundant labor continued through former slaves or poor immigrants from Europe and Japan.

The other expansion process was based on extractive economy, not agriculture: the rubber boom in the Amazon at the turn of the nineteenth to the twentieth centuries. A new system of socio-environmental interactions arose in the great Northern forest when the auto and bicycle industries outside of Brazil started producing tires from the latex of rubber trees (*Hevea brasiliensis*). The new system connected extensive areas at the heart of the forest, separated in private domains called *seringais*, to leading sectors of global capitalism. This flow occurred through a chain of exchanges – actually, a chain of debts – that went through international trade firms, international navigation companies, local trade agents who went up and down rivers exchanging rubber for consumer goods, *seringais* owners, their employees and, at the end of the chain, the *seringueiros*, workers extracting the latex from the rubber trees, spread throughout the forest without any legal protection, and highly exploited by their bosses, who paid very little for the rubber and charged dearly for anything workers needed to survive in the jungle (Weinstein 1983).

From a deforestation perspective, though, the whole process generated little damage. Rubber extraction did not require clearing the forest. Quite the contrary: to be reproduced for a reasonable time, the daily extraction of latex required maintenance not only of the rubber trees but also of their surroundings, which provided ecological support to them. In spite of rapid growth of a few cities, such as Manaus and Belém, followed by equally rapid decadence of exports from 1920 on, the environmental consequences were further diluted. By the early 1970s, when an intense deforestation cycle began, 99 percent of the Brazilian Amazon Forest original coverage remained untouched (Pádua 1997).

In the case of rubber, we are facing a clear bulk commodity. It is quite obvious, though, that its supply based on extraction from native trees spread in a forest would only be possible in a very early moment of the auto industry. Its conversion into agriculture was the only option, considering the industry's future growth, and the demand for rubber. It is true that there were ecological problems concerning planting rubber trees in the Amazon. Its introduction as exotic in the Asian tropics freed it from potential native plagues. Even though accommodation and parasitism of local elites deterred any continuous research efforts to do it in the Brazilian territory at the time (Dean 1987).

Thus, all things considered, it can be stated that Brazil had a quite modest presence in the first stage of the Anthropocene. As a supplier of raw materials to the industrialization of other countries, its importance was limited and occasional. Only the coffee production earned a longer permanence. The production of cotton, for example, that existed since pre-colonial times for local use, gained some export momentum in the last decades of the eighteenth century. Brazil could have become a significant supplier to the European factories. But that momentum proved short-lived given the inability to compete with the dominance of North American production in the nineteenth century (Dean 1995, 121). On the other hand, as a participant of the Anthropocene's production and consumption standards – of the world built on fossil fuels – as well as in the universe of its cultural trends and ideologies, Brazil's presence was very limited. Indeed, up to the first decades of the twentieth century, the country has been very distant from the vanguard of the contemporary process of industrialization and urbanization. It was a very unequal country, where a very small fraction of the society lived according to European consumption standards, and hovered over a poor and illiterate population, essentially rural, that lacked rights and opportunities. At the regional level, these upper classes based their power on farm production aimed at the internal market, such as meat and sugar, or on exports of rubber, cotton, tobacco, cocoa, and coffee (the source of the larger fortunes). Manufactured products were almost all imported. Such economic reality, though, was satisfactory to the dominant elites, as it reproduced a social order that cast them as masters.

It is important to mention cultural factors subjacent to that reality. According to some historians, in the early nineteenth century there was a substantial volume of financial resources in Rio de Janeiro, especially due to the slave traffic. It would have been theoretically possible to invest in manufactures and other activities closer to capitalist modernity. However, being imbued with the rural aristocratic culture as the utmost expression of social status, the mercantile elite invested instead in large slave properties, especially to produce coffee (Fragoso and Florentino 2001). Intellectual and corporate voices in favor of industrialization, which arose at the turn of the nineteenth to the twentieth century, found stern resistance in the dominant ideology of Brazil's "agricultural vocation" (Luz 1975). Even important abolitionist, reformist, and nationalist authors insisted in not prioritizing industrialization. The counter-image of European industrial cities, dirty and contaminated, was used to that end. They believed agriculture's modernization, with smaller properties and free labor, was the route to the country's progress (Pádua 2002, Ch. 5).

In any case, from an environmental point of view, the low availability of mineral coal reserves would have been a hurdle to industrialization. Energy's basic source was wood, still responding for 73 percent of primary energy consumption in Brazil even in 1941 (Wilberg 1974), and most essential materials of the economy were organic. Only from the 1920s on the national production of mineral coal showed some growth, based on findings in the South of the country. Hydroelectricity also started growing slowly, along with imports of fossil fuels. In 1915, for instance, Brazil consumed only 0.14 percent of the global mineral coal production, and 0.6 percent of the global oil production (Leite 2014, 58–61).

Thus, until the first decades of the twentieth century Brazil was essentially an extension of the territorial model consolidated in the nineteenth century. It is true that some significant changes occurred. Monarchy ended in 1889, making room for a federalist Republic that formalized local oligarchies' power. Urban spaces started to grow steadily, in spite of a clear dominance of two Southeastern cities: Rio de Janeiro and São Paulo. By 1940, those were the only cities with a population of around 1.5 million inhabitants, while major regional capitals still had only 200–300,000 residents. All along, though, the country was brewing the conditions that would result in a large growth of population, industry, and the economy in general from the mid-twentieth century on. As the next section will show, the growth of Brazil's presence in the world happened in the context of the Great Acceleration. Today, to think of its future in a profound way, it is fundamental to approach it in the context of the Anthropocene's dilemmas.

Brazil in the Great Acceleration, and beyond

Between 1900 and 2000, when it reached 170 million inhabitants, the Brazilian population grew tenfold. But the great transformation in the country's social and environmental reality happened from 1945 on. In 1950, Brazil had 51.9 million inhabitants, with a life expectancy of 43 years, and an illiteracy rate of 50.6 percent. Urbanization then was 36.2 percent. In 2014, the same indicators reveal the sheer velocity of the transformation: 199 million inhabitants, with a life expectancy of 73.4, and an illiteracy rate of 9.02 percent; urbanization of 84.3 percent (IBGE 2001, 2011).

The foundations for such change were in the making from the first decades of the twentieth century. The abolition of slavery accelerated immigration. Between 1884 and 1940, around 4.7 million people, especially Italians, Portuguese, Spaniards, Germans, and Japanese entered the country (Alvim 2001). Many of them were rural workers, and were directed to the coffee's new frontiers, or established themselves as small farmers in the Atlantic Forest regions in the South of the country. But there was also an immigration directed to cities, including European investors in search of industrial opportunities. At the same time, the logic of import substitution, supported by part of the revenues generated by the coffee exports, had been promoting a higher growth in manufacture in the first decades of the twentieth century, especially in the food and garment industries.

Cultural and political aspects must also be considered. New urban middle classes desired modernization. Intellectual and scientific life grew to be more vivacious, with vibrant artistic innovations and political debates that exposed the lethargy of rural oligarchies. That perception grew stronger with the coffee exports' crisis in the early 1930s, related to the 1929 crash of the New York Stock Exchange. The crash, as well as other previous episodes of coffee super production and the resulting drop in the international price of the commodity, highlighted the vulnerability of an economy dependent on export monocultures. The 1930 political revolution, with a strong military participation, generated a stronger and more centralized action of the national state in favor of urban-industrial growth. Progress was relatively slow, though. In spite of large iron reserves, for instance, steel industry based on blast furnaces and fueled by mineral coal only came to be in 1946 through a state company, the National Steel Mill Company (the CSN, in the Portuguese acronym, privatized in 1993).

From the end of the Second World War, even considering conjuncture oscillations, including moments of high inflation or economic stagnation, Brazil's history was marked by a growing connection to the Great Acceleration world. In the international context of the post-Second World War, the increased availability of foreign credit turned many statesmen toward the dream of developmentalism. President Juscelino Kubitschek, who governed Brazil from 1955 to 1960, had a slogan that puts in a nutshell the ideological seduction of the time: to advance fifty years in five! Brazil had natural resources, a growing consumer market (especially in cities), abundant and cheap labor, and a continental territory to be explored. The presence of transnational corporations, associated to State initiative and national companies, catalyzed an intense growing process. In the context of the military regime, from 1964 to 1984, when technocratic authoritarianism smothered previously existing political conflicts, the process grew substantially. Even after the dictatorship, though, in spite of the reduced scale, Brazil has been living significant waves of growth. In the world of the Great Acceleration, the Brazilian economy had average growth rate periods as high as 7 percent (1942–1962), 10.9 percent (1967–1973) and 3.52 percent (2003–2013) (Droulers 2001, 254; Gomes and Cruz 2015, 41).

The entrance of Brazil in the second phase of the Anthropocene, thus, is happening in the three dimensions previously mentioned. The dominance of the Anthropocene culture – the ideology of limitless growth, and urban-industrial transformation – is highly evident both in the right and the left of the political spectrum. Democratization of political institutions after the dictatorship did not essentially change the development model adopted from the Second World War on. In the last fifteen years, notably between 2003 and 2011, under the leadership of President Lula da Silva, of the Workers' Party (PT, in the Portuguese acronym), it is true that the country watched an acceleration of public policies aimed at reducing poverty and increasing access to consumer goods. Such initiatives took place as a result of social and union movements with roots in the resistance against military rule. However, in a broader historical perspective, one can understand that recent dynamic as an update, in the context of Brazilian political culture, of the classic social-democratic model that marked the history of the Great Acceleration. In other words, it was an expansion of public expenses in the direction of revenue distribution, more than redistribution. There was indeed a democratization of opportunities and public services. But the mass consumption increased without a reduction of the super-consumption of the wealthy. Such a model was only possible, in its materiality, through a strong increase in the extraction of natural resources, and Brazil was benefitted by a favorable wave in commodities prices.

In any case, differently from the Anthropocene's first phase, the insertion of Brazil in the Great Acceleration does not limit itself to supplying primary products for the growth of other regions on the planet. The change in its internal standards of production and consumption was also notable. Brazil never ceased to be an important exporter of primary products. However, population and urbanization growth generated a very attractive internal market, also for transnational corporations. Brazilian production of iron ore, for instance, grew from 9 million to 400 million tons between 1950 and 2014 (of which, currently, 344 million tons are exported). The steel production, on the other hand, also jumped from 788,000 to 33.9 million tons between 1950 and 2014 (for an internal consumption of around 28.5 million tons). The number of cars also increased notably, going from 650,000 in 1960 to 47.9 million in 2014 (IBRAM 2015; Instituto Aço Brasil 2015).

But there are clear signs that the aspect related to providing natural resources has again gained special relevance. The growth of international consumption of semi-manufactured and basic products exports, especially in Asia, has shown much more constancy and safety than the oscillations observed in the internal market or in the exports of manufactured products. The latter represented 57.3 percent of the country's total exports in 1994 but dropped to just 36 percent in 2014. As for the market for products such as iron ore, oil, gold, niobium, cellulose, ethanol, soybeans, and meats continue to grow even after the current crisis of the international economy (Gomes and Cruz 2015, 12–14; IBRAM 2015, 15–16). In other words, the advantage of a large territory and its ecological riches has been gaining more relevance, making room for what many experts call a move back to a Brazilian economy focused on primary goods. There is a great potential for this, as the territory still has large spaces with low population density and economic occupation, establishing considerable room for horizontal expansion of primary production. Even if, in the twenty-first century one cannot propose anymore, as in the mid-twentieth century, a limitless advancement toward the Amazonian Forest, there are, however, vast regions covered by other biomes that are much less valued by national and international public opinion.

The great irony, thus, is that the image of "open veins," which I considered inadequate to understand the colonial and post-colonial formation of Brazil, may become a reality nowadays. Moreover, the intensification of the primary exporter role may amplify a number of imbalances and social and environmental conflicts that can be observed today. Indeed, the axis of the analysis about Brazil in the context of the Anthropocene is not in the economy's numbers in themselves, but in what they represent in terms of biophysical materiality of the territory, and on national society's ecological basis. The numbers of exports of ore and steel do not immediately reveal the multiplication of mines and steel mills in different regions of the country, nor the territorial conflicts, pollution, and disasters related to them. It is enough to recall the colossal disaster in late 2015, when the dam holding rejects from the mining company Samarco, in the state of Minas Gerais, broke and devastated the whole valley of the Doce River. The same can be said of the export numbers of soybean, corn, and cotton, which do not immediately reveal the accelerated loss of native ecosystems within the Cerrado biome.

Since its inception, Brazil's participation in the Great Acceleration is characterized by important imbalances. The sheer speed of so-called "modernization" occurred in the context of a society with a large percentage of poor and vulnerable population. From the start, institutions and governmental entities were still held by the traditional elitism inherited from the country's formation. The great rural exodus mostly caused by the industrialization of agriculture, for instance, was not accompanied by policies directed toward welcoming to the urban world, in a minimally decent way, the mass of people dislocated from the fields. The logical result was the increase in informal communities (the *favelas*) on hillsides, mangroves and other devalued spaces in the cities. Furthermore, the *favelas'* urban chaos became the ideal locus for drug trafficking and criminality.

One can imagine the social and environmental consequences of some of the processes that took place since the 1950s: (a) expansion and remodeling of urban landscapes, with an increase in pollution, and the destruction of traditional architectonic complexes; (b) expansion of infrastructure, particularly highways and hydroelectric plants; (c) expansion of industrial areas and the storage of contaminants; (d) the opening of new frontiers for ranching and farming in regions previously covered with tropical forests, and other native ecosystems, and occupied by traditional (often indigenous) populations with low demographic density; (e) conversion of areas of longstanding traditional agricultural practices, with established rural populations, into large-scale operations based on the use of machines and agrochemicals.

One of the most notable features of the Great Acceleration in Brazil was precisely the aggressive opening of frontiers of economic occupation in the vast *sertões* that were a part of the country's territorial history. The "march to the West," one of the great goals of the national State since the 1930s, gained a powerful stimulus when Brasília, the new capital in the heart of the Cerrado, was inaugurated in 1960. That huge biome was traditionally considered unsuitable for agriculture, especially because of its soil acidity. In the last decades of the twentieth century, though, systematic agronomic research managed to turn the region into a gigantic frontier for agribusiness. It is highly revealing to check the Brazilian Institute for Geography and Statistics' anthropized areas map (Figures 1.1 and 1.2) in different ecological regions of the Brazilian territory in 1960 and 2000 (Torre 2010).

True, the concept of "anthropized area" is debatable, as it is based on modern patterns for demographic density, landscape changes, and socioeconomic occupation. Thus, the map presents as "empty" of economic life many areas, especially in the North and the West, which were occupied by indigenous populations and traditional communities that handled the ecosystems in a much lighter way, with much lower demographic density. In any case, precisely through the conceptual bias mentioned above, the map gives a good indication of the presence, on the territory, of a social dynamics articulated with the market economy, and the urban–industrial world. One can thus see that such "anthropized" areas in 1960 were still very limited, and concentrated on the

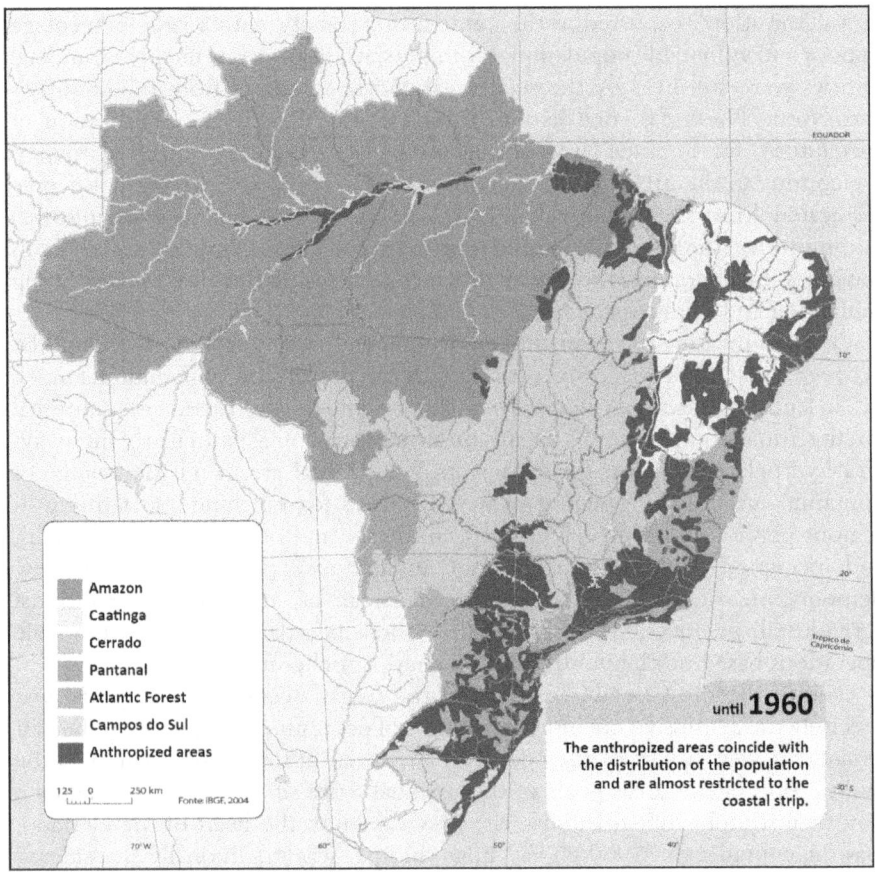

Figure 1.1 Anthropized areas until 1960

By William Torre. Based on the maps of the Brazilian Institute for Geography and Statistics.

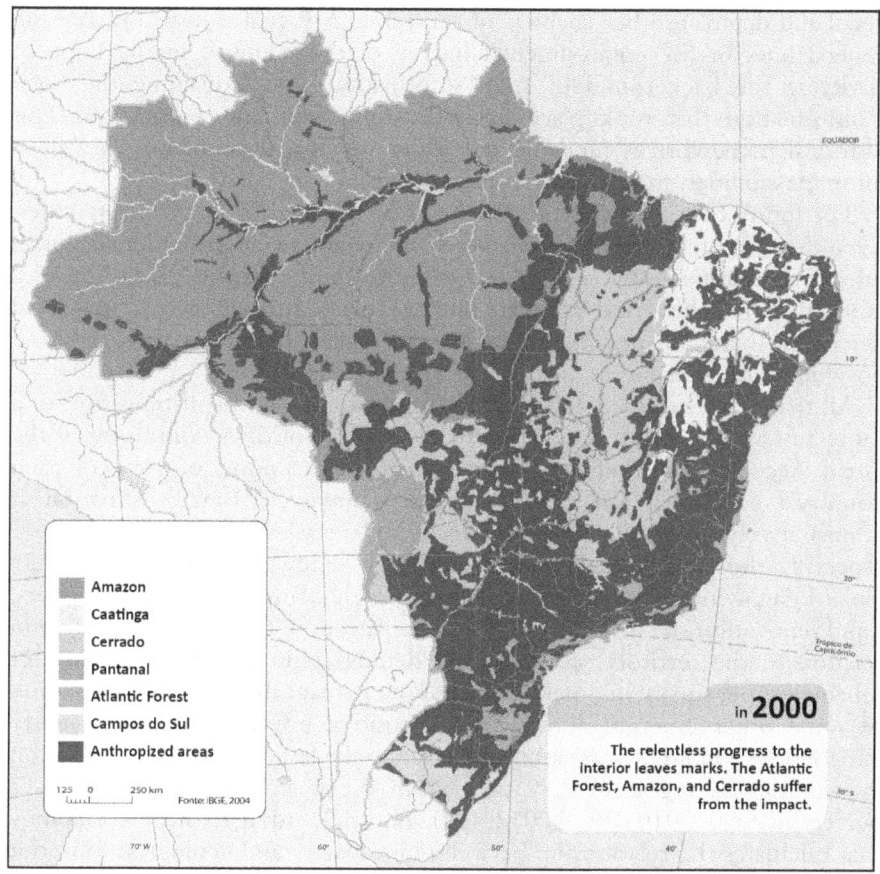

Figure 1.2 Anthropized areas in 2000

By William Torre. Based on the maps of the Brazilian Institute for Geography and Statistics.

Atlantic Forest region, close to the coast, or around the Amazon River. In the following decades, the occupation of such territories was impressive. The Atlantic Forest was almost completely converted, being reduced to around 13 percent of its original covering (if all fragments are considered). The Cerrado lost around 50 percent of its original covering, and the Amazon Forest, around 20 percent (Pádua 2015). In contrast, grain production, especially soybeans, increased from 39.4 million to 206.34 million tons between 1975 and 2014/15. It is true that such growth is becoming less horizontal and more vertical or technological, as the cultivated area grew in a smaller proportion in the same period, from 32.9 million to 57.9 million hectares (Contini et al. 2010; CONAB 2015). In parallel, Brazil became one of the largest consumers of pesticides in the world, using close to 1 million tons per year (Londres 2011).

I do not intend to develop here an analysis of many socio-environmental problems and dilemmas in contemporary Brazil, as those will be examined in detail and depth in other chapters of this book. My goal is just to situate the general lines of the environmental history of the country, in the hope of clarifying the background of Brazil's formation, and explaining the huge transformation that took place from the mid-twentieth century on. This historical framework is fundamental to discuss current problems, as well as future possibilities, from a broader perspective.

The future of Anthropocene, obviously, is wrapped in many uncertainties. Even the idea of a third phase, more self-aware and ecologically intelligent, is just a possibility. In any case, it is possible to think about the situation of a country in this historical context, but not in an isolated way. The global dynamics, including new technologies, will greatly influence the limits and possibilities of Brazil in the next decades.

All that was examined above is not promising in terms of transition to a more sustainable future, either in Brazil or the world. A continuity of the Great Acceleration model is dominant. From the point of view of each country's ecological and territorial basis, however, Brazil's situation is comparatively favorable. The territory is still very rich in ecological assets, especially those needed for an expansion of renewable sources of energy (solar light, winds, large reserves of fresh water, complex network of rivers, rapid renovation of biomass etc.). The population, in turn, is not excessive in relation to the territory, composing a demographic density of around 24 inhabitants per square kilometer, much more manageable than the population of some other emerging countries. A positive point is that the country experienced a rapid demographic transition in the last decades, the population growth rate falling from 2.99 percent in the period 1950–1960 to 1.17 percent in 2000–2010 (IBGE 2012, 134). Thus, according to the methodology that calculates the relationship between biocapacity and ecological footprint in different countries, Brazil is one of the few countries that still presents a biocapacity three times larger than its footprint. However, such biocapacity fell more than 50 percent between 1960 and 2011, as a consequence of intense devastation of native biomes (GFN 2013). Furthermore, one needs to consider that aggregated calculations of environmental consumption are very misleading in a society as unequal as the Brazilian society is in terms of wealth distribution.

From the point of view of environmental debate, though, one can say that Brazil is in a relatively good position. The territory's own ecological expression favors the presence of a strong debate concerning its future. In a more direct way, local consequences of rapid urban–industrial advance and the agribusiness' territorial expansion generated numerous problems and conflicts that had repercussions in national and international public opinion. Some of those conflicts – as in the case of the 1988 murder of Chico Mendes, a rubber tapper unionist, and a leader of the movement against the destruction of the Western Amazon Forest – became milestones in the global emergence of the so-called "ecology of the poor" (Alier 2003). The advance of the Amazon Forest destruction turned the country into a focus of the international ecological debate, especially with Rio de Janeiro hosting the 1992 United Nations Conference on Environment and Development.

There are indicators, thus, pointing to a considerable degree of environmental concern in Brazilian society. An international public opinion research made by the Pew Research Center in 2013, for example, shows Brazilian society in the second higher position (76 percent) in affirming that global climate change is "a major threat." As a comparison, Germany was at the level of 56 percent and Britain, 48 percent (Kopicki 2014). Nobody knows how many environmental associations are currently active in Brazil. In the National Roster of Environmentalist Entities, created and regularly updated by the Ministry of the Environment, there are 654 associations. The real number, though, considering all the small local groups, may be much larger. Beyond organized groups, environmental worries penetrated various spaces in Brazilian social and political life, including governmental agencies from the municipal to the federal levels. From the 1990s on, interacting with large international organizations recently arrived to the country, Brazilian environmentalism gained professionalism, and the capacity to produce and analyze data. In the context of academic research, the concern with the subject also grew considerably (Alonso and Maciel 2010; Pádua 2012).

In spite of progress on the degrees of knowledge and concern, concrete changes were too small in relation to the dimension of the environmental crisis in the country. Some developments occurred in the last decades, such as in legislation, sanitation and the increase of renewable sources of energy (especially wind power). The most successful policy, undoubtedly, was the reduction in yearly deforestation rate in the Amazon, which showed that a certain degree of real inversion of past trends is possible. Between 2004 and 2014, indeed, there was an 82 percent reduction in the Amazon yearly deforestation rate. Moreover, between 2003 and 2009, 73 percent of the total preserved areas created in the world were in Brazil. By the end of 2010, 43.9 percent of the Amazon Forest was protected by different types of reserves. The conservation policy for the Amazon, though, does not mean environmental policies in the country are strengthened. The Cerrado, for instance, is being destroyed, with little reaction from governments or public opinion. Facing the deforestation problem in the Amazon results mainly from concern with the country's international image, and with the search for political and diplomatic benefits. It also reflects a certain change in the geopolitical perception of the region by the national political elites. Indeed, there is more and more awareness about the Amazon forest conservation for future uses, including more sophisticated economic practices, now perceived as better than immediate destruction (Pádua 2015).

In any case, and in spite of positive aspects, the social and environmental situation in Brazil remains marked by very negative realities. Among other factors, one can mention increasing contamination in rural and urban spaces, *favelization* of cities, shortages in sanitation (especially in waste treatment), recurring urban-industrial accidents, chaos in the transportation system, and arrogance, corruption and violence of the elites. Despite a well-developed environmentalism, as seen, the report *How Many More?*, published in April 2015 by the British organization Global Witness, puts Brazil as the country with the largest number of environmentalists murdered in 2014, with 29 deaths. Globally, there were 116 murders, 87 of those in Latin America.

The short-term benefits of exporting primary resources may increase the pressure on the territory, and intensify the Great Acceleration model and its conflicts, reducing space for the search for options and the advance of social and environmental transformations required for a true transition toward sustainability. All things considered, beyond complex international interactions, the future will heavily depend on internal political struggles about what kind of country Brazilian society wants to build.

References

Alier, J.M. (2003) *The Environmentalism of the Poor: A Study of Ecological Conflicts and Valuation*, Edward Elgar Publishing, Cheltenham.

Alonso, A. and Maciel, D. (2010) "From Protest to Professionalization: Brazilian Environmental Activism After Rio-1992," *The Journal of Environment and Development* 19 (3): 300–317.

Alvim, Z. (2001) "Imigrantes: a Vida Privada dos Pobres do Campo," in Sevcenko, N. (ed.), *História da Vida Privada no Brasil*, vol. 3 Companhia das Letras, São Paulo.

Brannstrom, C. (2000) "Coffee Labor Regimes and Deforestation on a Brazilian Frontier, 1915–1965," *Economic Geography* 76 (4).

Chakrabarty, D. (2009) "The Climate of History: Four Theses," *Critical Inquiry* 35 (2).

CONAB – Companhia Brasileira de Abastecimento (2015) *Acompanhamento da Safra Brasileira – Grãos* Conab, Brasília.

Contini, E., Gasques, J., Alves, E. and Bastos, E. (2010) "Dinamismo da agricultura brasileira," *Revista de Política Agrícola*, Ano XIX n. 42.

Dabène, O. and Louault, F. (2013) *Atlas du Brésil*, Autrement, Paris.

Dean, W. (1987) *Brazil and the Struggle for Rubber*, Cambridge University Press, Cambridge.

——(1995) *With Broadax and Firebrand: The Destruction of the Brazilian Atlantic Forest*, University of California Press, Berkeley.

Droulers, M. (2001) *Brésil: une géohistoire*, PUF, Paris.

Fragoso, J. and Florentino, M. (2001) *O arcaísmo como projeto*, Civilização Brasileira, Rio de Janeiro.

Funai – Fundação Nacional do Índio (2015) Povos indígenas isolados e de recente contato (www.funai.gov.br/index.php/nossas-acoes/povos-indigenas-isolados-e-de-recente-contato), accessed January 22, 2016.

Galeano, E. (1997) *Open Veins of Latin America: Five Centuries of the Pillage of a Continent*, Monthly Review Press, New York.

GFN – Global Footprint Network (2013) *The National Footprint Accounts* GFN, Oakland.

Gomes, G. and Cruz, C.A. (2015) *Vinte anos de economia Brasileira – 1995/2014*, Centro de Altos Estudos Brasil Século XXI, Brasília.

IBGE – Instituto Brasileiro de Geografia e Estatística (2001) *Tendências Demográficas*, IBGE, Rio de Janeiro.

——(2004) *Mapa de Biomas do Brasil*, IBGE, Brasília.

——(2011) *Sinopse do Censo Demográfico*, IBGE, Rio de Janeiro.

——(2012) *Indicadores de Desenvolvimento Sustentável – Brasil 2012*, IBGE, Rio de Janeiro.

IBRAM – Instituto Brasileiro de Mineração (2015) *Informações sobre a Economia Mineral Brasileira 2015*, IBRAM, Rio de Janeiro.

IAB – Instituto Aço Brasil (2015) *Aço e economia* IAB, Rio de Janeiro.

Kopicki, A. (2014) "Is Global Warming Real? Most Americans Say Yes," *The New York Times*, June 1.

Kumar, K. (1986) *Prophecy and Progress: The Sociology of Industrial and Post-Industrial Society*, Penguin, Harmondsworth.

Leite, A.D. (2014) *A energia do Brasil*, Campus, Rio de Janeiro.

Londres, F. (2011) *Agrotóxicos no Brasil* ASPTA, Rio de Janeiro.

Luz, N. (1975) *A luta pela industrialização do Brasil*, Alfa-Ômega, São Paulo.

Mérenne-Shoumaker, B. (2015) *Atlas mondial des matières premières*, Autrement, Paris.

Moore, J. (2014) "The Capitalocene, Part 1" (www.jasonwmoore.com), accessed February 15, 2016.

Pádua, J.A. (1997) "Biosphere, History and Conjuncture in the Analysis of the Amazon Problem," in Redclift, M. and Woodgate, G. (eds), *The International Handbook of Environmental Sociology*, Edward Elgar, London.

——(2002) *Um sopro de destruição: pensamento político e crítica ambiental no Brasil escravista (1786-1888)* Jorge Zahar Editor, Rio de Janeiro.

——(2010) "European Colonialism and Tropical Forest Destruction in Brazil: Environment Beyond Economic History," in McNeill, J., Pádua, J. and Rangarajan, M. (eds), *Environmental History – As If Nature Existed*, Oxford University Press, New Delhi.

——(2012) "Environmentalism in Brazil: A Historical Perspective," in McNeill, J.R. and Stewart, E. (eds), *A Companion to Global Environmental History*, Wiley-Blackwell, Oxford.

——(2015) "Tropical Forests in Brazilian Political Culture: From Economic Hindrance to Ecological Treasure," in Vidal, F. and Dias, N. (eds), *Endangerment, Biodiversity and Culture*, Routledge, London.

Palsson, G., Szerszynski, B., Sörlin, S., Marks, J., Bernard Avril, B., Carole Crumley, C., Hackmann, H., Holm, P., Ingram, J., Kirman, A., Pardo Buendía, P. and Weehuizen, R. (2013) "Reconceptualizing the 'Anthropos' in the Anthropocene: Integrating the social sciences and humanities in global environmental change research," *Environmental Science and Policy* 28 (3–13)

Polanyi, K. (1944) *The origins of our time. The Great Transformation*, Farrart & Rinehart, New York.

Przeworski, A. (1986) *Capitalism and Social Democracy*, Cambridge University Press, Cambridge.

Ribeiro, D. and Moreira Neto, C. (1993) *A fundação do Brasil: Testemunhos (1500–1700)*, Petrópolis, Vozes.

Safier, N. (2008) *Measuring the New World: Enlightenment Science and South America* Chicago, University of Chicago Press.

Santos, T. and Câmara, J. (eds) (2002) *Geo Brasil 2002: Perspectivas do Meio Ambiente no Brasil*, IBAMA, Brasília.

Scarano, F. (2012) *Biomas brasileiros: retratos de um país plural*, Casa da Palavra, Rio de Janeiro.

Steffen, W., Sanderson, R.A., Tyson, P.D., Jäger, J., Matson, P.A., Moore III, B., Oldfield, F., Richardson, K., Schellnhuber, H.J., Turner, B.L. and Wasson, R.J. (2004) *Global Change and the Earth System: A Planet Under Pressure*, Springer, Heidelberg.

Steffen, W., Crutzen, P. and McNeill, J.R. (2007) "The Anthropocene: Are Humans Now Overwhelming the Great Forces of Nature?" *Ambio*, vol. 36, no. 8.

Steffen, W., Grinewald, J., Crutzen, P. and McNeill, J.R. (2011) "The Anthropocene: Conceptual and Historical Perspectives," *Philosophical Transactions of the Royal Society*, Vol. 369.

Steffen, W., Richardson, K., Rockström, J., Cornell, S.E., Fetzer, I., Bennett, E.M., Biggs, R., Carpenter, S.R., de Vries, W., de Wit, C.A., Folke, C., Gerten, D., Heinke, J., Mace, G.M., Persson, L.M., Ramanathan, V., Reyers, B. and Sörlin, S. (2015) "Planetary boundaries: Guiding human development on a changing planet" *Science*, vol. 347, issue 6223.

Steffen, W., Broadgate, W., Deutsch, L., Gaffney, O. and Ludwig, C. (2015) "The trajectory of the Anthropocene: The Great Acceleration," *The Anthropocene Review*, 1–18.

Théry, H. (2000) "Retrato cartográfico e estatístico," in Sachs, I., Wilheim, J. and Pinheiro, P. (eds) *Brasil – Um século de transformações*, Companhia das Letras, São Paulo.

Théry, H. and Mello, N.A. (2005) *Atlas do Brasil – Disparidades e dinâmicas do território*, EDUSP, São Paulo.

Topik, S. and Wells, A. (2012) *Global Markets Transformed – 1870/1945*, Harvard University Press, Cambridge.

Torre, W. (2010) *Almanaque Habitat*, Comdesenho, São Paulo.

UNCTAD (2015) *Review of Maritime Transport 2015* United Nations, New York.

Villela, A.A. (2011) "Exclusivo metropolitano, 'superlucros' e acumulação primitiva na Europa pré-industrial," *Topoi*, v. 12. n. 23.

Wallerstein, I. (1989) *The Modern World-System. Vol. III: The Second Great Expansion of the Capitalist World-Economy, 1730–1840*, Academic Press, San Diego.

Weinstein, B. (1983) *The Amazon Rubber Boom, 1850–1920*, Stanford University Press, Palo Alto.

Wilberg, J. (1974) "Consumo brasileiro de energia," *Revista Brasileira de Energia*, January–March.

2 Population, development and environmental degradation in Brazil

José Eustáquio Diniz Alves and George Martine

Brazil was one of the last countries in Latin America to achieve Independence from its European metropolis in 1822. It was the only country on the continent to implement a monarchy. It was also one of the last countries to eliminate slavery, in 1888, and to adopt a republican regime. The Brazilian Republic emerged in 1889 as the result of a military mobilization that deposed Emperor Dom Pedro II, while a bewildered population just looked on (Lobo 1889). Given the military's lack of a political project and the absence of any popular mobilization, positivist thinkers assumed greater relevance and, somewhat belatedly, put the ideals of Auguste Comte (1798–1857) into practice. The central theme of positivism reads, "Love as principle, order as basis, progress as end." Comte's Brazilian followers simply stamped "Order and Progress" on the national flag.

The Republican regime inherited a country having great territorial extension and enormous biodiversity but relatively few people. Its extension and heterogeneity stimulated an exploitative mentality towards nature, considering it simply as an obstacle to be conquered. Hence, it is unsurprising that the republican notion of progress involved population growth, economic development, domination of nature and the greatness of the Patria. The country's natural resources, which had been exploited during the colonial and imperial regimes through a succession of extractivist and agricultural cycles over different segments of its territory, were apparently unlimited.

Brazil thus entered the twentieth century as a rural and agrarian nation with an enormous interior emptiness. Understandably, Afonso Pena, the fifth Republican president (1906–1909), would proclaim that governing essentially involved peopling the interior. President Washington Luis (1926–1930) would expand on his predecessor's slogan, pointing out that settling the interior requires roads, therefore "to govern is to build roads."

A numerous population was considered essential for both national greatness and effective occupation of the national territory. Brazil's longest tenured president, Getulio Vargas (1930–1945 and 1951–1954), took power with the promise to focus more on the internal market and the development of the hinterland. He supported the extended family, population growth and westward migration. Vargas promoted various social policies aimed at fostering population growth, economic expansion and national greatness. Explicit policies to increase demographic growth during this regime included both positive and negative measures.

On the one hand, a "family salary" was paid to workers in the formal sector and, more generally, income support was provided to couples with children and to larger families (Fonseca 2001). International migration was actively supported for both economic (promotion of agriculture) and eugenic (whitening of the population) reasons. On the other hand, such policies were further bolstered by explicit measures against birth control, such as the prohibition of family planning support by doctors; the definition of marriage as indissoluble under the protection of the State, and the prohibition of "advertising processes, substances or objects destined to provoke abortion or avoid pregnancy."

Marshall Candido Rondon, a follower of positivist ideals, collaborated with the Vargas government and was given the responsibility of carrying out the political banner of interior growth, with the aim of occupying and improving agricultural production in more backward regions. The famous "March to the West" performed at this time intended to accelerate the occupation of open spaces in accordance with the conceptions of the "National Settlement Department." This type of policy obviously intensified the continuous process of decimating natural areas rich in biodiversity as anecumene or uninhabited areas progressively became ecumene ones.

The ideals of road-building and occupation persisted after WWII. The main achievement of the Dutra government (1946–1951) was the construction of a highway (BR 116) linking the country's two main cities, Rio de Janeiro and São Paulo. Subsequently, the second Vargas government intensified the industrialization process and initiated petroleum exploration with the creation of Petrobras. This was followed by the Juscelino Kubitschek presidency (1956–1960) which adopted the motto of "50 years in 5" and promised to accelerate modernization through the construction of hydroelectric plants, the promotion of basic industry, the production of automobiles and of consumer goods in general, and the conquest of the central Cerrado region. Throughout this process, Brazilian governments consistently viewed nature as a bottomless source of riches that needed to be exploited.

The military government that took over the country in 1964 gave continuity to the positivist ideals and took the lead in the unfettered exploitation of the environment and the expansionist population policy of a "Brazil Superpower." Despite the precarious living conditions of the majority and the lack of investment in the welfare of the population, the initial military governments supported a pro-natalist policy. Several documents, including the National Security Policy (1964–1970), the Strategic Development Program (1968–70), as well as the government's official reaction to the conservative Humanae Vitae encyclical (1968), reaffirmed the government's pro-natalist ideology as well as its concern with the occupation of remaining open spaces (Canesqui 1985).

After the military stepped down in 1985, subsequent governments (Sarney, Collor, Franco and Cardoso) did very little from 1985 to 2002 to revert the main thrust of environmental degradation. The debt crisis and hyperinflation that provoked "the lost decade" of the 1980s and the slow growth of GNP in the 1990s failed to eliminate the dream of economic growth as the mainstay of Brazilian progress. In the same manner, the two Workers' Party governments (Lula da Silva and Roussef) that have been in power since 2003 have adopted the neo-developmentalist line. They have supported mega projects such as the "pre-salt explorations," the transposition of the San Francisco River, the construction of hydroelectric power plants in the Amazon region and the promotion of agribusiness commodities that use chemical fertilizers and pesticides intensively, as well as the extraction of highly polluting minerals (iron ore, bauxite, niobium, gold and others). The use of mercury and cyanides in mineral exploration makes mining activities extremely polluting, contaminating fish and wild animals while also harming human health.

The positivist ideology of development at any price has almost become a religion or, at worse, a State ideology. Globalization in recent years has heightened this unfettered quest for economic growth. There are obviously detractors and "progress" has been questioned by many people and by social movements. Bishop Erwin Kräutler of Xingu, for instance, has emphatically criticized the way development has increased the genocide of indigenous populations and the ecocide of vegetable species in both the Cerrado and Amazon regions (Brum 2012). Nevertheless, such reactions have little impact and it is evident that national development continues to occur at the cost of the degradation of ecosystems and the loss of biodiversity.

Demo-economic growth and living conditions

To a greater or lesser extent, the original ideas of the positivists on progress have effectively been put into practice in Brazil over the last 125 years. The country experienced one of the highest rates of demographic and economic growth in the world during that period. Its population expanded by 14 times, going from 14.3 million in 1890 to 204.4 million in 2015, according to projections by IBGE (Brazilian Institute for Geography and Statistics). Meanwhile, the economy underwent a much faster growth, having expanded 211 times between 1890 and 2015, as shown in Figure 2.1. This means that per capita income has expanded by 15 times in the Republican period. In terms of purchasing power parity, Brazil has become the world's seventh largest economy and a featured emergent country, according to data from the IMF (International Monetary Fund 2015).

Figure 2.1 also shows that, since the proclamation of the Republic, the fastest demo-economic growth occurred in the so-called "thirty glorious years" (1950–1980), when the population grew at a rate of 3 percent per year and the economy at 7 percent per year. Despite social and regional inequalities, the Brazilian population experienced significant improvements in its living conditions during this period – which also coincided with the country's urban and demographic transitions, as explained below.

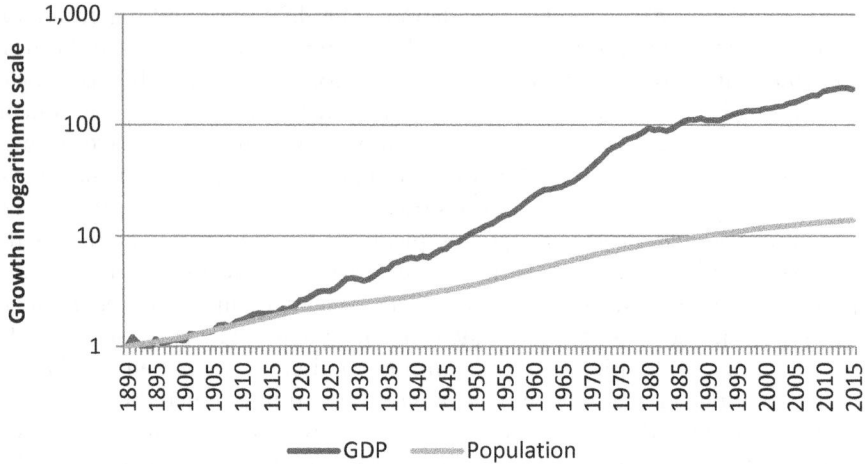

Figure 2.1 Growth of population and Gross National Product (GNP), Brazil: 1890–2015

Source: GNP data: Gonçalves (for years 1890 to 1900), IPEADATA for 1901 to 2014 (2008 to 2014, new series) and IBGE for population data.

Note: GNP for 2015 = -3.5%

Brazil's urban and demographic transitions

Brazil had an early and rapid urban transition. As of 1950, its population had grown to 52 million, two thirds of which lived in rural areas. By the second half of the 1970s, however, a majority of the country's inhabitants already lived in urban areas. The total population had increased to 191 million at the time of the latest census (2010), 84.3 percent of which lived in urban localities.

Two complementary trends, in both of which government policy played a significant role, explain this rapid urban transition. On the one hand, as shown earlier, the expansionist pretensions of successive governments had stimulated population growth through anti-birth-control policies, as well as aggressive pro-immigration measures. Yet, the main factor that accelerated population growth, as of the 1940s, was the reduction of mortality. Between 1890 and 1940, mortality decline had been slow and gradual but sanitation policies and internalization of the progress achieved in international medicine, together with improvements in public health and social security later accelerated the rate of reduction. In the context of persisting high fertility levels, mortality decline caused Brazil to experience a rapid rate of growth, especially during the 1950s and 1960s. A large manpower stock was thereby created, especially in rural areas, where fertility remained high over a longer period.

Meanwhile, in the mid-1960s, the military government established the institutional basis for an ambitious project aimed at the rapid modernization of agriculture that would also quell latent rural movements in support of land reform. It designed various instruments, chief among which was ample subsidized agricultural credit, in order to quickly modernize farming. The mechanisms utilized to distribute this credit inherently benefited larger landowners who could provide documentation of land ownership, and simultaneously excluded all classes of smaller farmers while favoring the concentration of land. Given that such measures were being adopted during a peak period of population growth, the number of people forced out of rural areas quickly multiplied. It is estimated that some 31 million people were expelled from rural areas between 1960 and 1980 (Martine and McGranahan 2010).

The consequences of this rapid transformation for urban areas were disastrous. Towns and cities were totally unprepared for the onslaught of mostly poor migrants; the resultant patterns of chaotic rapid growth have left a legacy of urban disorganization that persists to this day. The impacts of the rural exodus were also unexpectedly catastrophic for the country's major environmental treasure – the Amazon region. The military government, upon observing the dimension of the rural exodus and the resulting urban chaos, tried unsuccessfully to control migratory flows to urban areas, especially to the larger cities of the Southeast region. Failing this, they attempted to divert these flows to the occupation of the Amazon, coining the phrase – "Bringing landless men to a man-less land."

This initiative never accomplished its goal of diverting migrants from the cities of the Southeast to the Amazon, but it opened up the floodgates of ecological devastation. By clearing roadways and installing colonization projects that never really took root, the government facilitated the invasion of indigenous lands, large-scale deforestation, the spread of uncontrolled and inappropriate economic activities and the destruction of priceless biodiversity. Later, the expansion of monoculture and cattle raising on forestland greatly accelerated the rate of deforestation. Ironically, the major migration flows to the region were not directed to rural-agricultural areas but to the new and old urban localities of the Amazon region.

The Brazilian urban transition, catalyzed by the adoption of a conservative model of agricultural modernization, was also marked by the progressive concentration of the urban population in ever-larger cities and metropolis, as can be observed in Table 2.1. The number of urban localities having at least 20,000 inhabitants grew from 89 in 1950 to 870 in 2010, while the population residing in such localities rose from 24 to 132 million. Nevertheless, as can be observed in Table 2.1, the urban population became increasingly concentrated in a few large cities. The number of urban agglomerations having at least one million inhabitants showed only a small increase (from 5 to 16) between 1950 and 2010, but the number of residents in such localities escalated from 18 to 70 million in that same timespan. These 16 Metropolitan Regions accounted for 53 percent of all urban inhabitants in the 870 localities having at least 20,000 inhabitants in 2010, and for two-fifths of the entire growth of this group of cities between 2000 and 2010 (Martine and Ojima 2013).

The economic, social, demographic and environmental implications of this urban transition, especially of massive concentration, cannot be overstated. In

Table 2.1 Distribution of the urban population, by class of cities, Brazil: 1950–2010

City size	1950	1960	1970	1980	1991	2000	2010
1 million +	8,549	14,370	24,793	37,888	51,086	60,938	69,707
500 – 1 million	603	992	1,649	2,807	1,629	4,799	8,990
100 – 500,000	2,352	4,348	7,261	11,957	14,813	19,634	23,526
50 – 100,000	1,428	2,543	4,026	6,241	8,797	11,598	11,951
20 – 50,000	1,689	2,929	4,545	7,424	11,233	13,476	17,036

Source: IBGE, Demographic Censuses, *apud* Martine and Ojima 2013.

economic terms, the progressive concentration of people in urban areas allowed, with greater or lesser levels of adversity, the rapid expansion of the industrial activity that had emerged in the 1930s under an import-substitution model, and that had later expanded during the Second World War. The rural exodus multiplied the availability of a cheap work force and this became a key factor in the Brazilian industrial takeoff.

On the other hand, the multiplication of *favelas* provided a housing solution that reduced the cost of urban labor, but it promoted socio-spatial segregation and made it difficult for the majority of the population to access urban services and amenities, preventing them from exercising their right to the city.

From an environmental standpoint, large cities concentrate both consumption potential – and therefore greater environmental degradation – as well as the major levels of vulnerability, particularly in view of the progressive concentration of population on the cities' disorganized periphery, where basic services such as security, transport and others are typically lacking.

As elsewhere, Brazil's demographic transition was jumpstarted by mortality decline, a trend that had begun slowly in the first years of the Republic but remained relatively high until the 1930s, reflecting the period's socioeconomic under-development. Life expectancy at birth was less than 30 years when the Republic was proclaimed. By 1950, for reasons already described above, this had risen to 51 years and the crude death rate (CDR) was down to 16 per 1,000 inhabitants, while the infant mortality rate was 140 per thousand births. Recent statistics underline how far the country has come in more recent times: the CDR for 2010–15 has declined to its lowest-ever level (6.5 per 1,000), infant mortality is now 15 per thousand and life expectancy has risen to 75 years, according to IBGE data.

The Crude Birth Rate (CBR) remained at high levels for the first 70 years of the Republic and was estimated at an elevated 44 births per 1,000 inhabitants between 1950 and 1960. However, the CBR began to decline at an accelerated pace after the mid-1960s and reached a level of 14 per 1,000 in 2015. The Total Fertility Rate was still over 6 children per woman in 1960, but reached the replacement level (2.1 children per woman) in 2005 and dropped further to 1.9 children per woman in 2010.

Hence, against great opposition, despite government pronatalist policies as well as the efforts of religious institutions to restrict access to contraceptive methods and abortion, women are having fewer children. An ample literature explains this unexpected reduction in fertility as being the result of structural and institutional transformations that transpired at both the macro and micro level and that affected both inter-generational and gender relations and thereby reduced the demand for children (Carvalho, Paiva and Sawyer 1981; Merrick and Berquó 1983; Faria 1989; Alves 1994; Martine 1996). Because of this, the rhythm of Brazilian demographic growth has been declining since 1970 and the country will, in all likelihood, experience a reduction in its population size in the second half of the twenty-first century.

Socioeconomic impacts of the urban and demographic transitions

Brazil's demographic transition – which occurred concomitantly to its urban transition – produced a significant change in the country's age structure. After many decades in which the population pyramid was marked by its youth, it began its long-term trend towards ageing. As in other countries, a favorable demographic period known as the demographic dividend preceded the full transition to an older population pyramid. According to census data, only one-third of the population was economically active in 1970, with the other two-thirds being classified as dependents. This 2 to 1 relation adjusted to 1 to 1 in 2010 when 50 percent of the population was in the labor market, thereby reducing the dependency ratio and increasing per capita income.

The urban and demographic transitions, together with increased per capita income, made other social advances possible. Before 1960, 50 percent of homes had less than five rooms but, as conditions improved, the housing deficit was reduced and 70 percent of residences had five or more rooms in 2010. Moreover, the average number of persons per domicile declined from 5.2 to 3.3 between 1960 and 2010. Illiteracy was reduced while the average number of years of schooling went from 2 in 1960 to 8 in 2015. The Brazilian Human Development Index (HDI) went from circa 0.500 in 1980 to 0.750 in 2015.

Brazil remains one of the most unequal nations in the world, yet the fact that living conditions have clearly improved with demographic, economic and social progress during these 125 years of the Republic is undisputable. The consumer market expanded and broadened access to a variety of goods and services ranging from home ownership and sanitation to internet and cell phones. There are, however, clear signs that the continuity of progress within this model of development is limited, given its dependence on environmental regression, as seen in the next section.

The degradation of Brazil's main ecosystem

Brazil's territorial extension, with an area of 8,515,767 km², ranks fifth in the world. Its demographic density, as of 2015, is 24/km², much greater than Canada's (4/km²) and Russia's (8/km²), but well below that of China (146/km²) or of India (390/km²). However, this low density has not prevented the degradation of the country's main ecosystems. Brazil contains the largest absolute environmental reserve in the world, according to the Global Footwork Network (2015). As seen in Figure 2.2, the country's per capita ecological footprint was 2.9 global hectares (gha) in 2011, for a per capita biocapacity of 9.2 gha, yielding an environmental surplus of 6.3 gha. Although this seems acceptable, it should be noted that the gha surplus was three times larger in 1961.

The point is that while its human development indicators are improving, Brazil is losing biodiversity and degrading its ecosystems. The natural wealth of this immense country is threatened by negligence, degradation and unrestrained exploitation of its ecosystems. A large number of human activities are jointly contributing to the impoverishment of the country's natural capital through increasing pollution in cities, the destruction of rivers, the widespread use of fertilizers and pesticides, the construction of hydroelectric plants, the entire chain of industrial production, the network of commerce and services, the acidification of land and water, desertification, the expansion of agriculture and ranching activities, mining, deforestation, highway networks, slash and burn agriculture, the exploitation of biomass, and many other activities.

Reviewing data from the Ministry of the Environment's official site (MMA 2015), makes it clear that all of Brazil's ecosystems are being threatened to a greater or lesser extent by negligence and degradation, as summarized below.

The **Atlantic Forest** occupies an area of approximately 1,300,000 km² and extends over 17 Brazilian states. The remnants of this rich native vegetation

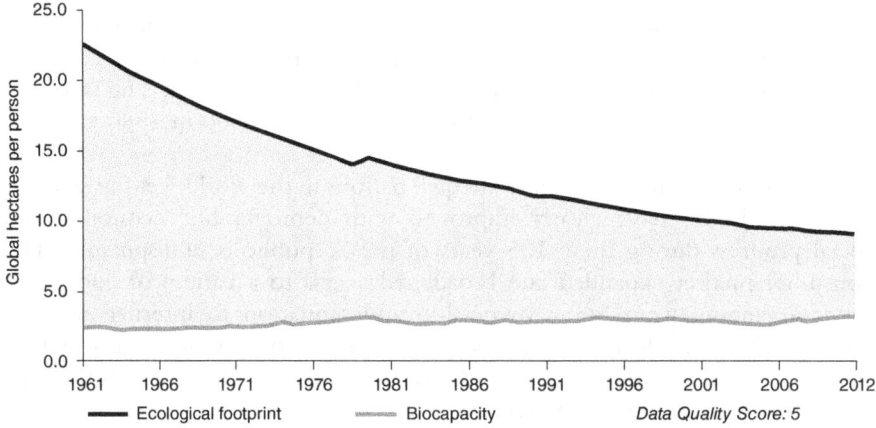

Figure 2.2 Ecological footprint and biocapacity, Brazil: 1961–2012

Source: 2016 Global Footwork Network. National Footprint Accounts 2016 edition have been reduced to 22 percent of the original forest cover and are found in various stages of regeneration. Only 8 percent of the area is well preserved in fragments of at least 100 hectares. In addition to being one of the richest biomes in the world in terms of biodiversity, the Atlantic Forest is of vital importance to some 120 million Brazilians who live in its domain, which also produces some 70 percent of the country's GNP.

The **Cerrado** is the second largest biome in South America, occupying an area of 2,036,448 km², equivalent to 22 percent of the Brazilian territory. It harbors the fountainheads of the three main hydrographic basins of South America (Amazon-Tocantins, San Francisco and Prata), thus, an enormous aquifer potential. The Cerrado also contains an extreme abundance of endemic species, harboring some 11.6 thousand species of catalogued native plants, 200 known species of mammals, 837 species of avifauna and numerous species of fish. However, a large number of plant and animal species are threatened with extinction. It is estimated that 50 percent of the Cerrado has been degraded by anthropic activites, making it second only to the Atlantic Forest in terms of alteration due to human occupation in Brazil.

Brazil's **Pantanal** (or "Swampland") is considered one of the largest continuous humid extensions on the planet. Its area covers approximately 150,000 km², occupying 1.76 percent of the Brazilian territory. The biome is influenced by rivers that drain into the Alto Paraguay basin. The Pantanal maintains 86.8 percent of its native vegetable cover. Despite its exuberant natural beauty, the biome is experiencing severe impacts from human activity, particularly from agribusiness. A mere 4.4 percent of the biome is currently protected in conservation units.

The semi-arid **Caatinga** occupies an area of 844,000 km² between the North of the state of Minas Gerais and the country's Northeastern region, equivalent to 11 percent of the national territory. With a rich biodiversity, the biome harbors 178 species of mammals, 591 of birds, 177 of reptiles, 79 of amphibians, 241 of fish and 221 of bees. Despite its importance, the Caatinga has been deforested at a high pace in recent years, mainly for pastures and agriculture, as well as for firewood.

The **Pampa** is located in the state of Rio Grande do Sul and occupies an area of 176,496 km², which corresponds to 63 percent of the state's territory and 2.07 percent of the national territory. It is estimated that it contains some 3,000 plant species, with a notable diversity of grass types. The fauna is still expressive, having more than 500 species of birds. However, the progressive expansion of monoculture and of pastures with exotic species has led to a rapid degradation and alteration of the biome's natural landscapes.

The **Forest of Araucarias** occupies an area of 231,000 km² spread out over four states – Paraná, Santa Catarina, São Paulo and Rio Grande do Sul. The forest is in danger of disappearing due to excessive exploitation, most of it illegal.

The **Mangues** (mangroves, or coastal and insular ecosystems) are some of the most complex and diverse ecosystems on the planet. They play an important role in transferring organic material to estuaries and contributing to the productivity of coastal areas, as well as in attenuating the effects of storms and waves, reducing erosion and maintaining coastal profiles. The great biodiversity of mangroves makes them natural nurseries for both the typical species of this environment as well as for the animals, birds, fish, crustaceans and molluscs who find in them ideal conditions for reproduction. Mangroves are naturally resilient, but they are threatened by clearing for land speculation, shrimp farms and tourism, as well as by the dumping of urban and industrial wastes, oil spills, silting and erosion. The destruction of mangroves involves huge losses, given the various important functions performed by these ecosystems.

The **Mata de Cocais** is situated in a transition zone between the Amazonian and Caatinga biomes in five states (Maranhão, Piauí, Ceará, Pará and Tocantins). The biome is naturally fragile and the combination of agricultural, mineral and logging activities, along with installation of industrial and commercial activities in the region is polluting the Tocantins-Araguaia aquifer and accelerating deforestation. The major section of this biome is in the process of becoming a savannah or desert.

The **Amazon Forest** is the largest biome in Brazil. It covers an area of 4,196,943 million km² and produces 2,500 tree species (or one-third of all tropical wood in the world) and 30,000 species of plants (of the 100,000 in South America). The Amazon basin is the largest hydrographic basin in the world, covering an area of some 6 million km², and has 1,100 tributaries. The Amazon River deposits 175 million liters of water into the ocean every second. The combination of water and forest makes up an unparalleled ecosystem. Unfortunately, data from Prodes/INPE[1] show that deforestation between 1988 and 2014 in this region reached 408,000 km². This is an area larger than the combined territories of the states of São Paulo, Rio de Janeiro, Espírito Santo, Alagoas, Sergipe and the Federal District. Another 400,000 km² were destroyed between 1965 and 1988. Recent estimates by the Ministry of the Environment indicate a 16 percent increase in deforestation in 2015 over previous year levels. Nobre (2014) warns that the forest is showing signs of wear and stress in its historical role of pumping humidity from the ocean into the interior of South America. That is, the forest's mission to serve as a "biotic water pump" is at risk and this affects the "flying rivers" and thus the quantity of water available in the highly populated Southeast region of Brazil while also aggravating the country's energy crisis.

The degradation of rivers and the energy crisis

Waterways have always been critical to Brazil's occupation and integration. Several large and small rivers that crisscross the nation have served a variety of functions from transportation to hydroelectric power, mineral extraction and irrigation. The recent story of Brazilian rivers, however, is an increasingly tragic one. This is best illustrated by a disaster that occurred on November 5, 2015. Tailing dams of the Samarco mining company, a joint venture between mining giants Australian BHP Billiton and Brazilian Vale do Rio Doce, broke up and unleashed a tsunami of toxic mud near the town of Mariana in Minas Gerais. Water and sediment from iron ore extraction sluiced down the mountainside. This sea of mud flooded the entire Rio Doce basin all the way to the Atlantic Ocean, destroying agricultural lands, killing several people and thousands of wild and domesticated animals, destroying hundreds of homes and, eliminating all forms of life in its path over thousands of kilometers due to its high content of heavy and toxic metals. Millions of people are struggling to find drinking water in the states of Minas Gerais and Espírito Santo while other critical issues related to health, employment, housing and living conditions will persist for many years.

At this writing, the ultimate dimensions of this tragedy – which is likely to evolve into the greatest environmental disaster of Brazil's history – are still difficult to fathom. However, the origins of the Samarco calamity are clear: they stem from lax regulation and vigilance by government and industry during a period of feverish mineral extraction detonated by increasing prices for iron ore. Negligence in the Brazilian mining industry is well documented and the Samarco disaster is only the latest and most catastrophic of the five dam breaks suffered by Minas Gerais in the last decade.

Negligence and misuse of waterways in development projects by government and industry also occurs on a large scale elsewhere. Attempts to explore hydroelectric potential in the Amazon region, such as in the Belo Monte dam, have received world attention and that experience would merit a full chapter by itself. However, other lesser-known incursions are significant. For instance, the São Francisco River – known as the river of national integration – is one of the waterways that has been most affected by deforestation, silting and the construction of dams for hydroelectric plants. The river, affectionately known as "The Old Chico," has a basin that embraces 500 municipalities having a population of 15 million inhabitants as it wends its way over 2,700 kilometers in seven states. Public neglect in the Parque Nacional da Serra da Canastra, the main fountainhead of the São Francisco River, caused it to dry up, for the first time in history, in 2014.

Yet, instead of proposing the recuperation and revitalization of the sources of the São Francisco River, the federal government intends to draw more water from the riverbed through a transposition. Unfortunately, the volume of water is decreasing, while silting is on the increase. Studies, such as those by Professor José Alves Siqueira from the Federal University of the Valley of São Francisco (Univasf) speak of the "inexorable extinction" of the São Francisco River. Ironically, the lack of water already threatens the great lakes of the hydroelectric dams on the river that prevent the free flow of water. Newspaper reports (Estado de Minas October 13, 2014) describe the partial paralysation of the Usina Três Marias, forced to function on only two of its six turbines at that time. According to a report from the National Operator of the Electric System (ONS), the lakes on the hydroelectric plants of the Northeast had less than 9 percent capacity, a level similar to that registered in 2001, when several blackouts occurred and the government was forced to adopt the rationing of energy.

The chronicle of the Paraiba do Sul River registers a more continuous process of degradation by different stages of regional "development." The river has an extension of 1,137 km and its basin covers an area of 56,500 km² that cuts through the country's most industrialized region in three states – São Paulo, Minas Gerais and Rio de Janeiro. It supplies most of the water used in the country's two major metropolitan areas – São Paulo and Rio de Janeiro. The Paraíba Vale formerly encompassed one of the lushest forests in the world but this has been progressively devastated, first by the coffee cycle during the nineteenth century, then by the increase of mineral extraction and agricultural activities and, more recently, by urban-industrial expansion. The progressive degradation of this river basin not only constitutes an instance of ecocide but it will inevitably aggravate the precarious water supply situation already existing in the most densely populated region of Brazil and also damage the economy of its most productive territory.

Neglect, misuse and degradation of waterways is also prominent in the locations where most of the Brazilian population now lives – its large cities. While the major Brazilian cities were growing, they were systematically degrading and burying their rivers and watersheds. Forests and natural vegetation were substituted by a "stone jungle," without consideration for the preservation of water. The story of rivers in the city of São Paulo provides a good illustration of how these critical assets have been ravaged by neglect and poor planning.

The Ipiranga River, cited in the national anthem and on whose banks the first Emperor, Dom Pedro I, proclaimed the cry of Independence, still runs from its spring in the Mata Atlantica through the city of São Paulo, but it presents a depressing scenario. Filth, stink and sewerage mark the "placid margins" cited in the anthem, while the area of the Monument to Independence in the central Ipiranga district reeks of a powerful stench. The quality of the river water in this stretch is classified as "terrible" by the state environmental company (Cetesb). Of the 9.5 kilometers in the river's course through the city, the water is considered as clean in only one point – its spring in the district of Jardim Botânico. Beyond that, the sewerage dumped into the river destroys any basic signs of life.

The degradation of the Ipiranga is typical: the rivers in the city of São Paulo have been debased, buried alive and transformed into "black tongues." The Tamanduateí and Anhangabaú rivers that fed the Patio do Colegio – the initial landmark of the state's capital – no longer exist as the source of clean and free waters. The Tietê River has become an open sewer, while the riparian forests on its margins have been destroyed, its floodplains drained and its margins restricted. The Pinheiros Rivers were channeled and transformed into sewers. At the meeting point of the Tietê and Pinheiros – the two main rivers of the city – a complex of overpasses and constructions was built, instead of a much-needed park for ecosystemic protection. Drinking water has to be supplied from ever more distant sources such as the Cantareira complex and the rivers that originate in the south of Minas Gerais. In recent times, these sources have been having their own well-publicized problems due to recent and ongoing "development" processes in other ecosystems, as described above. Serious periodic and lasting water shortages in the São Paulo Metropolitan area have been chronicled as a result.

Much the same kind of story could be told for Rio de Janeiro, Belo Horizonte and other major cities since the degradation of rivers and the water crises are recurrent phenomena in Brazil's urban growth. The situation of the Carioca River in Rio de Janeiro is particularly distressing, given the contamination of the waters of Guanabara Bay wherein Olympic athletes will compete in 2016. In general, the reiteration of inappropriate local urban practices is now combining with global processes involving climate change to jeopardize the supply of water.

Much has been told in recent times of the impending energy crisis in Brazil. This is closely connected with the water crisis and both result from negligence, deforestation, destruction of watersheds, silting as well as the disrespect and overuse of free-flowing waters. Thermoelectric plants using fossil fuels have been activated to compensate for the decline of hydroelectric production. The myth of a "clean energy matrix" is dwindling as plans for construction of hydroelectric plants in the Amazon region are meeting with resistance from the Forest People and the growing awareness of their high environmental costs. Undoubtedly, Brazil will need to increment the production and efficiency of its energy matrix. However, current evaluations indicate that hydric availability will drop significantly in coming years. Brazil would urgently need to promote and invest in renewable energies, diversifying energy sources and reducing its dependence on hydroelectric energy. However, vested interests and administrative inertia are making it extremely difficult to alter traditional solutions and to capitalize on the potentialities of renewable energy.

De-industrialization, re-primarization of the economy and the fiscal crises

Without question, the development model adopted during the entire Republican period resulted in important social advances while also boosting Brazil's participation in the world economy. Belatedly, we is learning that it has also caused enormous environmental stress that, in turn, are likely to hamper future development efforts. But the country's problems are not only environmental: Brazil's productive structure has actually been on a regressive track over the last 35 years and it is experiencing an economic crisis in the second decade of the twenty-first century that could be fatal for its ambition of becoming a fully developed country.

The process of acute, rapid and profound de-industrialization that has been going on since the mid-1980s is a major sign of Brazil's economic regression.

Over this period, Brazil's economy has been undergoing a process of "regressive specialization." That is, the more dynamic segments of the transformation industry are losing ground in both the internal and external markets, while agribusiness and mineral extraction have become the main paths to GNP growth. The country is thus gradually reassuming the primary-exportation style of the Old Republic (1890–1930).

Re-primarization signifies not only economic regression but also heightens concern over the global environmental consequences of Brazilian economic activity. Agribusiness expansion, for instance, has flourished on a cycle of deforestation and monoculture, both of which promote the invasion of indigenous and anecumene lands as well as the massive destruction of biodiversity. Large-scale mining operations, especially those using open-pit mining techniques, result in destruction of mountaintops and in deforestation, in the contamination of rivers and food chains with toxic compounds and in devastating spillage from tailing dams. Expansion of Brazilian meat-producing activities to meet growing per capita demands is portrayed in highly optimistic terms, completely overlooking the fact that they are one of the largest sources of greenhouse gases and a leading factor in the loss of biodiversity. In brief, re-primarization also signals a growing direct and indirect contribution to global environmental problems by Brazil.

A recent (May 2015) report by FIESP (Federation of Industries in the State of São Paulo) shows that the participation of the transformation industry in GNP fell from 22 percent in 1985 to 10.9 percent in 2014, the lowest level registered since the systematization of National Accounts. De-industrialization is a well-known phenomenon in developed countries who have, in recent times, farmed out heavy industry to less-developed regions and focused economic activity increasingly on the services' sector. However, Brazil's process of de-industrialization is precocious and regressive since it still has a long way to go in terms of human development, and since its movement is actually towards a more backwards primary sector. That is, its exports portfolio is increasingly slanted towards commodities, primary goods or manufactured goods having little added value and low technological content. Massive inflows of external capital from primary sector financing and exports have also resulted in an overvaluation of Brazil's real exchange rate, and this, in turn, has negatively affected the competitiveness of its manufacturing and agricultural sector in a typical case of the so-called "Dutch disease."

Since the 1980s, Brazil has been investing considerably more in the construction of shopping centers than in the erection of factories. That is, it has prioritized consumption over investment. Low rates of investment prevent the country from incorporating new technologies, improving infrastructure and increasing work productivity. Job opportunities also fail to accompany the growth of the economically active population. Wrong choices in long-term strategy are also making matters worse. A prime example of this is the government's commitment to the exploration of petroleum and gas in pre-salt layers. So much faith was placed in this initiative that it was publicized as

Brazil's "passport to the future." Petrobras would be the company that would guide national salvation, distributing petroleum royalties to many municipalities, while also guaranteeing an upgrade in the nation's health and education systems. On another level, the royalties would be used to support "national champions," the designation applied to selected companies that would be assisted in the process of becoming giants capable of competing globally in their respective sectors.

However, the high expectations and precocious spending of expected profits were soon shattered by a combination of extreme corruption in Petrobras, the State-owned company, with the practical realities of the oil market. As repeatedly demonstrated over months of investigation by the Public Ministry and by the "Lava-Jato" operations, Petrobras squandered uncounted billions of dollars in a variety of frauds and money-laundering maneuvers. These involved high-level functionaries, politicians, political parties, and the federal government itself in a web of corruption that spawned hundreds of tentacles and that instantly disintegrated Petrobras' vaunted high level of credibility and efficiency. However, the fact is that, even without these scandals and under the best of circumstances, the high costs of extraction in the pre-salt layers and the drop of international prices for petroleum would inevitably have wrecked this strategy of transforming fossil fuels into a national redemption strategy.

The policy of "national constituent" aimed at expanding national participation in the production process at all levels also went awry as the entire production chain went into crisis in that sector. The companies and shipyards contracted during the height of the debt crisis by Petrobras are dismissing manpower and run the risk of downsizing and foreclosures. Petrobras itself, once the largest Brazilian company, is deemed the most indebted company in the world and it will inevitably be forced to reduce investments in coming years.

In the light of this crisis, Petrobras' Plan for Business and Management for 2015–2019 forecasts investments of US$130.3 billion for the period, a reduction of 37 percent in comparison to the previous plan for 2014–2018. Less investment by Petrobras inevitably means smaller investments all along the production chain and, thus, a worsening of the recession.

Brazil's second largest national company, the Vale do Rio Doce company, one of the largest mining companies in the world, was privatized in 1997 through a negotiation process that is still under scrutiny. Since then, as a private company, it has also accumulated its share of scandals and investigations. However, it may have suffered its major blow when Samarco, a subsidiary of the Vale do Rio Doce caused a widespread environmental disaster, as discussed above.

The recent tragedy in Mariana reflects the continuing promotion of un-economic and environmentally disastrous activities, but the lessons appear hard to learn. Thus, the main axis of infrastructure concession in Dilma Rousseff's second presidency (2015–2018) is the development of a Transcontinental Railway Project (or Bi-oceanic Railway), in partnership with

China. Its object would be to channel Brazilian commodities to the Pacific. However, the impacts of this project on the environment and on the indigenous populations in the regions affected by the railway would be extremely negative. Such a railway would, in some ways, be the coronation of the "March to the West" ambitions of earlier positivists, but it would come with catastrophic impacts in both Brazil and Peru.

Altogether, the Petrobras crisis, the re-primarization of the country's economic structure and the reversion of macroeconomic indicators have thrown Brazil into a recession of serious proportions, perhaps the greatest in the history of the Republic. Both internal and external debts have soared. The Minister of the Economy promised a primary surplus of 1.2 percent of GNP in 2015, but was unable to approve any of the measures required to achieve this goal in Congress; as a result, the country will actually have a primary deficit of 1 percent of GNP. The nominal deficit is nearing 10 percent of GNP. Brazil finds itself in a situation described by economists as "Fiscal Dominance." This refers to a situation in which the disorder of public accounts creates its own dynamics and where monetary and income policies become incapable of containing inflation by increasing interest rates. Indeed, high interest rates only aggravate public finances, increasing the debt size and making fiscal adjustments ineffective. In brief, Brazil has fallen into a situation of "chronic primary deficit" and since it lacks the governance to alter this situation, the economic crisis will only get worse.

Low rates of economic growth and fiscal crises are not exclusive to Brazil; practically all Latin American countries are experiencing this situation now that a decade-long super commodity cycle and favorable terms of trade have vanished. Over this period, the leftist governments that took over power in many countries have eschewed their traditional opposition to "enclave economies" in order to defend a "new extractivism" supported largely by state-owned companies whose products were exported mostly to China. Gudynas (2009) has been a main critic of this shift, arguing that the new extractivism is similar to the old, inasmuch as it repeats the same entrepreneurial strategies based on competitivity, cost reduction, increased profitability and tolerance of negative social and environmental costs. Along the same lines, Martinez-Alier (2010) considers that the new extractivism increases social costs, environmental injustice and causes increasing conflicts in the process of resource extraction and waste disposal.

Population ageing and the precocious expiration of the demographic bonus

The positivists' development project is not the only one at risk under the present economic situation. The job market is facing a serious crisis and the prospect of attaining the goal of "full employment and decent work" proposed by the International Labour Organisation seems increasingly remote in Brazil. During the 1970–2010 period, the country took advantage of its most favorable demographic moment with increasing levels of occupation, greater formalization of employment and increasing educational levels of the economically active population. However, this scenario has been inverted since 2012, although the demographic dividend window has not yet shut completely.

Although unemployment rates remained low for some time, occupation rates declined and the absolute number of occupied persons stagnated. The job market, which had been stagnant since 2012, collapsed in 2015, causing millions of workers to become unemployed and increasing the instability of working conditions. Economic and demographic conditions have completely changed for the worse in comparison to previous decades in Brazil. This increases the probability that the country may be falling into the "middle-income trap," with little probability of achieving or maintaining the progress dreamt about by earlier positivists.

Given the significant reduction in aggregate demand, 2015 and 2016 are sure to witness an aggravation of the social situation. Subsequent recuperation will undoubtedly be slow and all signs point to a new "lost decade" (Benjamin 2015). Given the lack of economic dynamism, precious "human capital" is being squandered. In the context of low rates of savings and investment, a precocious expiration of the demographic bonus may spell the conclusion of Brazil's development trajectory (Alves 2015).

Even under normal conditions, population ageing tends to transform the demographic dividend into a burden. Table 2.2 shows that, in 2001, the working age population (WAP) of 15 to 59, increased by 2.2 million people, while the older population (60 and over) increased by 373,000, according to Census data. However, projections show that the WAP decreases rapidly from then on, while the older-age group increases steadily until mid-century. In 2019 – only three years from now – the older age groups will grow faster than the WAP (1.1 million to 972,000) for the first time in history. As of 2032, the WAP will begin to decrease in absolute terms, while the older age groups will continue to add over a million people every year. In 2043, the WAP population will decrease by 1.1 million people while the older age group will increase by 1.4 million people.

These changes in the age structure will obviously have an enormous impact on the labor force, on social security and on the health system. First, the number of people paying taxes and contributing to social security will diminish, thereby reducing the production of goods and services, in view of the low and stagnated productivity of labor (Negri and Cavalcante 2014). Concomitantly,

Table 2.2 Annual variation of population by age groups, Brazil: 2001 to 2060

Age group	2001	2011	2021	2031	2041	2051
15–59	2,182,158	1,544,540	757,624	4,279	–921,944	–1,074,071
60+	372,753	780,855	1,183,629	1,203,366	1,408,436	895,742

Source: Population projections by IBGE (2013 revision).

population ageing, especially the segment aged 80 and over, will significantly increase the costs of health systems and also increase pressure on family systems to take on part of the responsibility for care of the elderly.

Ultimately, the major impact of ageing will be on the actuarial equilibrium of social security. Tafner et al. (2014) show that social security costs as a proportion of GNP are particularly high in Brazil, similar to that of more advanced societies. The system already faces a large deficit, but this will only tend to increase until mid-century. Given the chronic fiscal deficit, increasing imbalances in social security will aggravate economic growth and employment, impeding the eradication of poverty and the improvement of living conditions for the Brazilian population. Brazil faces the trap of low growth and high public deficits, thereby downgrading historic trends of GNP growth. Yet, even lower growth rates would not guarantee less environmental degradation given the historical and deeply entrenched values involving exploitation of nature and public negligence of environmental assets.

Final considerations

The developmentalist ideology derived from the "Order and Progress" motto of the positivists who participated in the construction of the Brazilian Republic has materialized over the long run through the construction of an urban-industrial society and the diffusion of material progress. Despite the persistence of social and regional inequalities, it is unquestionable that living conditions for the entire population have improved significantly over the past 125 years. Yet, today, this ideology is facing a triple impasse of economic, social and environmental problems that result from its subordination to the logic of a globalized capitalism that is itself losing legitimacy. This development model now lacks economic dynamism, redistributive justice and ecological sustainability.

The country is squandering the last years of a favorable demographic situation. It is getting old before becoming rich, marching rapidly towards an age structure typical of developed countries without having solved its economic and social bottlenecks. The economy's low productivity and international competitivity hamper increases in per capita income and the sustainability of an effective social security model. The country still has not been able to formulate and adopt adequate policies in education, health and housing. Worse, it may be discarding a last window of opportunity to takeoff in human development with environmental preservation.

Brazilians are currently witnessing an unexpected interruption of long-term economic growth – long before having overcome its historical weaknesses – and entering a period of submergence, diminution and impoverishment. It is stagnating in the middle-income trap without having made proper use of the country's demographic bonus. Viewed in international terms, Brazil may be seen as a country that never really emerged from under-development and which will remain needy and unequal in social terms.

Current crises in Brazil's two huge national companies, Petrobras and Vale do Rio Doce, companies that symbolize its extractivist and polluting model of development, reflect and encapsulate the country's impasses. Yet, the international context offers little hope since the entire world is facing an impasse in relation to sustainable development. Deep social, economic and environmental dangers loom on the global horizon. Explosions on the Deepwater Horizon platform in the Gulf of Mexico, the nuclear accident in Fukushima and the more general transgression of planetary boundaries all warn of environmental risks that face the planet. The process of financialization (the growing dominance of capital market financial systems and the increasing political and economic power of the rentier class) absorbs the gains of economic growth, while the concentration of wealth reaches scandalous levels, with the richest 1 percent of the pyramid owning as much as the other 99 percent taken together (Credit Suisse 2015). Severe obstacles preclude the sustainability of development. According to Martine and Alves (2015):

> The "hegemonic system of production and consumption" (whether capitalist or socialist), does not have the ability to be simultaneously socially just and environmentally sustainable. Consequently, it is impossible for the model of development that we know to simultaneously maintain and promote the Three Pillars of sustainability; in practice, they have become humanity's main trilemma in the 21st century.

Brazil's difficulties in resolving ongoing economic, social and environmental crises are compounded by the deficiencies inherent in a political system of "presidentialism by coalition," which makes it incapable of planning or monitoring "sustainable" development activities. For instance, the Brazilian Government had committed itself to reducing greenhouse gas emissions by 37 percent as of 2015, and by 43 percent as of 2030, in comparison to its 2005 patterns, as part of its Intended Nationally Determined Contributions (INDCs) for COP21. This is undoubtedly a positive intention, but one that is unlikely to be implemented in practice since, as noted by Viola and Basso (2015) – "The Brazilian record in the implementation of reduction emission goals is poor and the evolution of national policies in the various sectors that contribute to climate change is heading in an opposite direction to de-carbonization." Currently, Brazil is going through a perfect storm, confronting simultaneous crises in the economic, social and environmental domain, while also traversing a particularly complicated political period. The relative successes enjoyed during the first 125 years of the Republic are unlikely to be repeated in upcoming decades. Perennially considered to be the country of the future, Brazil may face serious problems in pursuing the course of progress in coming years, denying the slogan depicted so optimistically on the national flag.

Note

1 Projeto de Monitoramento do Desmatamento na Amazônia Legal, do Instituto Nacional de Pesquisas Espaciais (Prodes/INPE).

References

Alves, J.E.D. (1994) Transição da fecundidade e relações de gênero no Brasil. Tese Doutorado, Centro de Desenvolvimento e Planejamento Regional, UFMG, Belo Horizonte.

Alves, J.E.D. (2015) "O fim do bônus demográfico e o processo de envelhecimento no Brasil", Revista *Portal de Divulgação* 45, Ano V. Jun/jul/ago 6-17, São Paulo.

Benjamin, C. (2015) "É pau, é pedra, é o fim de um caminho," Revista *Piauí* Edição, 103.

Brum, E. (2012) "Dom Erwin Kräutler: Lula e Dilma passarão para a História como predadores da Amazônia," Revista *Época*, Rio de Janeiro.

Canesqui, A.M. (1985) "Planejamento Familiar," *Revista Brasileira de Estudos de População* 2(2), 1–20, Campinas.

Carvalho, J.A.M., Paiva, P.T.A. and Sawyer, D.R. (1981) "A recente queda da fecundidade no Brasil: evidências e interpretação," Cedeplar/UFMG (Monografia, 12), Belo Horizonte.

Credit Suisse (2015) "Global Wealth Report 2015," Research Institute, Switzerland.

Faria, V.E. (1989) "Políticas de governo e regulação da fecundidade: conseqüências não antecipadas e efeitos perversos" in *Ciências sociais hoje* ANPOCS, São Paulo.

FIESP (2015) "Perda de Participação da Indústria de Transformação no PIB," FIESP, São Paulo.

Fonseca, A.M.M. (2001) *Família e política de renda mínima*, Cortez, São Paulo.

Global Footwork Network (2015) Global Footwork Network, Available in: www.footprintnetwork.org/en/index.php/GFN/page/trends/brazil/

Gonçalves, R. (2010) "Evolução da renda no Brasil segundo o mandato presidencial: 1890-2009," IE-UFRJ, Rio de Janeiro.

Gudynas, E. (2009) "Diez tesis urgentes sobre el nuevo extractivismo," in *Extractivismo, política y sociedad*, CAAP, CLAES, Quito, Ecuador, pp. 187–225.

IBGE (2013) Projeções Populacionais (Revisão 2013), Rio de Janeiro.

IMF–International Monetary Fund (2015) World Economic Outlook (October 2015), IMF Data Mapper, Washington, available at www.imf.org/external/datamapper/index.php

Lobo, A. (1889) "O povo assistiu àquilo bestializado," *Diário Popular*, Rio de Janeiro.

Martine, G. (1996) "Brazil's Fertility Decline, 1965-95: A Fresh Look at Key Factors," *Population and Development Review*, 22(1) 47–75.

Martine, G. and Alves, J.E.D. (2015) "Economia, sociedade e meio ambiente no século 21: tripé ou trilema da sustentabilidade?" *R. bras. Est. Pop.*, 32 (3), Rio de Janeiro.

Martine, G. and McGranahan, G. (2010) "Brazil's early urban transition: what can it teach urbanizing countries?" IIED/UNFPA Urbanization and Emerging Population Issues, Working Paper # 4, IIED and UNFPA, London.

Martine, G. and Ojima, R. (2013) "The challenges of adaptation in an early but unassisted urban transition," in Martine, G. and Schensul, D. (eds), *The Demography of Adaptation to Climate Change*, UNFPA, IIED and El Colegio de Mexico, New York, London and Mexico City.

Martinez-Alier, J. (2010) "Environmental Justice and Economic Degrowth: an alliance between two movements," ICTA, Universitat Autònoma de Barcelona, Coimbra, 20–22 October 2010.

Merrick, T. and Berquó, E. (1983) "The determinants of Brazil's recent rapid decline in fertility," National Academy, Washington.

MMA, Biomas brasileiros. (2015) www.mma.gov.br/biomas (accessed on 21 May 2015).

Negri, F. and Cavalcante, L. (eds.) (2014) "Produtividade no Brasil: desempenho e determinantes," ABDI, IPEA, Brasília.

Nobre, A. (2014) "O futuro Climático da Amazônia," Relatório ARA, São Paulo.

Rocha, M.I.B. (1987) "O parlamento e a questão demográfica: um estudo do debate sobre o controle da natalidade e planejamento familiar no Congresso Nacional," Texto Nepo 13, Campinas.

Tafner, P., Botelho, C. and Erbisti, R. (2014) "Transição demográfica e o impacto fiscal na previdência brasileira," in Camarano, A.A. (ed.), *Novo Regime Demográfico: uma nova relação entre população e desenvolvimento?*, IPEA, Rio de Janeiro.

Viola, E. and Basso, L. (2015) "Dá para acreditar nas metas do Brasil?" Observatório do Clima, São Paulo.

3   The Amazon before the Brazilian environmental issue

Violeta Refkalefsky Loureiro

Throughout history, the Brazilian government has had a changing and ambivalent attitude toward the Amazon region within the national scenario. It has sometimes perceived the Amazon as a paradoxical ecosystem (both superabundant and fragile at the same time) with a wide range of problems requiring solutions (the need to protect its extensive land borders, the purchase of large tracts of land by foreigners, criticism from international organizations for the deaths of members of indigenous communities and environmental damages), and as a challenge to development with part of its territory occupied both by local communities, considered uneducated and primitive by the ruling elites and economic organizations, and indigenous peoples divided into approximately 180 ethnic groups, speaking over 100 different languages. At other times, however, the Amazon region was seen as the prime location for national exploration and a solution to Brazil's problems. Periods exemplifying the latter are: a) the first years of the 20th century, when the region emerged as a producer of rubber, a raw material extracted from the native forest and exported to Europe and the US; b) during World War II, when the region was called upon to meet the allied troops' demand for rubber, because the Axis nations blocked the access to the Malaysian supply of this product; or c) in the 1980s when Serra Pelada began to produce gold and the military regime planned to use this commodity to pay off the Brazilian national debt: gold was extracted by thousands of people working under degrading conditions who risked their lives on a daily basis, in an insane venture encouraged by the Federal Government.

In contrast to other periods, from the 1980s onward, the Brazilian government decided on the permanent exploration of the Amazon region's natural potential. Instead of being the site of occasional bursts of exploration, the region was seen as a permanent solution to improve the country's international trade balance and solve structural problems such as the shortage of electricity in more developed regions. As a result, it prioritized two main activities:

1   Mining: the region contains the largest and most diversified mineral deposits in the world. Since 1985, when Albras – the world's eighth largest

aluminum company – started operations in the state of Pará the government has been encouraging the installation of several new mining ventures. However, these businesses have been extremely harmful to the Amazon environment, in particular because they resulted in 30 years of charcoal extraction from the native forests to supply local mills and because of dams containing discharged material from explored mines.

2 Electricity production: the 1987/2010 National Energy Plan was subsequently extended. This plan establishes that by 2030 there will be 26 hydroelectric plants in the Amazon region, six of which are already under construction. This plan sets out two strategies, namely, to supply energy to mining companies and, prioritize hydroelectric power as the main source of electricity in the country to supply the more developed Center-South regions with energy generated by the voluminous Amazonian rivers. Once the decision to implement this plan was taken, the Amazon no longer existed for its own sake, but became an important asset for the rest of the country; its river basins and nature in general left the human sciences and geography textbooks to be transformed into low value raw materials on the international stock exchanges.

At first glance, this government policy could be considered a development plan for the region, but it is not. It is a permanent plan for exploring the area for the benefit of more developed areas to the detriment of the Amazon. First of all, this decision did not take into account that the Amazon plains require the building of very large reservoirs to accumulate sufficient water to generate enough energy, resulting in an enormous loss of land and biodiversity; hydroelectric complexes also lead to the displacement of indigenous communities, the so-called forest people, as well as riverside populations. Furthermore, "development" policies established by the Federal Government for the Amazon region since the 1960s have attracted millions of immigrants. The result has been uncontrolled population growth which put an enormous pressure on natural resources and local public services such as health, education and security, further reducing the poor living standards evident in the region. The population of the seven Amazon states grew from 2,579,442 in 1960 to 15,864,402 in 2010 (IBGE 2010). In addition, this growth proved to be a way of transferring poverty from other regions to the Amazon, instead of implementing public policies aiming to reduce poverty in the migrants' places of origin and the country as a whole. Throughout the twentieth and twenty-first centuries there has not been a global development plan for the region, but only projects to exploit its natural resources to benefit the Federal Government and other Brazilian regions. The 1987/2030 Energy Plan symbolizes this attitude, revealing the government's intention to explore the Amazon on a permanent basis.

The export of pig iron, aluminum, alumina, copper, manganese, cattle and other products stimulated by Brazilian policies over the last decades consolidated the position of the Amazon region as a producer of commodities. This could be called the *re-primarization of the Brazilian economy* in so far as the production of these commodities has been promoted not only in the Amazon region, but in other parts of the country. The profile of Brazilian exports today is similar to that of the 1950s when Brazil was an agrarian society. Just five commodities (iron ore, crude oil, soybeans, sugar and meat) currently account for 49 percent of Brazilian exports (Ministério do Desenvolvimento Indústria e Comércio Exterior – Ministry of Development and Foreign Trade 2015). Moreover, iron ore (14.3 percent) and cattle meat (6.58 percent), responsible for over 20 percent of exports, are mainly produced in the Amazon. Incentive policies for the production of primary goods represent an enormous setback in terms of Brazil's status, reinforcing its place as a peripheral economy in the Western world. They also cause immeasurable loss to the two richest Brazilian biomes situated in the Amazon and the Center-West (home to the *Cerrado* biome) regions. When large tracts of an area that is probably the richest and most biodiverse tropical forest in the world are transformed into barren pastureland, it is, at the very least, controversial.

The Amazon entangled in internal neocolonialism

After Brazil's independence, the Amazon economy remained unaltered until the end of the nineteenth century with the beginning of the rubber cycle. Natural rubber filled the coffers of the Federal Government, foreign banks and those of a very small regional elite, while fueling the already significant social inequality. The Brazilian government consistently ignored and abandoned the local, incipient and fragile rubber goods manufacturing sector, while benefiting from the arduous work of rubber tappers which also involved thousands of poor immigrants attracted to the forest where they became sick and often died.

In Brazil, history textbooks register the exploration of "herbs from the forest" as a typical example of the economy of the old colonial system and the "rubber cycle" as an important but short-lived historical event in the Amazon's history, restricted to the few years at the end of the nineteenth and the beginning of the twentieth centuries. However, local social history shows that both cases are examples of the typical vigorous form of exploration that has persisted to this day, albeit involving new products.

The Manaus Free Zone, situated in the capital of the state of Amazonas, is the only industrially-based economic space in the region. It could be analyzed in terms of a rupture from the primary commodity exports–oriented economic model predominant in the region. However, it is not an exception to neocolonialism, since the only beneficiaries are multinational companies. In my understanding, the greatest benefit the Manaus Free Zone has brought to the State of Amazonas cannot be measured in monetary terms. It is the fact that it has spared the state from the environmental damages other Amazon states have suffered from for over half a century. For example, between 1970 and 1980 the population of Rondônia grew by 324 percent and by 2000, deforestation amounted to 28.5 percent of the state's area and continued to grow in the following years. In the other Amazon states deforestation grew at a lesser pace. However, only the state of Amazonas managed to reduce its deforestation rate (7 percent). By 2013, total deforestation in the region amounted to 762,979 km^2, corresponding to between 18 and 20 percent of forested areas (Nobre 2014).

The Amazon's primary-export oriented model has very clear characteristics: it is enduring, promotes income concentration and its main bases lies outside the region (markets in the Brazilian Center-South and abroad). In addition, it neglects local populations. Another feature is that the Federal Government and large economic corporations have been in charge of this process since the end of the nineteenth century. The current economic model based on the production of electricity, iron and other commodities, such as beef cattle and soybeans, consists in a new form of internal colonialism. Although this model is the continuation of a perduring phenomenon, it presents marked and distinctive phases.

The beginning of recent transformations

From the 1970s, the 'Amazon frontier' became a constant and significant topic of academic and political studies and debates in Brazil. The various human groups which began to occupy the region from the mid-sixties onward were predominantly small rural producers involved in family farming. The 'frontier' was thus a place with sufficient land to allow for the number of peasants to continue to grow. Thus, in the 1970s, poor peasants and financial capital, stimulated by the authoritarian government of the military dictatorship, relentlessly pushed back these 'frontiers', taking advantage of the roads opened to reach new lands.

For the poor peasant migrants who came to the region, the Amazon was a place where they could pursue the utopic dream of constructing and reconstructing their family history under better conditions than in their places of origin. Nonetheless, in their view, it was also, and still is, a place of conflict – where the land as *workplace* clashes with the interests of those who consider it an *enterprise* and a capital reserve. The economic groups, businessmen and unscrupulous exploiters who settled in the region exerted strong pressure both over nature and over the various government spheres, in direct proportion to their greed for profit and the size of the territories they appropriated. There was less pressure when pioneers only aspired to a better life, as in the case of family farming. Nonetheless, all segments are a threat to the land, the way of life, culture and rights of the region's indigenous population.

In the 1970s and 1980s, the Amazon served some very specific functions, for example, the Federal Government used the region (via stimulated migration) to diffuse tensions resulting from social discontentment, regional crises (such as droughts in the Northeast), and the lack of income distribution policies. To excuse government planners and authorities, during the 1970s and 1980s, by arguing that they lacked scientific knowledge about the fragility of the Amazon ecosystem is to be extremely condescending: their disregard for the intrinsic values of nature did not rest on (their lack of) economic or scientific knowledge. The reason for attributing zero value to the richest and most biodiverse forest in the world (as evidenced by the federal and state agencies' concessions of land and incentives for cattle raising and logging which lasted the entire military dictatorship period – 1964/1985) were cultural: the elite's prejudice against the "jungle." This wild and untamed habitat provided unquestionable proof of Brazil's condition as a primitive society.

The Brazilian government's disastrous choices and the advance of the primary export oriented economy

The loans taken out in the 1970s led to the debt crisis of the 1980s and part of the 1990s, resulting in approximately zero GDP growth for the next 15 years (1980/1994) and uncontrolled inflation that began to surge from the mid-70s onward. In 1975, annual inflation reached 29.89 percent; in 1976, 38.06 percent; in 1977, 41.51 percent and in 1979, 79.42 percent (Baer 1987). During the following years, the inflationary scenario worsened: between 1981 and 1990 GDP growth was 1.6 percent and inflation reached 224 percent in 1984. Despite several economic plans, inflation continued to grow and, even with the onset of the democratic regime in 1985, it remained uncontrolled (above 200 percent per year) until 1994, when the Real Plan established a new currency, stabilizing the economy. In the middle of a crisis, in order to boost the economy, the military government decided to attract foreign investments in mining and the metal sector and permitted the use of wood charcoal. Moreover, they decided to construct the fourth largest hydroelectric plant in the world (Tucuruí/Pará) to support new mining ventures, while at the same time encouraging cattle raising and logging.

In addition to the Central Government's economic errors and misjudgments, long-standing and widespread cultural misconceptions also prevailed, such as the idea that nature in the Amazon region is extremely abundant, resistant, limitless and self-regenerating. As a result, disregarding the fact that the Amazon is the largest genetic bank in the world, businessmen and adventurers of all kinds separated the forest (that they exhaustively exploited) from the idea of biodiversity (considered exclusively a topic for scientific studies). In the 1970s and 1980s the average annual deforestation rate in the region was 21,130 km^2, reaching an annual average of 29,059 km^2 by 1992/94; it started to decline in 2004, dropping to 5,831 km^2 in 2015. Nevertheless, the economic model of commodity production continues to be an environmental threat (MMA 1998, INPE/IBAMA s/d, INPE/PRODES 2016).

The Amazon in the current context

Today, the Amazon region can be understood as a *commodities frontier* and exporter of electricity. It is important to consider at least two essential points related to this *commodities frontier* phase:

1 In the current form of neocolonialism, the leading role in terms of public policies concerning the Amazon is played by Brazil itself, in contrast to colonial times when Portugal decided the forms of occupation in the region. Both in previous decades, when the Amazon functioned primarily as a *frontier for expansion of peasants and capital* (1966/1985), and in the current phase, which started in 1985, when the region started to play the role of *commodities frontier*, it is the Federal Government that stimulates the regional economic model. In other words, although the world today is much more globalized than at any other time in history, the main structures that make up this phase of neocolonialism and their connections with global commodities markets depend far more on internal decisions than on any external conditions. The simplistic theory, which has predominated for decades, that the economy of peripheral countries is defined by the rich countries, cannot be considered absolute: there are innumerous productive activities that can be implemented through internal social and economic policies, creating an economy that depends very little on the fluctuations of foreign markets, promoting solidarity while respecting local knowledge and culture. This is especially viable in the Amazon region which has the largest genetic bank in the world and in Brazil which has an internal market of almost 200 million people.

2 The role of the Brazilian government as promoter and inducer of the commodity export oriented model is somewhat paradoxical, given that this has only increased regional inequalities, to the detriment of the Amazon region.

It must be emphasized that the commodity export oriented model that has marked the Amazon since colonial times has changed and been converted into a form of regional neocolonialism that is just as imbricated in social exclusion as during the colonial period, with the aggravating factor of also containing the seeds for the depredation of the environment. The public policies implemented by the Federal Government in the region since the 1970s have produced more negative than positive effects. The highways that opened up the forest in the 1970s and 1980s provoked the uncontrolled occupation of land, produced chaos in terms of land ownership and violations against one of the most diverse environments in the world. Incentives to extensive cattle raising in the region resulted in a herd of 70 million heads of cattle, occupying an area of approximately 70 million hectares. In the 1960s, logging in the Amazon region was responsible for a very small portion of the national production as there were no roads connecting the region to the rest of the country. Almost 30 years later (1989), logging in the region accounted for 27 percent of national timber production (Meirelles Filho 2014; IBGE 1989). Illegal felling of the native forest continues as before: at least 78 percent of logging in the State of Pará between 2011 and 2012 was illegal and mainly exported to Europe and the United States (Imazon 2013). Large concessions of land to logging groups settled alongside the highways were, and continue to be, followed by land conflicts. In these areas, big business clashes with the poorer riverside communities and ethnic minorities who suffer all sorts of human rights violations: the General Secretariat for the Pastoral Land Commission reports that 622 land conflicts were registered in the Amazon in 2009 and 489 in 2012 (CPT 2010; CPT 2012).

Governmental policy options were not decided outside Brazil; they did not derive from international pressures or conjectures. On the contrary, policies implemented in the Amazon region were introduced by the Federal Government in order to solve the problems of other regions, the Brazilian Treasury and the country's international trade balance. That is why after the military dictatorship, these policies continued and were further developed as part of the Federal Government's actions for the region. It could have chosen less economically damaging and more democratic options such as promoting the cultivation of local fruits, of which there is an enormous range still unknown in developed countries, fish-farming (considering that the region contains about 1,800 fish species, 100 of which are classed as edible) and the promotion of the chemical and pharmaceutical use of native plants.

The Amazon region's population started to grow in the 1970s, caused by the migration to the area of poor inhabitants of other regions and by businessmen and big companies, and continues to this day, becoming a permanent phenomenon and completely changing the region's traditional occupation patterns and increasing the pressure on natural resources. Over and above its status as a *commodities frontier* and exporter of electricity, the Amazon region acts as an internal colony, explored internationally, with the acquiescence or consent of the authorities and national elites. From a geo-economic standpoint, the Amazon region could be said to be peripheral, subordinate and dependent.

The current model is harmful to the environment

Commodities production (cattle, steel and other metals, and charcoal from the native forest) is responsible for alarming rates of deforestation. Deforested areas began to be measured in the Amazon region from 1988. Although it is well-known that rates have been very high since the 1970s, 1990s data provide an example of the scale of the disaster. In 1994/95, 29,059 km^2 of forest was cleared and deforestation remained high until the beginning of the twenty-first century (in 2003/2004, 26,130 km^2 of forest was cleared (INPE/PRODES 2016)). Since 2004, deforestation in the Amazon region has been decreasing on a yearly basis, although with some fluctuation. In 2012/13 the area of forest felled was 5,891 km^2 and this dropped to 5,012 km^2 in 2013/2014 (although deforestation grew again in 2015 to 5,831 km²). Much less, therefore, than in the previous four decades of the twentieth century. Although this is welcome, the environment is just as much of a concern as before for the following three reasons:

1   The production of commodities, in particular soybeans, has been displaced to a different region, in particular, the Brazilian Center-West, where it is advancing into the biome known as *Cerrado*, where fragile riparian vegetation and water sources have been destroyed in order to plant soybeans. The clearing of the *Cerrado* has caused inestimable environmental damage to the country, as it is the source of many important rivers, forming six major Brazilian hydrographic basins. The hydric potential of the *Cerrado* is such that Brazilian geographers have named it the *Cradle of Rivers*. Thus, it is this habitat that is now being destroyed instead of the Amazon. International pressure on the Brazilian government has resulted in increased actions to protect the Amazon biome, forcing large soybean farmers to turn to the *Cerrado*.

2   Studies have shown that to produce 1 kg of beef, 20,000 liters of water are needed and 2,000 liters of water are necessary to produce 1 kg of soybeans (ANDA 2009). This means that it would be more profitable and rational, though absurd, for Brazil to export water to countries lacking this resource, preventing deforestation, soil degradation and river pollution.

3   Both the Amazon and the *Cerrado* continue to produce an enormous environmental liability, that is, there has been no recovery of deforested areas, and in the few cases when this has occurred, the species planted were exogenous, such as pine or eucalyptus, further eroding the fragile soil environment of both biomes.

A new factor in the region's process of exploration of natural resources – power generation

The Amazon has been a *commodities frontier* since the 1970s with the construction of the Tucuruí hydroelectric plant. It was inaugurated in 1985, representing the start of the second phase of the recent capital expansion. It also symbolizes the new transformations which took place between the two phases (the Amazon's transition from a *peasant frontier* to a *commodities frontier*). Its implementation was mainly motivated by the mining ventures that started in the region. Once in operation, 50 percent of its highly subsidized energy was sent to the mining companies in the states of Pará and Maranhão. Given their need for vegetable charcoal and cattle ranching, mining companies were responsible for vast areas of deforestation in Southeast Pará and Western Maranhão which are currently known as the "Arc of Deforestation." Until recently, these areas were covered by pristine, dense forest, but now 256 municipalities are established here, 52 of which have been seriously deforested (MMA 2013).

Changes in the transmission regime only occurred once hydroelectric power capacity increased and the electricity produced is now supplied as divided thus: one-third to mining companies and mills, one-third to the national electricity grid and one-third to the Amazon region. In short, while the Tucuruí plant was unable to reach maximum capacity, mining companies and mills had priority over the local population. Today, the highways are no longer responsible for the arrival of venture capital or migrants into the Amazon; they have been replaced by mineral prospecting and exploration and government infrastructure works to support the commodity export oriented economy (such as waterways to transport soybeans and other products), as well as opportunities for major investments. Economic activities have changed. However, whether they are the result of the first phase (1970s and 1980s) – logging, cattle raising and settlement projects – or the second phase – mining, steel and other metals production, grain and palm oil cultivation and power generation – these activities all have common characteristics: They evolved regardless of the interests and ways of life of the local populations; involve large investments; receive government support; and have very serious impacts on the environment.

The *commodities frontier* model leads to income concentration and violates the Constitution

The primary export oriented economic model in force in the region since the old colonial times to the mid-twentieth century and based on forest products (export of brazil nuts, forest essences, oils and seeds, among others) did not involve the destruction of the forest but its conservation, because plant-based extractive activities required as much forest as possible. If the military government defined their development plans for the Amazon based on environmentally harmful activities, the post-1985 democratic government showed no better understanding of Amazon issues than their predecessors. The current model is also underpinned by two lines of activities which are harmful to the environment and whose origins also lie in the military regime. The first was aimed at the foreign market: mining and the metals industry in their primary phases. The second, power generation, not only supports the above activities, but also meets the demands of other Brazilian regions.

What emerges from the continuity of these two economic models is that the issue is not only democracy – to a greater or lesser degree – but a permanent lack of understanding of the Amazon as a biodiverse and fragile region and its potential both within the national and international contexts. Moreover, the Federal Government imposed on the Amazon region the status of 'power generator' for the rest of the country without providing compensation for a loss of profits in carrying out this activity. This argument could also be applied to commodities production, given that in the Amazon the states are legally forbidden to levy local taxes on exports, as set out by the Constitutional Amendment n. 87, 1996, known as the Kandir Law. At the time, the Brazilian trade balance was heavily negative and in order to make Brazilian commodities more attractive, prices were lowered with the approval of this law in Congress, exempting commodities and energy from local taxes. This measure seriously affected the Amazon region by reducing its tax revenue base. Furthermore, the Federal Government didn't introduce any measures to compensate for the consequences of its incentive policies, namely, a loss of resources, violence caused by social unrest and a lack of adequate basic services due to the resultant increase in population and environmental damages.

The situation described above raises issues and violations of different orders. The first is moral: policies introduced result in the re-establishment of internal neocolonialist relations where the region is in a subordinate position and unable to define its own history and destiny; in addition, the authorities do not take into consideration the rights and ways of life of the local population when deciding to develop activities and projects. The second relates to a breach of the Brazilian Constitution considering that it states in more than one article, and in particular in Article 3, that one of the government's main objectives is to "eradicate poverty and exclusion and reduce social and regional inequalities" and, in Art. 43, which refers to regional public policies and states that the Government "must focus on the reduction of the country's social and regional inequalities." However, as we have observed, economic policies underway serve only to reinforce regional inequalities to the detriment of the Amazon states.

The current economic model has been unable to raise the region's GDP

One of the most serious shortcomings in the social sciences dedicated to the study of the Amazon region relates to its inability to address the issue of as to why Brazil has acquiesced to ventures that produce so little benefit? Any first year Economics student learns that such ventures lead to an increase in GDP but that benefits incurred are not retained for a number of reasons:

1   Given these enterprises form economic enclaves, companies are not rooted in the local economy and do not participate in all stages of the productive chain which ends with the production of final goods, as a result, they create few jobs and little wealth in the region.
2   Enclaves produce only raw materials or semi-finished products with low market value (thus, companies are only interested in producing large quantities). When comparing the regional GDP *per capita* of the various Brazilian states in relation to the National GDP, it is possible to observe the unfavorable position of the Amazon/North Region, and in particular, that of the state of Pará, the location of mining companies and mills producing semi-finished products and the fourth largest hydroelectric plant in the world. In 2010, the state of Pará's GDP was only 52 percent of the average GDP of Brazilian states. Furthermore, the enclaves-based model does not present future possibilities for improvement, with forecasts

of 62.93 percent of the average GDP of Brazilian states for 2022 (IPEA). Among the five Brazilian states with the smallest GDP in 2010, four are found in the Amazon region, with the exception of Piauí in the Brazilian Northeast (0.6 percent), namely, Tocantins (0.4 percent), Amapá (0.2 percent), Acre (0.2 percent) and Roraima (0.2 percent) (Terra 2016).

3   These enclaves damage the environment.

4   Since enclaves produce minerals, their exports are classed as tax exempt. Therefore, even if states producing raw-materials and semi-finished products had a high GDP, they would not be able to improve the social welfare of their populations, given that the income generated by these companies is not socially distributed but sent directly to the coffers of economic conglomerates.

The most serious cases of income loss in the Amazon region occur in the states of Pará and Maranhão, where the Carajás mining complex is located. During the first quarter of 2012, Vale (the world's fourth largest mining company) produced 83,900 metric tons of iron ore, approximately 340,000 metric tons/year, the equivalent of $1.5 billion Brazilian reais of export sales, only for that year. In addition to iron, Vale is also the largest producer of gold – 18 t/year – and other products in Latin America. Despite its mineral wealth, the state of Pará has one of the worst GDP *per capita* among Brazilian states, equivalent to less than half of the national *per capita* GDP average. In 2013, only five Pará municipalities out of 144 had a *per capita* GDP above the Brazilian average – R$26,445.72 (US$13,225). While the GDP *per capita* of the more industrialized states was R$38,262.13 (US$19,131), in Pará it was only R$15,176.18 (US$7,588) (IBGE 2015). Both in Maranhão and Pará poverty is visible everywhere. Income concentration, a result of the commodity export oriented economy, in particular because of new enclaves and the Federal Government's extortion mechanism (imposed by the Kandir Law), is one of the most serious obstacles to improving the quality of life of the region's population.

Theories that explain the state of economies based on the export of primary goods are highly repetitive in that they insist on the simplistic argument that the international division of labor assigns peripheral countries a status of exporters of raw material and semi-finished goods. These are indeed valid explanatory factors, but not the only ones. This type of analysis disregards the conditions, obligations and internal public policy decisions of countries, as well as the decisions of national and international interest groups and their alliances with international corporations. In the case of the Amazon region, even after unsuccessful experiences like those of the rubber boom, manganese, timber and other products, the Federal government continues to insist on imposing on the region a raw materials and semi-finished goods export model. In Brazil, the deeply rooted culture of corruption and public resources misuse in public administration are factors that drain the vitality out of policies, reducing potentially positive impacts.

The decision by the Brazilian government to consolidate and boost the export model (at least in its recent form) was voluntary. It was the sole funder of the Tucuruí Hydroelectric Dam, exempting the Japanese government from these expenses, even though the first contract signed by the interested parties stipulated that Japan should bear part of the infrastructure costs. Japan's exemption was all the more absurd given that this occurred in the 1980s when the national economic situation was so critical – due to high inflation, economic stagnation and Brazil's large public debt – that economists called it the "Lost Decade." Within the global context, assigning the region to a position of secondary economic importance, of periphery and of exporter of primary products deepened its dependency on foreign markets and weakened the already fragile autonomy the region enjoyed until then.

Large projects are now commonplace in the Amazon landscape

The *commodities frontier* was boosted through innumerous new ventures: the production of grains and palm oil is advancing in the Amazon region, in areas that have been previously deforested, as well as in the *Cerrado*, where it is taking over riverside and water spring areas.

> As a whole, the mining companies will invest US$ 24 billion dollars between 2012 and 2016 to increase production of iron ore, bauxite and other metals found in the Amazon basin, according to the Brazilian Mining Institute – IBRAM. Brazil has already received one fifth of the investments in the world, and the Amazon represents, for many people, the largest unexplored potential in the Country. According to Fernando Cara, president of IBRAM "The Amazon will be our California".
>
> (*The Wall Street Journal* 2012)

Furthermore, the so-called "green economy," established in areas deforested in recent decades, has been transformed into monocultures with the predominance of pine trees. These are commercial forests that destroy, rather than recover, regional biodiversity which could have been achieved through the cultivation of economically profitable endogenous crops and respecting, at least to some degree, the diversity of species, as would have been expected from the recovery of degraded areas in the Amazon region. "Green municipalities," the pride of some mayors, destroy one of the key features of nature in the region – its valuable and abundant biodiversity. Meanwhile, the region is slowly losing the characteristics for which it is known across the world – the most biodiverse system of rivers and forests on the planet (Vieira et. al. 2001). Indifference and lack of understanding of the scale of the disaster is also present among researchers and specialists, as observed when, in 2015, Embrapa-Pa (the Pará division of the Brazilian Agricultural Research Agency) suggested the "recovery" (EMBRAPA 2015) of degraded pastureland in the Amazon region through the cultivation of eucalyptus and soybeans.

Table 3.1 Urban population growth in the Legal Amazon: 1960–2010

Situation	1960	1970	1980	1991	2000	2010
Urban	1,041,213	1,784,223	3,398,897	5,931,567	9,002,962	11,664,509
Rural	1,888,792	2,404,090	3,368,352	4,325,699	3,890,599	4,199,945
Total	2,930,005	4,188,313	6,767,249	10,257,266	12,893,561	15,864,454

Source: IBGE – Demographic Census for 1960; 1970; 1980; 1991; 2000; 2010.

In addition to these activities, there is the expansion in mining and the building of new hydroelectric plants which, just as in the 1980s, are implemented without prior consultation with the local population, in particular the indigenous communities. This disregards the fact that Brazil is a signatory to the International Labor Organization's Convention 169 that ensures ethnic minorities the right to express their view on projects that affect them.

Social breakdown becomes all the more evident when analyzing the growing migration from rural areas to the cities. The 1960, 1970, 1980, 1991, 2000, 2010 demographic censuses (see Table 3.1) provide a good explanation as to why the urban peripheries of cities in the Amazon region have the lowest social indicators in Brazil: there are no local funds to provide these cities with basic public services.

An inevitable consequence – increase in inequalities

In this scenario, it is not surprising that the Atlas of Human Development reveals that out of the 48 Brazilian municipalities with the lowest living standards, 35 are located in the Amazon region. In fact, the municipality with the worst living standards among all 5,565 Brazilian municipalities is in the state of Pará (Melgaço), the Amazon state with the largest number of mining and metal processing companies (PNUD/IPEA/FJP 2013). The Atlas of Human Development showed that among the 30 municipalities with the lowest Human Development Index (HDI) in the country, 18 were located in the Amazon region (Época Negócios 2013). When adding together the percentage of people living in extreme poverty (monthly income of less than one quarter of the minimum wage) and in poverty (monthly per-capita income between one quarter and one half of the minimum wage) it can be observed that in 1990, the poor and extremely poor made up 45 percent of the regional population; in 2002 they were 44 percent. In 2009 the two categories – poverty and extreme poverty – encompassed 53 percent of the population of the state of Maranhão, 40 percent of Acre, 45 percent of Pará, 40 percent of Amapá, 42 percent of Amazonas, 37 percent of Tocantins, 39 percent of Roraima, 30 percent of Rondônia and 24 percent of Mato Grosso. According to IBGE/ PNAD (2009), the average number of people living in poverty and extreme poverty in the Amazon region was 42 percent, in comparison to 29 percent for the rest of Brazil.

In 2010, in Pará, the Amazon state where the largest mining companies are located and one of the largest timber exporters in the country, 53 percent of the families had a monthly income lower than the national minimum wage. This means that this economic model has only aggravated regional and social inequalities (IBGE 2010). In over 20 years (1990/2010), the Northern Region's participation in the GDP has only improved by 0.6 percent. The economic model established is not only destroying the richest ecosystem in the world by exposing the region and the country to the risk of severe and harmful environmental changes, but it is also making its population the poorest in the country (Vieira 2012).

In the areas of education and culture, the North states are the most seriously affected in Brazil. According to the Ministry of Education and Culture (MEC/INEP/IDEB 2013), the state of Pará had the worst national performance in terms of secondary school education (2.9) and performed equally badly in primary education (3.6). This demonstrates that these large economic ventures cannot be transformed into social benefits. Although the other Amazon states have shown slightly better results than Pará, they still present some of the lowest indicators in Brazil, alongside the Northeastern states (on average this region performed better than the North) (ibid.).

Modernization gone wrong

Therefore, are these investments in the Amazon region worthwhile? There are many answers and they require reflection:

1   It is undeniable that the federal government has made heavy investments in the region for almost five decades. Nevertheless, the dubious manner in which these investments were implemented – confronting cultures, ignoring local knowledge and destroying nature – in addition to wrong decision-making – cancelled out positive and strengthened negative effects. Consequently, results have not corresponded to expectations.

2   The population's living standards were neither modernized nor improved due to a number of reasons already mentioned in this text. Large corporations produce and export goods which are in the main raw materials; there is no product diversification and a few dozen large companies form "modern enclaves" and control the entire production and trades systems. They are considered modern because they use very efficient and sophisticated machines and equipment that increase production and productivity. Since these companies do not process their goods internally or finish them locally, they do not create a chain of small- and medium-sized businesses. The result is that these large investments neither develop the regional economy nor distribute income. On the contrary, they increase income concentration, create few jobs and pay low salaries. Finally, governments continue to invest in available resources to strengthen this destructive economic model: as already stated, this is *modernization gone wrong* (Loureiro 1993).

Modernization gone wrong did not eliminate arduous, degrading work, or even slave labor which is still found with some frequency in areas which have been deforested to create pastureland and in the exploration of charcoal to supply mills. On the contrary, these new economic activities bring back to the region old forms of work (Guimarães 2005). In 2004, as result of repeated accusations made by the press and institutions defending workers' rights which revealed the existence of slave labor both in the metal industry and in charcoal production, the Attorney General's Office (PGR) and the Ministry of Labor, together with companies and institutions representing workers signed an agreement in which the mills committed themselved not to employ slave labor in the production of charcoal. However, by 2012, there had been little change. The weekly magazine *Revista Época* (03/27/2012) reported on the mills' illegal practices in the states of Pará and Maranhão, based on a survey conducted by *Instituto Observatório Social* (Social Observatory Institute). At the time, several mills were found using charcoal produced by slave labor and encouraging poor settlers to clear the forest in their small plots to guarantee the cheap supply of charcoal the mills needed.

3   It is important to stress that the current economic model violates the Constitution in several aspects, especially with regard to Art. n. 170 which states that:

> the economic order, based on the value of human labor and free initiative, aims to ensure a decent life for all individuals, according to the precepts of social justice, observing the following principles.
>
> ...
>
> VI – defense of the environment;
> VII – reduction of regional and social inequalities

4   The national programs for the Amazon region have revealed the inability of the Brazilian government to understand, articulate and develop the region's potentials. The foundations upon which the current Amazon development model are based – in particular the perpetual pursuit of progress and modernization – reveal the contradictions within the Brazilian government in relation to the Amazon. This is no longer a simple regional question, and has not been for a long time, it has become a national, and even an international, question. The complexity of the Amazon was underestimated during the dictatorship and this is still the case. Whenever the region is seen through a simplistic lens, it undermines the country's governance capacity. During an informal conversation on the region, the noted Brazilian sociologist, Prof. Octávio Ianni, once said to this author: "For the Brazilian people, the Amazon is an enigma that still needs deciphering."

5   Governments and the economic elites have always considered the region's populations to be an obstacle to development, though they have been conserving, protecting and caring for the environment for centuries. Their

interests are exclusively considered within a market logic, often resulting in conflicts. However, the same perspective is not taken when giant projects squander national resources and collective assets, such as the Amazon's natural environment.

6   As Western developed countries evolved, levels of democracy and the respect for constitutional rights increased. However, this has not been taking place in relation to the Brazilian Federal Government's modernizing project for the Amazon (neither during the authoritarian period, nor the phase that followed). The link between modernization and universalization of benefits, a feature of modernization in the central countries, has been internally broken. This is why real democracy and social justice in the Amazon region were transformed into diffuse and abstract ideas of democracy, confined to legal texts, and remain distant or absent from people's lives. This reinforces a paradox: the Federal Government has for decades proclaimed a modernization that does not materialize in people's daily life, that is, a more inclusive and diverse justice, capable of fulfilling the expectations of the poorer and socially excluded population or of those who are culturally different. Several recent programs have willingly consolidated the pseudo "vocation" of the region as an exporter of primary goods. When I refer to the voluntary capacity with which groups in the government have to act, I do not disregard the existence of a highly competitive global market where central economies generally sell advanced technologies and peripheral countries export raw materials. Nevertheless, I want to stress that, despite this global context, there are internal choices which are entirely ignored and disregarded by both central and regional governments. For more than a century, government actions have clearly demonstrated that the Amazon's potential is not used in a truly modern and rational way. In many cases, governments made the wrong decisions, in others, political groups have clearly chosen to assist international capital to acquire economic advantages.

Final considerations

Market logic and the ideology of progress established a form of seeing the regional populations as if they were marching in the opposite direction to history. They have forged a form of seeing economic growth, "development" and "progress" in opposition to, or irreconcilable with minorities and the poorer and voiceless populations in society. These groups are considered inferior, undesirable, unexpressive or are simply an obstacle. Moreover, in recent years, environmental issues have relegated the violation of rights and the harmful effects of this economic model on local populations to a secondary position. These groups, in particular indigenous communities, have not had a right of say with regard to any of the hydroelectric and mining projects of recent decades, despite the fact that this requirement is enshrined in legislation, although, in some cases, they have openly and blatantly expressed their opinions against these projects. But these populations' requests that original projects be altered to conform to their just demands generally fall on the deaf ears of economic elites and those in positions of power. There is no doubt that within the national context the Amazon region has been forced into a neocolonial position.

It is interesting to note the consensus that the classical colonial system is outdated and "antiquated," while the current model is considered modern. Even though, in essence, the two systems are identical, as they are both forms of colonialism. Nevertheless, historical colonialism differs from the current form of neocolonialism in the Brazilian Amazon on the following: while the old colonial system was imposed, the current neocolonial model was a definite and voluntary choice of governments which decided on the same commodity export-oriented economic model that had been so criticized by history and economics textbooks in schools and universities. This voluntary decision can be evidenced by the innumerous incentives, attractive options, facilitations, consortiums and joint projects involving the government, as well as other actions whose aim is to attract new primary goods export businesses to the region.

"Modernization gone wrong" is manifested in everyday life. Take any "modern" activity, for example the iron industry. It does not have any traces of modern civilized Western culture. On the contrary, this activity implies slave labor, the destruction of native forest for charcoal production, environmental degradation, the violation of laws, disregard for collective welfare rights and wealth evasion. This is the same old economic model dressed in new robes and it continues to sustain economic enclaves which are now considered "modern" because they use efficient and sophisticated machines and equipment to increase production and productivity. However, they maintain the same old forms of labor exploitation and disrespect for the environment.

When governments and companies showcase development policies, funding and giant infrastructure projects as symbols of progress and the region's modernization (in particular big mining companies and powerful hydroelectric dams) and promise the future improvement of the region's living standards, they omit some background information – that the GDP has increased by means of economic enclaves which have enormous negative impacts on the environment and on working conditions. A question arises in terms of the objectives of development: although material gains are implicit in the concept of development, they cannot be reduced to economic results which, when converted into numbers, become tokens and are the only possible expression of progress and modernization.

By stressing the value of export sales – grain production in 2014 attained 204.5 million tons (Ministério da Agricultura 2015) – or pointing to a future power crisis, the Federal Government, the media and the elites mobilize the national sentiment in the name of progress and better future living standards for the country's population. However, what sort of peculiar policy is this? It does not take into account environmental damages. It does not care for other illegal practices such as the violation of the human rights of the poorer population who are subjected to degrading working and living conditions. It also does not consider the rights of minorities (indigenous populations) who are forcibly removed from their territories, without any regard for the law and international treaties to which Brazil is a signatory. Daily life and statistics have consistently contradicted the promise of improved living standards for the local population. No utopia, whether progress or development, justifies the violation of rights of poorer people and other groups widely known to be living in abject poverty and unable to confront economic groups and the government.

It is important to highlight that the future is constructed on today's utopias and not on a continuity of the past and that, between the past and the future, there is a present that leads us to reflect and incites us to break with the past and plan a better and more compassionate future. Breaking with the past means to experiment with new routes, test possibilities and take advantage of opportunities, based on regional wealth and local knowledge, with the indispensable support of science and technology. This break demands courage to invent new forms of production and social coexistence which are less excluding and more compassionate. This is undoubtedly a hazardous path to take, but every new venture is hazardous. The perpetuation of the past and its projection toward the future means continuing on a path of inequality and ever growing violence: a path we already know and reject. In order to change our route, first and foremost, the Brazilian government needs to change its attitude toward the Amazon region. A modern country is more democratic when it serves the interests of its people, when it protects its poorer populations and minorities against social violence, when it defends the environment against excessive depredation (as in the case of the Amazon region), and maintains it for the survival and enjoyment of future generations. Finally, a modern country is more democratic when it conciliates real development with respect for human rights.

References

ANDA (2009) Impacto da pecuária bovina – Parte 1 (www.anda.jor.br/29/06/2009/impacto-da-pecuaria-bovina), Accessed 13 February 2016.

Baer, W. (1987) "A retomada da inflação no Brasil: 1974-1986," *Revista de Economia Política*, 7 (1), 2–972.

CPT (2010) Dados 2009 (www.cptnacional.org.br/index.php/publicacoes/noticias/articulacao-cpt-s-da-amazonia/199-dados-2009-release-amazonia-concentra-maior-numero-de-conflitos) Accessed 13 February 2016.

CPT (2012) Conflitos no Campo no Brasil (www.cptnacional.org.br/index.php/downloads/viewdownload/43-conflitos-no-campo-brasil-publicacao/316-conflitos-no-campo-brasil-2012) Accessed 13 February 2016.

EMBRAPA (2015) Soja e eucalipto para recuperação de pastagens no sudeste paraense (www.embrapa.br/busca-de-noticias/-/noticia/2673414/soja-e-eucalipto-para-recuperacao-de-pastagens-no-sudeste-paraense#) Accessed 30 May 2015.

Época Negócios (2013) As 30 Cidades do Brasil com o melhor e o pior IDH (http://epocanegocios.globo.com/Informacao/Resultados/noticia/2013/07/30-cidades-do-brasil-com-o-melhor-e-o-pior-idh-do-brasil.html) Accessed 30 May 2015.

Guimarães, E. C. (2005) *Trabalho Cativo por Dívida na Amazônia Paraense*. Instituto de Ciências Jurídicas, Universidade Federal do Pará.

IBGE (1989) Anuário Estatístico do Brasil, 1989, R.J., v.49, 716 p, p.335. (http://biblioteca.ibge.gov.br/visualizacao/periodicos/20/aeb_1989.pdf).

IBGE (2010) Instituto Brasileiro de Geografia e Estatística. Sinopse do censo demográfico 2010 (www.censo2010.ibge.gov.br/sinopse/index.php?dados=8) Accessed 13 February 2016.

IBGE (2015) *Contas regionais do Brasil: 2010-2013* – Rio de Janeiro n. 47 (http://biblioteca.ibge.gov.br/visualizacao/livros/liv94952.pdf) Accessed 16 February 2016.

IBGE/PNAD (2009) *Pesquisa Nacional por Amostra de Domicílios* www.ibge.gov.br/home/estatistica/populacao/trabalhoerendimento/pnad2009/ Accessed 16 March 2016.

IMAZON (2013) Boletim Transparência Manejo Florestal Estado do Pará 2011-2012 (http://imazon.org.br/publicacoes/boletim-transparencia-manejo-florestal-estado-do-para-2011-2012/) Accessed 13 February 2016.

INPE/IBAMA (s/d) Amazônia Desflorestamento 1995-1997 (http://www.obt.inpe.br/prodes/Prodes1995-1997.pdf) Accessed 13 February 2016.

INPE/PRODES (2016) Monitoramento da Floresta Amazônica Brasileira por satélite (www.obt.inpe.br/prodes/index.php) Accessed 13 February 2016.

IPEA (2014) "Dinâmica recente dos PIBS *per capita* regionais: quanto tempo para chegar em 75% do PIB *per capita* nacional?" *Boletim Regional, Urbano e Ambiental* 9 (www.ipea.gov.br/portal/images/stories/PDFs/boletim_regional/141211_bru_9_web_cap9.pdf) Accessed 16 February 2016.

Loureiro, V. R. (1993) "Préjugés et modèles: les plans d'intégration économique de 1'Amazonie brésilienne ou la modernisation à rebours" *Cahiers du Brésil Contemporain*, n.21.

MEC/INEP/IDEB (2013) *Escolas estaduais do Pará têm pior Ideb de ensino médio da Região Norte* (http://g1.globo.com/pa/para/jornal-liberal-2edicao/videos/v/escolas-estaduais-do-para-tem-pior-ideb-de-ensino-medio-da-regiao-norte/3611540/) Accessed 16 February 2016.

Meirelles Filho, J. C. (2014) "É possível superar a herança da ditadura brasileira (1964-1985) e controlar o desmatamento na Amazônia? Não, enquanto a pecuária bovina prosseguir como principal vetor de desmatamento," *Boletim do Museu Paraense Emílio Goeldi. Ciências Humanas*, 9 (1) 219–241.

Ministério do Desenvolvimento Indústria e Comércio Exterior (2015) *Análise da balança comercial brasileira 2014* (www.mdic.gov.br/arquivos/dwnl_1399060864.pdf) Accessed 16 March 2016.

Ministério da Agricultura (2015) Produção brasileira de grãos é de 204,5 milhões de toneladas (www.agricultura.gov.br/vegetal/noticias/2015/06/producao-brasileira-de-graos-e-de-204-milhoes-de-toneladas) Accessed 16 February 2016.

MMA – Ministério do Meio Ambiente (1998) Primeiro Relatório Nacional para a Convenção sobre Diversidade Biológica (www.mma.gov.br/estruturas/chm/_arquivos/cap2c.pdf) Accessed 13 February 2016.

MMA – Ministério do Meio Ambiente (2013) Plano de Ação para prevenção e controle do desmatamento na Amazônia Legal (PPCDAm): 3ª fase (2012-2015) pelo uso sustentável e conservação da Floresta / Ministério do Meio Ambiente e Grupo

Permanente de Trabalho Interministerial, Brasília. www.amazonfund.gov.br/FundoAmazonia/export/sites/default/site_pt/Galerias/Arquivos/Publicacoes/PPCDAm_3.pdf Accessed 1 February 2016.

Nobre, A. D. (2014) *O Futuro Climático da Amazônia*, Relatório de Avaliação Científica. Patrocinado por ARA, CCST-INPE, e INPA. São José dos Campos, Brasil.

PNUD/IPEA/FJP (2013) *Índice de Desenvolvimento Humano Municipal Brasileiro; Índice de Desenvolvimento Humano Municipal Brasileiro* PNUD Brasil, Brasília.

Terra Estados com maior PIB do Brasil (www.terra.com.br/economia/infograficos/pib-do-brasil) Accessed 16 February 2016.

The Wall Street Journal (2012) Mineradoras aceleram seu avanço na Amazônia brasileira (http://br.wsj.com/articles/SB10001424127887324660404578201912498326692) 26/12/2012 Accessed 16 February 2016.

Vieira, I. C. G., Cardoso da Silva, J. M., Conway Oren, D., D'Incao, M. A. (2001) *Diversidade Biológica e Cultural da Amazônia* Museu Paraense Emílio Goeldi, Belém.

Vieira, I. (2012) *Norte, Nordeste e Centro-Oeste aumentam participação no PIB nacional*, Agência Brasil 23/11/2012 (www.fetecpr.org.br/oito-estados-brasileiros-concentram-77-das-riquezas-nacionais-13-e-de-sao-paulo/) Accessed 1 February 2016.

4 Deregulation, relocation and environmental conflict

Considerations on the control of social demands in contemporary Brazil

Henri Acselrad and Gustavo Neves Bezerra

The term 'ecological modernization' is currently used in the literature for the process by which environmental variables have been internalised by institutions to promote the market economy, political consensus and technological adjustment. Yet what can account for the resistance to adapting the technical and material basis of the accumulation of capital – production model, energy mix, etc. – beyond what the market itself determines? Has the ecological self-criticism of capitalism proved unjustifiable and unconvincing for the agents of capitalism themselves? The critical interpretation associated with some social movements is that no changes have actually been adopted to date in the technical-spatial practices and patterns of capitalist intensive accumulation to the requirements of utilitarian ecosystem reproduction – except to the extent permitted by mercantile dynamics – because of the prevailing socio-spatial divide in environmental degradation. The environmental damage caused by capitalistic development is systematically apportioned to the dominated social and ethnic groups, either through expropriation of the territorial bases of non-hegemonic forms of social production, or through deterioration of the reproductive bases of social groups that are not integrated into the capital circuit. The social mechanism used to achieve the unequal imposition of risks within or between countries in liberalized capitalism is the threat of company relocation, or investment location blackmail, forcing workers to compete not only for their pay but also for the legal rights and conditions designed to ensure social and environmental protection. This chapter will discuss the operation of this kind of social mechanism in Brazil based on case studies of a steel works and a pulp producer.

The environment and spatial mobility differentials

Isabelle Stengers sees capitalism today as a system that paralyses and traps social actors in its 'infernal alternatives' – situations that seem to leave no other option than resignation or impotent complaints about the inescapable economic war (Stengers and Pignarre 2005, 39–40). The imperative of acceptance replaces politics by submission, and the 'infernal alternatives' become the standard, a rule of judgement that produces the common measure of things, and a mechanism of regulation and control (Stengers and Pignarre 2005, 39–40) that leads people to find themselves caught by the imperatives of competitiveness and by the need to be able to attract market investment to themselves and to their own locations. But how are these 'infernal alternatives' generated and imposed? They appear to be produced through the constant reorganization of the way the system works, neutralizing the power of those who think differently. Through the reorganized forms of the system, the movement of investments inculcates the standard (Ewald 1993, 104) – the governing attributes that justify the undertaking's location.

It is well known that the main agents of global capital accumulation – the large corporations, multilateral development banks and state apparatuses – have sought to 'green' their discourses. They generally trumpet environmental protection as one of the main goals that their 'development' policies aim to achieve. Yet there is an obvious contradiction between the growing consensus that 'something needs to be done' and the half-hearted measures taken to actually replace the damaging techniques that characterize the current capitalist development model. In this chapter we will discuss how specifically the relocation threat stratagem may be related to the social processes of creating environmental 'sacrifice zones' that penalize the poorest (people who have 'less to lose' from risky activities since they own little), thereby allowing capitalism to carry on regardless and maintain its damaging production system.

The model created by Boltanski and Chiapello (1999) is useful in analysing this relationship between capitalist discourse and practice. In their view, the success of capitalism in terms of its longevity and geographical spread may be accounted for by its ability to neutralize social critique. Capitalism has been able to disarm critique in two ways. First, it may incorporate part of the critique into the repertoire of ideologies that justify capital accumulation, for example by claiming that the profits of capital will provide the money to protect the environment. Boltanski and Chiapello call this set of effective discursive justifications, which vary from one moment to the next, the 'spirit of capitalism'.[1] Second, it may disable critique by changing some of the material grounds for it and ensuring that the previous critique no longer applies to the new circumstances; teachers' strikes, for example, cannot effectively target an educational system that is in the process of being privatized; and the new competitive environment forces small soya producers to support the operations of the genetically modified soya multinational. The authors call this kind of reaction to critique 'displacement':[2] capitalism subtly shifts in relation to the moral postulates that society attempts to impose on it by means of certain devices (such as the law and collective bargaining[3]) or it relocates geographically[4] to places where the social critique is weaker.

Boltanski and Chiapello point out that the formation of the spirit of capitalism[5] at any given time paradoxically depends on the emergence of social critique: capitalism needs to incorporate part of the critique in order to strengthen its 'domination' – in other words, to ensure that both workers and capitalists are engaged in the accumulation process. Critique also helps to establish social procedures that limit accumulation by making it prove that it is keeping its promises.

Capitalism has always been criticized for damaging not only 'natural' areas but also its workers' health. Yet, insofar as the fruits of capitalist development began to be partly shared with the workers in the mid-twentieth century (including in some peripheral countries), the more 'qualitative' critiques of the accumulation process (those not referring specifically to the revenue of subsistence) tended to die down. However, after the 'world revolution of 1968',[6] there was a proliferation of activist movements that became known as 'new social movements', one of which was environmentalism. Since then, capitalism has been under pressure to justify itself in response to the ecological critique.

There have been countless attempts to reconcile capital accumulation with environmental protection in theory. The most powerful in terms of social penetration both inside and outside the managing techno-bureaucracy of capitalism was carried out by the Brundtland Report (United Nations 1987), which established the idea of 'sustainable development'. It argued that growth was 'necessary' to wipe out poverty in less developed countries, and technical progress would ensure that growth would be based on a more economical use of raw materials. In the wake of this conservative modernization of the spirit of capitalism, new areas of social life were also incorporated into the accumulation process – so-called 'clean' technologies, genetic information on biodiversity, etc.

It is a curious fact that before coming up with the idea of sustainable development the accumulation imperative had been challenged by a critique from within, from a group of entrepreneurs. In the early 1970s the group known as the Club of Rome pointed out that after 30 years of continuously high levels of economic growth the spirit of capitalism needed to wake up to the concern that the materials on which accumulation was based might run out. Their proposal to set limits to growth was, as might be expected, not very well received by the agents of a system that is defined by the addition of more value to value – capital as expanding value. This portrayal of capitalism as needing to limit its own expansion therefore proved incompatible with the reproductive urge of capital in its current form.

The equation has only been partly solved through the rhetoric of 'ecological modernization' (inspired by the principles set out by the Brundtland Report), in which the 'environment' variable has been internalized by institutions to promote the market economy, political consensus and technological adjustment. However, no arrangement seems yet to have emerged in the sense of admitting that this technical adjustment (including spatial practices of adding, removing and transferring materials and energy) might be directed and coordinated by political bodies, the so-called instruments of 'command and control' that are constantly condemned by the agents of the environmental economicism that dominates ecological modernization strategies. The model that has prevailed is 'market environmentalism', which holds that the market is the solution and not the problem for the ecological adjustment of capitalism.

Why, then, has there been so little change in the forms of production and waste disposal even though there is an evident concern to use resources sparingly? As the sociologist Robert Bullard (1990) points out, no measure aimed at altering the impact of capitalism on its material bases will ever be put into practice so long as the environmental damage caused by accumulation can be dumped largely on the most dispossessed − those who are least able to influence locational decisions and who have the least spatial mobility, since they operate entirely within circles of risk. Therefore, just as the term 'local production systems' is used in the economics literature to mean 'production arrangements whose substantial interdependence, interconnection and linkage result in interaction, cooperation and learning, enabling innovation in products, processes and organizational formats and generating greater business competitiveness and social enablement,' (CNPq/FINEP/SEBRAE 2002, 13) here we might suggest the existence of 'local pollution systems': production arrangements whose interdependence and linkage result in a spatial interconnection of the 'negative externalities' of production so as to optimize investment by distributing environmental risk to the agents with the least economic and political resources (Acselrad 2006).

If we can accept that this 'law' (the law of overcoming the environmental strangulation of the system by transferring risk to the poorest) has held true over time, governing technical and locational choices within national territories, we have to recognize the fact that its operation has gradually been hindered as some countries have adopted environmental regulations and as certain social movements have reduced the leeway available for the conventional socio–spatial division of environmental degradation. A symptom of this is the recurrent campaigning by business interests against environmental regulations, which are presented as 'bureaucratic obstacles to development'. It has thus become more difficult to dump the risks on the most dispossessed simply by playing on the different levels of regulation and social organization within the country. The solution has been to use what has come to be known as the 'Summers rationale',[7] which, based on global economic efficiency, justifies the internationalization of the socio–spatial division of environmental degradation by claiming that the cost of human lives is lower in less developed countries, and so forth.

What is the mechanism that has put the 'Summers rationale' into effect since then, by systematically arranging to transfer the environmental damage caused by capital accumulation to dominated social groups in peripheral countries? The increased socio-environmental risk is systematically allocated to the most deprived populations under a logic of so-called freedom of choice − a choice between the 'infernal alternatives' of precarious, hazardous conditions of work and no work at all. Therein lies the key to this political economy of demobilization: the concentrated apportioning of risk to deprived social groups is generally achieved through mechanisms that to a large extent depend on the acquiescence of the populations targeted by this 'location blackmail'.

The threat of relocation imposes an enhanced burden of risk at the receiving sites. The social risks (i.e. the unstable conditions of 'social insecurity' and the constant threat of job losses) have already been pointed out, but the receiving sites are also subject to a variety of environmental risks. These include risks relating to the hazardous sanitary, physical, chemical and geotechnical conditions of undertakings located in densely populated areas and those arising from the expropriation of the resource base in regions into which the capitalist frontier is advancing.

Richard Sennett points to the appearance in flexible capitalism of a unique factor that may be called 'compulsory risk-taking', whereby individuals feel obliged to play against themselves: not taking the risk, he says, means being left out (Sennett 2001). We could say the same about the 'voluntary' acceptance of environmental risks and damage – not accepting them could make a bad mobility situation worse. Consent is given not in the hope of gaining anything from capitalism, but in the expectation of avoiding further losses.

Locational mobility and stripping of rights: empirical evidence in Brazil

What has been the basic logic behind capital location since the new cycle of exporting capital began in 1980? Has it been to completely ignore territories endowed with rights while focusing instead on socially and environmentally vulnerable areas? Is there a trend to switch entirely from northern countries to new southern 'tigers'? The fact that flows do not work according to this over-simplified logic has led many people to contest the argument that relocation blackmail has deleterious effects. However, a more detailed analysis of the territorial dynamics of each specific group of production activities reveals two things: a) there is a strong trend in the main sectors of goods production towards the respatialization of activities, although relocation in some sectors tends to be to areas relatively close to former units; b) as mentioned above, relocation does not actually have to happen, since it may be enough for companies or local politicians to show that relocation is a possibility for demands for greater democracy to be curbed in a particular area. Here we will seek to illustrate these scenarios with some empirical examples.

According to Chesnais (1996), the locational decisions made by an oligopoly obey not only the imperative of seeking low production costs but also a strategy of remaining close to customers and suppliers. He suggests that it would not be rational for companies only to look for sites with the lowest wage costs, since the internationalization of capital is accompanied by new rationalizing practices that can save work anywhere in the world (in so-called 'lean' companies). The relocation of undertakings should therefore systematically take into account in their calculations the costs of work (and, we might add, environmental costs) with their proximity to elastic and 'retainable' consumer markets. Chesnais believes the issue of customer loyalty leads global businesses to try everything possible to keep their production sites close to their points of sale, while taking costs into account. Essentially they explore differentials in social (and environmental) rights within a particular 'region' (which may often be a continent); thus, for example, they produce goods in Mexico to be consumed in the United States, or in eastern Europe for consumption in Western Europe.

> The combined effect of the industrialization of countries with very different wage levels within a totally liberalized Common Market, the freedom of foreign investment, and Thatcherite neoliberal policies, which have also been adopted by other countries, means that *currently there are considerable wage differences within the EEC*[8] (a differentiation that will only increase with the 'association' of certain ex-socialist countries). No industrial group needs to relocate its production outside the EEC and a few neighbouring countries to the east in order to find cheap skilled labour.
>
> (Chesnais 1996, 131)

Global companies thus set up regional strategies[9] that take account of accentuated differentials in rights between territories that are not very far apart. Similarly, it is easy to find examples of differentials in rights (or in the political ability to fight for rights) that make all the difference even within countries. In the case of the US automotive industry, for example, there is a flagrant move away from old 'brownfield' areas with intensive union activity, such as the Great Lakes region, to virgin or 'greenfield' car production sites within the United States. Chesnais also points out that location strategies vary with the type of undertaking. He argues that there are two kinds of pattern: whereas technology- or machine-intensive world oligopolies try to site their production bases fairly close to high-income consumer markets, other sectors tend to outsource their activities to areas far from their headquarters (these include 'network corporations', such as Nike, and department store and supermarket chains).

Both these trends may be observed in Brazil. We have found both a set of intra-national inequalities (in environmental policies, tax policies and wage levels), which make a great difference for companies' relocation strategies, and well-delimited sector-specific relocation patterns. There is empirical evidence that the search for lower costs and government benefits (tax incentives and public grants of various kinds) had a major impact on the respatialization of employment in the country in the 1990s (Azevedo and Toneto Júnior 2001, 173–175.). Labour-intensive sectors and sectors highly dependent on natural resources showed significant inter-regional relocation (from the south-east to the north-east[10] and the mid-west,[11] respectively), while capital-intensive sectors confounded sceptical expectations[12] by showing some degree of inter-regional relocation (from the South-East to the South, especially to Paraná state, since it lies next to São Paulo) and, particularly, significant intra-regional relocation (mainly from metropolitan São Paulo to the interior of the state).

Overall, this respatialization of production in Brazil reveals that wage levels were higher than the national average for the industry in question in the states where jobs were lost, but lower than the national average in the states where new businesses opened (Azevedo and Júnior 2001, 173–75), which would appear to confirm the preference for lower costs. However, that is not the whole story. It begs the question of why certain states characterized by low wage levels (such as the north-eastern state of Ceará) received a mass of new undertakings, while others where labour costs are as low or even lower (such as the neighbouring state of Piauí) did not become major foci of investment. The answer may be that market forces alone did not determine where new companies would relocate, but local (state and municipal) authorities may have played an important role by offering tax incentives and other benefits.[13]

The importance of state action shows that relocation processes do not take place in a vacuum, free from political agendas. Just as some authorities see capital mobility as offering 'opportunities' and try to put forward policies that will help to attract new investment, other social actors (including other governmental actors) have joined the public debate by denouncing inter-regional blackmail and the framework of deregulation that has made it possible. Some social aspects of location blackmail, in which various social actors enter into conflict because of their differing views about whether to accept risky undertakings or not on environmental and/or economic grounds, are discussed in detail in the empirical case studies that follow.

Eucalyptus monoculture projects: relocation blackmail and relocated opposition

The pulp and paper sector has long been seen as an important vehicle for exports designed to ensure the country's ability to make payments in foreign currencies (for imports, profit remittances, international loans, etc.). Pulp in particular came to take pride of place in the balance of trade, especially in the form of kraftliner, the outer layer of corrugated wrapping paper, which has less value added and makes more intensive use of natural resources. In general, the Brazilian pulp and paper industry is average in terms of the intermediate goods sector in Brazil in that it is large-scale, strong in exports, unsophisticated in its product range and has a high environmental impact (Schlesinger 2001, 53–56).

The spread of eucalyptus monocultures for pulp and paper production in Espírito Santo and Bahia dates back to developmentalist times, especially the authoritarian period, which sought to occupy areas considered 'empty'[14] and undervalued in both states.[15] Northern Espírito Santo and the southern tip of Bahia had flat areas near the coast with good rainfall conditions, suitable for eucalyptus plantations. Moreover, land was cheap, since it was considered 'empty' or was overgrazed. With the idea of bringing much-vaunted development to the region, the government granted considerable tax incentives to logging, cattle ranching and, later, eucalyptus planting activities (Fanzeres 2005). This convergence between state action and business interests in these 'empty' lands with no market value led to a change in the way the state recognized land ownership, favouring occupation by agri-business undertakings.[16]

Nonetheless, these lands that had historically been treated as empty were in fact inhabited by communities, including *quilombos* and indigenous groups that were able to provide for their subsistence entirely through traditional economic practices. The spread of eucalyptus plantations was made possible through 'dispossession' mechanisms, to use Harvey's (2005) terminology. For example, the notarial offices that sprang up everywhere in northern Espírito Santo and southern Bahia at this time were essentially tools for seizing land from peasant communities, *quilombos* and indigenous villages.[17] It was very common in Espírito Santo for land ownership to be given to third parties who later transferred it to the pulp company.

> When Barra [Conceição da Barra] got its notarial office they started making the documentation. And they got what they wanted. In '68 they transferred a lot of black people's land to the pulp company ... Some time later many owners of notarial offices moved out of the region.
> (*Quilombo* resident from Sapê do Norte, Espírito Santo,
> *apud* Acselrad 2006)[18]

The companies' agents used a variety of arguments to persuade *quilombo* families to sell their land. One was to sell up in exchange for jobs with the company, but before getting the jobs they would have to leave the land to go to school in the city. Coercion was also part of the strategy – if people had sold land to the pulp company, all their neighbours would 'have to' sell up as well, since the company would not be a 'good neighbour'. Promises of employment for small farmers and their children were also used as strategies to persuade families to give up their land. Initially, many small farmers and *quilombo* inhabitants worked at eucalyptus planting and then building the factory.

For the indigenous populations, the company offered work on the condition that they gave up their indigenous identity. In order to get jobs, many indigenous people began to deny their origins.[19] Anyone who sold their land for the sake of better living conditions was soon to regret it. The sum paid was purely symbolic and insufficient to start a new life in the city; moreover, the *quilombo* residents' lack of skills – since they were unschooled and had only ever worked their smallholdings – did not enable them to find good jobs in the city.

In Minas Gerais state alone, an area of 2 million hectares (almost the size of Belgium) in the Jequitinhonha valley region and the Cerrado in the north of the state is used to supply charcoal for the state's pig-iron furnaces and major steel works. Further plantations are planned for the southern tip of Bahia and Espírito Santo, associated with new pulping plants (Aracruz's third factory and Veracel Celulose) and the expansion of existing installed capacity (Bahia Sul Celulose). Near the border with Bahia, almost 70 per cent of the Espírito Santo municipality of Conceição da Barra is covered with eucalyptus, and neighbouring São Mateus has 50,000 hectares planted with the species. In Aracruz municipality, according to the local indigenous movement, the factories are located precisely on the site of the former main Tupiniquim village of Aldeia dos Macacos.

It was only in the late 1990s, after Aracruz tried to expand its plantations in Espírito Santo, that any real organized opposition appeared in the region, consisting of Guarani and Tupiniquim people, charcoal burners, students, teachers, lawyers, parliamentary advisors and some NGOs, who formed the Rede Alerta contra o Deserto Verde (the Green Desert Alert Network). New actors and issues began to come together to oppose the monoculture at the end of 1998. It is not clear exactly when the Network came into being in Espírito Santo, but it soon reached southern Bahia and in 2002 was organizing in Minas Gerais. Its repertoire of action has expanded since 1998 and includes mobilizations, demonstrations, marches, occupations, pressure groups, open letters, meetings, publications, films, public hearings and attendance at parliamentary commissions of inquiry, disputes about licensing, certification of plantations and carbon credits, reporting to international networks and forums and lawsuits, etc.

In 2002 the Network showed its capacity to oppose the way businesses take advantage of differences in spatial regulation. Aracruz had been forced into a corner in Espírito Santo (having been scrutinized by parliamentary commissions of inquiry and banned from increasing its plantation areas) and was trying to expand into economically depressed parts of Rio de Janeiro state. The Network then 'exported' its expertise and capacity for action to the neighbouring state, allying itself with INCRA-RJ, rural workers' movements, various environmental bodies and researchers. Their joint pressure resulted in Law 4063/2003, under which any large-scale eucalyptus plantation project in the state could only be implemented after thorough ecological-economic zoning had been carried out. The pollution of the River Pomba by the Cataguazes pulp company and the resulting water crisis in the city of Campos and the north of the state of Rio de Janeiro helped their opposition.

With the increasing visibility of the actions taken against them, company representatives have often tended to respond by threatening to cancel their investment plans or relocate their undertakings. That happened during the recent episodes involving the MST (Movimento dos Sem Terra – the Landless Movement) in Barra do Riacho in Rio Grande do Sul state, and the Tupiniquim people in Aracruz municipality in Espírito Santo, as the two news reports below illustrate:

> The MST's violent action took place a few weeks before Aracruz specified where its new investment would be sited. The federal government is disputing with Espírito Santo and Bahia over the construction of a US\$ 1.2 billion factory to produce 1 million tonnes of pulp, which could provide 50,000 jobs directly and indirectly. Despite the destruction of its research laboratory, the company is still prepared – at least officially – to invest in Rio Grande do Sul ... The company has carried out a study to determine the social risk. Ponte [the state's development secretary] has

mentioned two important questions in relation to this kind of investment, which is wrapped in controversy over its environmental impact: compliance with the law, and an atmosphere of good community relations. The subject is so important that last year Aracruz, still smarting from the frequent invasions of its lands in the states where it has production units (Bahia and Espírito Santo), had already hired a company to analyze the social risk to the institution if it set up a new unit around Porto Alegre. Compared to its rivals, the Metropolitan Region had come out well since society there is highly politicized.

(Cruz 2006)

Espírito Santo, which was a strong candidate to attract Aracruz Celulose's fourth factory, lost the investment of over US$ 1.3 billion to Rio Grande do Sul. The state has not yet been ruled out as a site for future company projects, but at the moment it is not seen by Aracruz as a priority location for new investments. The conflicts with indigenous people who claim they own the land, the fact that groups supporting the indigenous people have been trying to discredit the company with major customers abroad; the attempts by the Legislative Assembly to prevent new eucalyptus plantations, and the committees of enquiry set up to investigate Aracruz have all sounded the alert for the company's directors and shareholders. When the time came to choose where to site its new factory, Aracruz weighed the support and solidarity it had been given by civil society, politicians and business people in Rio Grande do Sul against the countless problems it had been facing in Espírito Santo in recent years. In the end, even with the logistical and infrastructure benefits offered by Espírito Santo, Rio Grande do Sul was the winner. 'All that is leading us not to leave Espírito Santo but to look for other options,' said the company president, Carlos Lira Aguiar. 'One shouldn't put all one's eggs in one basket,' he explained, as he emphasized the strategy of distancing the company from its problems. Problems that have occupied 40% of the executive's agenda. There are days when Aguiar, other directors and managers in the company spend 80% of their time solving the indigenous land issue. 'I envy executives who can devote all their time to their businesses,' he complained.

(Bridi 2006)

By counting on social contexts favourable to the ideology of 'development' at any price (first because the military regime wanted to legitimize itself, and later because of the social landscape of economic depression and the need to generate hard currency), the companies promoting eucalyptus monocultures have managed to take over large areas. To do so they have not shrunk from dispossessing traditional communities by exerting economic pressure or using fraudulent and violent means to get land. Conversely, social opposition has been gathering strength, not least through the emergence of political initiatives that are both pluralist (in that diverse groups are involved) and innovative in terms of their organization and objectives. By operating as manoeuvrable 'networks', these movements try to inhibit the ability of undertakings to induce competition between localities and force down all their social and environmental standards. By tracking companies as they relocate, the movements have, in contrast, managed to export and raise the standard of the rights that should be associated with economic activities.

Differential regulation spaces in a steel works on the Brazilian–Bolivian border: the case of EBX

In the middle of the first decade of the new millennium, the Brazilian EBX group tried to establish a mining and steelworks site in the Pantanal region on both sides of the Brazilian–Bolivian border. The idea was to mine ore in Mato Grosso do Sul state and process it at low cost in a steel works on Bolivian territory, located in the free-trade zone of Santa Cruz department. However, the Bolivian Government was highly critical of the MMX[20] plant for tax and environmental reasons. Evo Morales's new government resented the enterprise setting itself up in an area where the use of Bolivian natural gas was subsidized. Gas would make up 20 per cent of the fuel used by the steelworks (about 340,000 cubic metres per day). Even more controversial was the fact that 80 per cent of the remainder would come from charcoal (Pedra 2006). Fearing that the Pantanal 'would be turned into charcoal' (FOBOMADE 2006), Bolivian environmental groups also called for an embargo on the undertaking.

From the start the company had been planning to mitigate criticism of its potential devastation of natural resources by including in the project the goal of replacing the use of native forest with planted eucalyptus forests (which some environmental bodies considered equally bad because of their possible impact on the Pantanal's water resources). But the company did not even wait for the Bolivian government to decide whether or not the measure was appropriate. Two blast furnaces were quickly built before it had even received the environmental licence to install the steelworks. Some environmental bodies felt that such an action could be a strategy to present a *fait accompli* and overcome social opposition to the undertaking. EBX had recently suffered a major setback in another undertaking in Santa Cruz, when it was prevented from bidding for a contract to mine ore in Mutún because of reports that the company had been unduly favoured by the state.[21] Therefore it had hurriedly built the blast furnaces in Porto Quijarro on the Brazilian border, not least since it feared environmental legislation might be changed to prohibit the use of charcoal.

In April 2006 the Bolivian government published a decree requiring that the illegal works be stopped. It said that to be built the steelworks would need not only the environmental licences, but also a specific law, given that the area in question lay on the border. The embargo on the works ended up exacerbating the conflict between the political opposition in Santa Cruz and the central government. To put pressure on the government to allow the works to continue, demonstrators in the region held three ministers for 13 hours. They declared that even if the work on the steelworks was illegal, the embargo might jeopardize further investment in the region. The company was promising to provide 900 jobs directly and 5,500 indirectly (Otávio 2006).

A week after the kidnapping, the Bolivian president announced on a Brazilian television channel that the company was considered to be in an illegal situation in the country and that 'EBX is trying to split Bolivians apart and blackmail them. Therefore, I reiterate, there are two options: either it leaves voluntarily or we will end up forcing it to leave.' The chairman of EBX, Eike Batista, stated the following day that he would be taking the equipment out of Bolivia and he would no longer be building two thermal electric power stations that would have used Bolivian gas, one in Corumbá (Mato Grosso do Sul) and the other in Bolivia. The candidate areas for receiving the blast furnaces already built in Bolivia were the Brazilian states of Mato Grosso and Amapá (where the company was mining iron ore) and even Paraguay, where the steelworks project could be completed with two more blast furnaces (Reuters 2006). He also tried to explain away the Bolivian government's position by saying that it was the result of ill feeling against the previous government, with which Batista had made the deal.

EBX was back in the news a few days later because of its bid to construct a pig iron steelworks in Corumbá, not very far from the Bolivian territory that it had just left. The company maintained that the Corumbá steelworks had already been in its plans before the clash with the Bolivian government. The project was, in any case, very similar: the main source of energy was to be charcoal, first from native timber and later from its own 35,000 hectare eucalyptus plantation, which would supply the works with 225,000 tonnes of charcoal a year. The project was again controversial because of the risk of environmental damage. The public prosecutor's office for Mato Grosso do Sul state had asked a group of researchers to review the project's environmental impact assessment (EIA), and they concluded that the operation of a steelworks in the city would threaten the Pantanal ecosystem; Sônia Hess, one of the authors of this review, found that the EIA did not make it clear where precisely the 25 truck-loads of charcoal needed to supply the steelworks every day would come from (Yafusso and França 2006). The review also found that the project ignored the health risks caused by this kind of industry and the EIA failed to detail the types of particulates that would be emitted into the atmosphere during pig iron production.

A public hearing on the licensing process was held on 4 May that year. However, trouble surrounded the event from the outset. Many of the authors of the review of the EIA preferred not to attend the public hearing for fear of being verbally and physically attacked (Agência Popular 2006). On the day of the event, cars fitted with loudspeakers, some of them marked with the logo of the mayor's office, are said to have driven round the town making threats, ranging from 'Environmentalists out!' to 'Let's gun them out of town!' (Nassif 2006). There were also demonstrations by the members of the homeless movement, who were concerned that establishment of the steelworks would mean a number of poor rural producers would be forced off their lands. They wore white t-shirts with the question 'Steelworks what for?'. They also took the opportunity to ask reporters 'Why do we have to shut up and take what Bolivia has thrown out?', and 'Well, if Bolivia didn't allow charcoal to be used in the steelworks, why should we allow that here in the middle of the Pantanal, with the natural gas pipeline running past our doors?' (APUFSC 2006).

Sônia Hess, the researcher, said she had been intimidated at work (in the Federal University of Mato Grosso do Sul) by the rector[22] and also by Senator Delcídio Amaral and Governor Zeca of the Workers' Party (PT). The EBX manager dealing with environmental matters tried to minimize the reprisals against the researchers and the demonstrators protesting against the steelworks, but also stated that he was sure the company had an 'infernal alternative' scenario on its side: 'There was no disrespect shown to anyone at the public hearing. What happened was that all of society in the region was mobilized. That is a very deprived area, which was hit even harder recently with the foot-and-mouth disease crisis. Our project is very well received there' (Yafusso and França 2006).

Despite the situation of the Corumbá undertaking, EBX did not give up its mining, steel and energy projects in Bolivia, on account of the support promised by the Brazilian government, which tried to negotiate a way for the company to get back into Bolivia (Yafusso and França 2006). In September 2006 Eike Batista restated his plans to return to Bolivia on two conditions: 'the Bolivians must let us supply iron ore from Corumbá, and local tax regulations must remain stable' (Reuters 2006). Until the middle of 2008 he was still saying he wanted to invest in Bolivia. The main reason why EBX never tried to sue the Bolivian government was its hope that it might develop new projects, especially thermal power stations, in the country (Rios Vivos 2007). The fact is that even though it had been expelled from Bolivia, EBX never stopped contributing to deforestation in the country because it was buying up charcoal there for the Corumbá steelworks even before it was finished. The company defended itself by saying that it was from legally felled timber. In March 2007 the Bolivian government allowed the company to remove the two blast furnaces that it had built in the country (Rios Vivos 2007).

The Corumbá undertaking was finalized in August 2007 after winning a court battle. The steelworks was set up on a 60-hectare plot granted to the company by the state government, which had previously belonged to an export processing zone (Agência Popular 2006). Just 14 months later the concerns voiced by the researchers and activists opposing the installation of the charcoal-powered steelworks proved to be grounded, as the company had already been fined a total of R$ 29.4 million for using timber with no proof of origin (Rodrigues 2008). Batista's reaction was to threaten to close the previously celebrated plant: 'This has got to be looked into and we are querying these fines. [The steelworks] is giving us a lot of trouble and *we are even thinking of closing this operation down.* In the context of the company's business, it is a very small item that is causing us so much trouble' (Rodrigues 2008). While querying the fines, the tycoon tried to legitimize his business approach in the eyes of governmental and non-governmental bodies by donating R$ 11.4 million to the national parks of Fernando de Noronha (Pernambuco), Lençóis Maranhenses (Maranhão) and Pantanal Matogrossense (Mato Grosso).

The limits of a large project aimed at exporting slightly processed goods (and benefiting from ample government resources and low levels of environmental and labour protection) became clear at the end of 2008. This time, the threat to close down the controversial plant was not part of an attempt to force an even lower level of collective regulation but the result of cash-flow difficulties arising from the slump in international demand for commodities. About two weeks after the steelworks stopped production (on 25 November 2008) due to lack of orders, there were already some 2,000 charcoal producers working for accredited EBX suppliers who were out of work (Vargas 2008). The Corumbá Extractive Industry Workers' Union managed to negotiate a two-month suspension of the contracts of 350 workers directly employed by EBX (and a further 270 in the sector in the same region, in companies such as Vale), and only then would the steelworks assess whether it needed to actually get rid of its workers. However, the mayor of Corumbá, who had been one of the main supporters of the project (to the point of being accused of intimidating demonstrators at the public hearing in 2006), stated that he did not believe the company would operate again, unless it was bought up by a larger multinational.

In short, the case of EBX in Corumbá was yet another instance where, for the sake of development at any price, a company making intensive use of natural resources managed to get direct public funding.[23] However, as usual, it failed to benefit the territory in social and environmental terms since it aggravated deforestation (as shown by the heavy fines handed out by the Brazilian environmental protection agency IBAMA), created few direct jobs, generated extremely low-quality and unstable indirect jobs, and declined rapidly because it was tied to the unstable international commodities market. In Bolivia the government tried to denounce the weakness of the project in terms of encouraging a desirable form of development, but was 'punished' by the political radicalization of social groups trapped in the logic of tacit consent for the market, which is powerfully produced by the current stage of highly mobile capitalism and collective under-regulation.

Final considerations

This chapter has sought to assess the extent to which social technologies producing 'relocation blackmail' – threats to relocate investment and, consequently, jobs and public revenue – might be contributing to the weakening of the ability of society to curb the adverse effects of the market on social and environmental well-being. Our hypothesis is that the control/

containment of demands for democratization (of the environment, labour relations, land, income, etc.) is no longer resulting directly from the action of an authoritarian state but is undertaken through the action of market deregulation, which enables free capital to impose a 'pedagogy' of political conformity on society.

This deregulation is widely known to have occurred at a global scale, and does not only affect countries where civil society has become stronger with the overthrow of authoritarian regimes (as in Latin America and Eastern Europe). In fact, the countries at the heart of capitalism, which have generally been characterized by major social struggles for wider rights and the denunciation of capitalism (civil rights and pacifism in the United States, and the denunciation of unbridled industrialization and imperialism in Western Europe) have been experiencing a reversal of demands for increased democracy in favour of submission to the requirements of ultra-mobile capitalism. A zero-sum game therefore appears to be operating in relation to the raising of social rights on a global scale, in which some social spaces 'win' by attracting capital hostile to increased democracy, because other spaces have not been 'flexible' enough to make democracy retreat or become neutral.

Deregulation and the pruning of social rights are clearly not the only mechanisms for attracting capital that are used in the contemporary capitalist world (since some investments require social spaces with highly qualified professionals or nearby consumer markets, for example). Our question is why poverty, deprivation of social rights and environmental degradation are being reproduced and intensified across the globe despite the rise of an increasingly productive capitalist system. Our starting point is the idea that political docility and privation of rights are prerequisites for accumulation in this form of capitalism.

For adherents of what has come to be known as the 'sociology of global flows', the idea of fixed boundaries in a society or nation-state has been replaced with 'borderless global fluids' (Moll and Spaargaren 2003). It is not exactly a case of analytically replacing places with flows, but of knowing how to analyse the new relationship established between places, including by means of such flows. The space of flows, which Castells considers the 'dominant spatial manifestation of power', could be seen differently with regard to relocations, as the space through which power is redistributed between points and us. In our case, the flow of relocated capital forges a link between two places/moments in a flow of value and in an 'environmental flow'. Here the notion of 'environmental flow' acquires a specific meaning other than the idea of a flow of matter and energy (which could be expressed in the notions of 'ecological footprint' (Wackernagel 1996), 'ecological rucksack' and 'environmental space'). It is the movement of the international transfer, particularly in a north–south direction, of environmental conditions associated with productive investment (e.g. steelworks waste, chemical waste or electronic waste) or of the transfer of assets in search of more favourable environmental conditions (e.g. sunlight, available land) to the detriment of the non-hegemonic forms of social production of small farmers, fishermen, former slave communities (*quilombos*), indigenous communities, the landless, etc. Increasingly today, given the restrictions required through social pressures associated with the construction of the 'environmental question', every technological decision contains a political dimension consisting of the socio-spatialization of its environmental impacts. Examples include the planned obsolescence of electronics products, which internalizes the exporting of the E-waste resulting from that rapid obsolescence to poor communities in India and China, which have taken on the recycling stage and the resulting self-contamination with hazardous substances (Pucket 2002); and the calculation of pulp and paper production costs, which increasingly internalizes marketing and 'community relations' activities designed to demobilize local societies that might listen to the arguments of social movements opposed to the planting of vast tree monocultures.

Such processes of imposing environmental risks on the weakest groups will not go unopposed, because the struggles for spatial democratization will also include demands for locational and environmental justice. Opposition of this kind to discriminatory land use decisions is a relatively new phenomenon associated with a rethinking of the environmental question, which now incorporates concerns about the way impacts are distributed as a result of the spatial dimension of activities. Instead of environmental education and lobbying, in many countries these struggles have involved barricading streets, sit-ins, mass demonstrations and boycotts. What they have in common is their protest against the mechanisms of dualization: a) there is a disconnection between locational decision-makers and the victims of the undesirable, risk-bearing aspects of these decisions (political power is said to be used to keep pollution away from the powerful); b) so long as there are areas of less resistance, any decision restricting the environmental damage done by undertakings will be followed by the transfer of the harmful activities to poor urban residential areas. In the case of Brazil, a number of episodes suggest that this kind of opposition to the unequal imposition of environmental risk is beginning to spread; they include the cancellation of a project to locate a coal-burning power station in Itaguaí, Rio de Janeiro state (Ferraz 2004), and the stopping of chemical waste transfers from Cubatão to Camaçari at the instigation of ACPO (Associação de Combate aos Poluentes Orgânicos Persistentes – Association to Combat Persistent Organic Pollutants) (Malerba 2004). The same may be said of the initiative by the Rede Brasileira de Justiça Ambiental (Brazilian Environmental Justice Network) to require Petrobras to use the same criteria when siting its undertakings in Ecuador as it does in Brazil. Because of that the Yasuní people felt empowered to demand that Petrobras stop its operations on indigenous lands classed as Unesco Biosphere Reserves. In the words of a young *quilombo* (black community member) resident in Espírito Santo state, Brazil, faced with the inequitable expansion of modernizing undertakings into their lands, it is a case of social actors who, despite all the uncertainty imposed by big business, 'have learnt to say no'.[24]

Notes

1 Boltanski and Chiapello therefore propose a dual model for the structure of capitalism. On one hand are those economic practices that reproduce capital accumulation 'insatiably' (i.e. without any value judgement), and on the other there are the ideational productions (which may be materialized in the operation of various social institutions, especially the provisions of law) that justify and give meaning to people's engagement in this never-ending process of accumulation, which even its main beneficiaries can see is absurd (Boltanski and Chiapello 1999).

2 For Boltanski and Chiapello these dislocations are neither fully planned by conscious actors nor the outcome of an unconscious process without a subject, but the result of the collective formation of critiques by think tanks, consultants, management experts, journalists, etc. (Boltanski and Chiapello 1999)

3 For example, by replacing collective forms of benefit distribution contained in job and wage plans with mechanisms for performance measurement and individual distribution of the same resources, which divide workers politically.

4 'These displacements may be geographical (relocation to regions where the workforce is cheap and labour law is undeveloped or accorded little respect) if, for example, the firms do not wish to improve the division between wages and profits in the way demanded by the critique. (The same could be said of new environmental requirements)' (Boltanski and Chiapello 1999, p. 34).

5 The idea of the spirit of capitalism is a reformulation of Weber's concept of the same name. Weber (2009) described how a certain set of motivations inspired capitalism in its early stages. Boltanski and Chiapello argue that all of the subsequent development of capitalism remained dependent on the validity of a 'spirit' that justified practices aimed at accumulation. They also observed that the set of justifications varied over time. In their view, we are today witnessing the emergence of a third historical set of justifications.

6 'The details of the 1968 uprisings were different in the various arenas of the world system, but such uprisings did occur everywhere [...] In all of them, however different the local situation, there was a recurrent double theme. The first was opposition to US hegemony and to Soviet collusion with that hegemony [...] And the second was disillusionment with the Old Left in all its forms (communist, social-democrat, movements of national liberation).' (Wallerstein 2000, p. 255).

7 In a 1991 memo distributed only to World Bank executives, Lawrence Summers made the following proposal: 'Just between you and me, shouldn't the World Bank be encouraging MORE migration of the dirty industries to the LDCs [Least Developed Countries]?' See 'Let Them Eat Pollution', *The Economist*, February 8 1992 p.82 UK Edition.

8 European Economic Community, now the European Union.

9 Chesnais disputes the idea that global companies as a rule have 'global factories'. There is instead a multi-regionalization of production, with production units responsible for supplying the markets closest to them. He maintains that the fact that world production is increasingly being relocated towards Asia has in most cases less to do with establishing global factories there than with strategically preparing to supply the rapidly growing consumer markets in that region.

10 Footwear manufacturing saw the disappearance of former clusters (such as Franca) and the establishment of new clusters, especially in the North-East. The textile industry, which tends to be more dispersed, seems to have relocated to several regions, especially Paraná and the north-east. São Paulo state's share of textile jobs in the country fell from 47% in 1986 to 34% in 1998, with most relocations taking place between 1994 and 1998.

11 In the food and drink sector, most growth was seen in the mid-west and south (especially Paraná state, where transport costs were the lowest in the country),

whereas timber and furniture industries left the south-east mainly for Mato Grosso state in the mid-west. The metropolitan region of São Paulo saw its share of jobs in the timber and furniture sector fall by half between 1986 and 1998, with 1994 again marking a rapid increase in the relocation rate.

12 Azevedo and Júnior began their study expecting that capital-intensive sectors (the metallurgy and mechanical industries) would be spatially stable. They were surprised to find a high rate of relocation, albeit generally to areas not too far from the main consumer markets in the south-east.

13 Azevedo and Júnior note that, unlike the compensatory regional policies of the past, this kind of local government action was merely 'confirming the movement that might be expected from the independent action of companies in search of lower production costs' (Azevedo and Júnior, 2001-179). Moreover, the process cannot be said to have induced a spatial deconcentration of income, since the jobs that moved most were those with a low technological content, while higher-ranking positions in design and marketing continued to be concentrated in the south-east.

14 Land occupied by traditional communities was also considered idle, since it was not being used for market-driven purposes.

15 Espírito Santo was regarded as the least 'developed' state in the south-east, 'where the South-East was not the South-East', alongside the southern tip of Bahia, a huge, almost forgotten area sidelined by progress.

16 'Few people had documents proving they owned the land: ownership was ensured by working and occupying it. The land was divided up among families, with streams marking the boundaries. It remained like that until the large private companies arrived in the region' (Salomão 2006)

17 The 'Diagnóstico de conflitos sócio-ambientais em relação à plantações de árvores' (Diagnosis of socio-environmental conflicts regarding tree plantations) commissioned by the Ministry of the Environment states that almost 10,000 hectares used by the Tupiniquim indigenous people in Aracruz municipality – more specifically the Caiera Velha indigenous reserve – had been occupied in 1942 by the iron and steel company Companhia Ferro Aço de Vitória (COFAVI). In the 1970s the pulp and paper company Aracruz established itself on the steelworks land and thereby 'the socio-environmental debt relating to the occupation of indigenous land in the region was transferred to it.' The same document claims there are reports that a former mayor of Primo Bitti and a land-grabber known in the region as Captain Orlando expelled indigenous families from their land, so that the mayor's real estate company – Bitti Imóveis Ltda – could register the land and later sell it to Aracruz (Fanzeres 2005).

18 Another interviewee stated: 'The company arrived and said, "You have to get out of here, because this land has a new owner"' (Indigenous inhabitant of Três Palmeiras village, *apud* Acselrad, 2006).

19 It should be noted that 'that strategy is still in use today, since ... the main argument used by the company to deny that it is taking land occupied by traditional populations is that there are no indigenous or *quilombo* people in that region' (Malerba and Schottz 2006).

20 MMX is the mining and metallurgy branch of EBX.

21 There were complaints that the government under Carlos Mesa, Morales's predecessor, had changed the tendering rules in favour of EBX, since it had made the use of charcoal a condition of the contract, as requested by the Brazilian company. The newspaper *Folha de São Paulo* reported that 'the change prompted criticism from environmentalists because, instead of a clean source of energy that was plentiful in Bolivia, a fuel was being favoured that required the intensive exploitation of forestry resources in an ecologically sensitive area, the Pantanal'

(Maisonnave 2006). The complaints led the company to be excluded from the tendering procedure even before the Morales government took office.

22 'I was ashamed to be subjected, as a scientist, to a veritable inquisition, alongside other colleagues of mine, simply because we were fulfilling our role in society, which is to inform it about what we study, the area we specialize in: the environment … I am ashamed to find out that a university rector prevents his staff from freely expressing their scientific knowledge, the philosophical and ethical basis of a university. What is the world coming to?!' (Hess, *apud* Nassif 2006).

23 Apart from donating the land to EBX, the state government promised to invest R$ 163 million in restoring the old North-West Brazil Railway (Estrada de Ferro Noroeste do Brasil), which would serve the company's interests (Agência Popular 2006).

24 Interview with the authors, June 2003, Porto Seguro, Bahia.

References

Acselrad, H. (2006) 'Tecnologias sociais e sistemas locais de poluição', in *Horizontes Antropológicos* 12 (25), Porto Alegre.

Agência Popular (2006) 'Pólo siderúrgico inaugura novo ciclo econômico em Corumbá', in *Diário MS*, 26 July 2006. Available at http://diarioms.com.br/polo-siderurgico-inaugura-novo-ciclo-economico-em-corumba/ (accessed 10 March 2016).

APUFSC – Associação de Professores da Universidade Federal de Santa Catarina (2006) 'Siderúrgica EBX será instalada em Corumbá', in *Notícias APUFSC*, 15 May 2006.

Auer, P., Besse, G. and Méda, D. eds. (2005) 'Délocalisations, normes et politique d'emploi', La Découverte, Paris, 2005.

Azevedo, P.F. and Toneto Júnior, R. (2001) 'Relocalização do Emprego Industrial Formal no Brasil na Década de 90', in *Pesquisa e Planejamento Econômico* 31 (1), 153–186.

Boltanski, L. and Chiapello, E. (1999) *El Nuevo Espíritu Del Capitalismo* Akal S.A., Madrid (*The New Spirit of Capitalism*. Transl. Gregory Elliott. Verso, London, 2006).

Bourdieu, P. (1998) *Contrafogos*, Jorge Zahar, Rio de Janeiro.

Bridi, R. (2006) 'Espírito Santo perde novos investimentos da Aracruz', in *A Gazeta On-line*, Vitória, 30 July 2006.

Bronfembrenner, K. (2000) 'Uneasy Terrain: the impact of capital mobility on workers, wages and union organizing' US Trade Deficit Review Commission. New York, mimeo.

Bullard, R.D. (1990) *Dumping in Dixie – Race, Class and Environmental Quality*, Westview Press, Boulder, Co.

Chesnais, F. (1996) '*A Mundialização capitalista*', Xamã, São Paulo, 1996.

CNPq/FINEP/SEBRAE (2002) 'Interagir para Competir – promoção de arranjos produtivos e inovativos no Brasil', Brasília, 2002.

Cruz, T. (2006) 'Ameaça ao investimento de US$ 1,2 bilhão', *Zero Hora*, 9 March 2006.

Engels, F. (1975) *A Situação da Classe Trabalhadora em Inglaterra*, Edições Afrontamento, Porto.

Ewald, F. (1993) *Foucault – a norma e o direito*, Vega, Lisboa, p.104.

Fanzeres, A. ed. (2005) *Diagnóstico de conflitos sócio-ambientais em relação à plantações de árvores*, Ministério do Meio Ambiente, Relatório Final.

Fassin, D. and Bourdelais, P. (2005) *Les Constructions de l'Intolérable – études d'Anthropologie et d'Histoire sur les frontières de l'espace moral*, La Découverte – Recherches, Paris.

Ferraz, I. (2004) 'O fim do projeto da usina termelétrica a carvão mineral de Itaguaí', in Acselrad, H. ed., *Conflito Social e Meio Ambiente no Estado do Rio de Janeiro*, Relume Dumará, Rio de Janeiro.

FOBOMADE (2006) 'EBX: Legalizar lo ilegal? Los bosques de Santa Cruz en peligro de convertirse en carbón', available at www.fobomade.org.bo/mutun/doc/EBX_LEGALIZAR_LO_ILEGAL.pdf, (accessed 25 October 2015).

Fontagné, L. and Lorenzi, J-H. (2005) *Désindustrialisation, délocalisations*, Conseil d'Analyse Économique, La Documentation Française, Paris.

Foucault, M. (1976) *La Volonté de savoir*, Gallimard, Paris.

Hardt, M. and Negri, A. (2001) *Império*, Record, São Paulo.

Harvey, D. (2005) *O Novo Imperialismo*, Loyola, São Paulo.

Lipietz, A. and Leborgne, D. (1990) 'Flexibilidade Defensiva ou Flexibilidade Ofensiva: os Desafios das Novas Tecnologias e da Competição Mundial' in Valladares, L. and Preteceille, E. eds, *Reestruturação Urbana: Tendências e Desafios*, Nobel, São Paulo.

Maisonnave, F. (2006) 'Governo Morales mantém decisão de expulsar EBX', in *Folha de São Paulo*, 25 April 2006.

Malerba, J. (2004) 'Meio Ambiente, classe e trabalho no capitalismo global: uma análise das novas formas de resistência a partir da experiência da ACPO' in Encontro da ANPPAS, mimeo., Indaiatuba.

Malerba, J. and Schottz, V. (2006) 'O movimento de resistência à monocultura do eucalipto no Norte do Espírito Santo e Sul da Bahia', Relatório de Pesquisa de Campo.

Mauss, M. (1948) 'Les Techniques et la Technologie', in Meyerson, I. ed. *Le Travail et les Techniques*, PUF, Paris.

Moll, A.P.J. and Spaargaren, G. (2003) 'Towards a sociology of environmental flows: A new agenda for 21st century environmental sociology', International Conference on Governing Environmental Flows, Wageningen, 2003, pp. 43–97 (www.researchgate.net/profile/Arthur_Mol/publication/40114932_Toward_a_Sociology_of_Environmental_Flows_A_New_Agenda_for_Twenty-First-Century_Environmental_Sociology/links/0a85e5354f506997c4000000.pdf, (accessed 8 September 2015).

Nassif, L. (2006) 'A EBX é coisa nossa', *Folha de São Paulo*, 7 May 2006.

Otávio, C. (2006) 'Petrobrás vira objeto de xenofobia na Bolívia', *O Globo*, 30 April 2006, available at www2.senado.leg.br/bdsf/handle/id/399854 (accessed 28 October 2015).

Pedra, D. (2006) 'EBX desiste de siderúrgica na Bolívia e de termelétricas na fronteira', *Correio do Estado*, 26 April 2006, available at http://riosvivos.org.br/a/Noticia/EBX+desiste+de+siderurgica+na+Bolivia/9010 (accessed 28 October 2015).

Pottier, C. (2003) *Les Multinationales et la mise en concurrence des salariés*, L'Harmattan, Paris.

Puckett, J., Byster, L., Sarah Westervelt, S., Gutierrez, R., Davis, S., Hussain, A. and Dutta, M. (2002) 'Exporting harm – the high-tech trashing of Asia' Basel Action Network, mimeo, 50p., available at www.ban.org/E-waste/technotrashfinalcomp.pdf, (accessed 28 October 2015).

Reuters (2006) 'Siderúrgica EBX sai da Bolívia' 25/04/2006, available at http://noticias.uol.com.br/ultnot/2006/04/25/ult29u47463.jhtm (accessed 28 October 2015).

Rios Vivos (2007) 'EBX compra carvão boliviano para siderúrgica no Brasil', 19 March 2007. Available at http://riosvivos.org.br/a/Noticia/EBX+compra+carvao+boliviano+para+siderurgica+no+Brasil/10357 (accessed 28 October 2015).

Rodrigues, L. (2008) 'Eike pode fechar siderúrgica em Corumbá por problemas ambientais', *Folha de São Paulo*, 14 October 2008, available at www1.folha.uol.com.br/mercado/2008/10/456168-eike-pode-fechar-siderurgica-em-corumba-por-problemas-ambientais.shtml (accessed 28 October 2015).

Salm, C. 'Sobre a recente queda da desigualdade de renda no Brasil: uma leitura crítica', available at www.centrocelsofurtado.org.br/adm/enviadas/doc/17_20070512201723.pdf (accessed 20 February 2009).

Salomão, J. (2006) 'O movimento de resistência quilombola à monocultura do eucalipto no Norte do Espírito Santo', Relatório de Pesquisa de Campo, ETTERN/IPPUR/UFRJ, Vitória.

Schlesinger, S. (2001) *Indústria no Brasil: produção sustentável, consumo democrático* FASE, Rio de Janeiro.

Sennett, R. (2001) *A Corrosão do Caráter: conseqüências pessoais do trabalho no novo capitalismo*, Record, Rio de Janeiro, 5th edn.

Stengers, I. and Pignarre, P. (2005) *La Sorcellerie capitaliste*, La Découverte, Paris.

The Economist (1991) 'Let Them Eat Pollution', February 1991.

United Nations (1987) Report of the World Commission on Environment and Development: Our Common Future. UN Documents.

Vargas, R. (2008) 'Setor carvoeiro demite 2.000 funcionários no MS', *Folha de São Paulo* 09/12/2008, available at www1.folha.uol.com.br/mercado/2008/12/476955-setor-carvoeiro-demite-2000-funcionarios-no-ms.shtml (accessed 28 October 2015).

Wackernagel, M. (1996) 'La Huella Ecológica de las Ciudades – Como Asegurar el Bienestar Humano dentro de los Límites Ecológicos?' Encuentro internacional hábitat, Medellin, Colombia, March 11–15, mimeo, 9 pp.

Wallerstein, I. (2000) 'Globalization or the Age of Transition? A Long-Term View of the Trajectory of the World System'. *International Sociology*, 15(2), 251–267.

Weber, M. (2009) *The protestant ethic and the spirit of capitalism*, W.W.Norton & Company, New York.

Yafusso, P. and França, M. (2006) 'Projeto da EBX em Corumbá causa polêmica', *O Globo*, 21 May 2006.

5 Markets or the Commons?

The role of indigenous peoples, traditional communities and sectors of the peasantry in the environmental crisis

Jean-Pierre Leroy

Brazil's economic policy that prioritizes development based on the production and export of commodities, accelerates the destruction of ecosystems and the marginalization of indigenous peoples, traditional communities and sectors of the peasantry, which see their rights to the environment and to their own future denied and are trapped and even exterminated by the expansion of cattle ranching, mining and large infrastructure projects in their territory. These activities are politically and economically supported by the federal government and carried out by the legislature that seeks to reverse the recognition of these people's rights. In the end, the socioenvironmental issue that should be central is concealed or captured by private interests. We are miles away from realizing that these populations are part of the solution not only to the problems brought on by economic growth that cannot absorb the available labour or provide decent living conditions in the cities but also in terms of managing ecosystems through the traditional common use of their territory, which they have historically preserved. In the context of the commodification and privatization of nature, and in the face of the huge environmental and social crisis, the academic and political debate on common goods and the study of its underlying practices are, as such, at the top of the agenda. Is there any possibility that alternatives to the current catastrophe can be imposed or at least considered? Is a movement for the 'common goods' credible? Is there still time to consolidate the Commons (goods) of indigenous peoples, traditional communities and the peasantry? These are some of the questions addressed in this chapter.

The construction of territories

Why are the Commons related to indigenous peoples, traditional communities and sectors of the peasantry important? What do these individuals, families, peoples and communities teach us? Through their struggles and strategies of resistance, they are not only affirming that it is worth fighting for their rights but also that it is possible; that the market has not invaded all of the territories and areas of life, that they are living proof that the Commons are still a reality; and that, therefore, there are other ways of organizing the economy than through the capitalist market; a form of organizing social and cultural life that is not subordinate to the dominant mode, demonstrating that it is possible to politically act in favour of the public interest rather than under the injunctions of private interests; that the destruction and homogenization of territories is not inevitable; and that they hold valuable knowledge for our future.

As of now it is important to delimit what I understand to be the Commons (goods) involving indigenous peoples, traditional communities and sectors of the peasantry. By bracketing the word 'goods', I should point out that emphasis is placed not on the resource base of these social groups, but on the groups themselves.[1] In the midst of so many definitions (see for example Helfrich 2008), I highlight two contributions in particular: Alain Lipietz (2010, 22), reflecting on the Commons in France, states that 'common goods are not things, but social relations' and David Bollier (2008, 30) that 'the common goods concept… refers to a wide variety of social and legal systems for managing shared resources in fair, sustainable ways'. In this sense, I will often use the word 'Commons' as a noun.

Brazil has a long tradition of common use related to land and natural resources derived from indigenous peoples, European migrants, particularly the Portuguese, and from African descendent populations. This tradition is registered in numerous denominations that qualify these forms of land use and in the legislation itself (Almeida 2008) without this implying, however, that the political and economic forces or society as a whole recognize the peoples and social groups that occupy these spaces. With the end of the military dictatorship, indigenous peoples, traditional communities and the peasantry, both with and without land, started becoming visible and characterizing the spaces in which they lived in as 'territory'. The contemporary political history of this 'new wave of territorialization' or 'territorial turn' is best summarized in the words of Henri Acselrad. The interaction between the direct struggles of different peoples, communities and sectors of the peasantry, the creation of related organizations, the approval of legislation and the elaboration of academic papers, led to the consolidation of a political-organizational field around these territories. As analysed by Acselrad (2010, 14),

> The demand for territory, [unlike a simple demand for land] evokes questions of power, identity, self-management and control of natural resources. A territorial claim seeks to impose a new territorialization that, within the national space and based on territorial citizenship, attempts to redefine the relationship between groups, the state and the nation.

The following events can be highlighted as a few of the milestones in the history of this struggle in recent decades: the first 'blockade' (*empate*) against the clearing of forests in the 'Carmen' rubber exploitation (*seringal*) in the Amazonian state of Acre in 1976; the assassination of the president of the

Brasileia Rural Workers Union, Wilson Pinheiro, in 1980 and of the rubber tapper and union leader Chico Mendes in 1988, both also from Acre; the Meeting of the Forest Peoples in Altamira in the Amazonian state of Pará in 1989 and the public hearing to discuss the hydroelectric project in the *Xingu* River, when the indigenous leader Kaiapo Tuira pulled her machete to the face of the representative of *Eletronorte*, the energy company responsible for the project; the drama of the indigenous people *Xavante* to return to the *Marawatsede* Indigenous Land in 2004; the long battle for the recognition of the *Raposa Terra do Sol* Indigenous Land in Roraima; and the assassination of Sister Dorothy Stang in 2004.

To support their struggles, these social sectors were able to organize themselves and create their own institutions: the National Council of Rubber Tappers (*Conselho Nacional dos Seringueiros – CNS*), created in October 1985 during the first National Meeting of Rubber Tappers; Extractive Reserves in 1986; the Coordination of Indigenous Organizations of the Brazilian Amazon (*Coordenação das Organizações Indígenas da Amazônia Brasileira – Coiab*) in 1989; the Amazon Working Group (*Grupo de Trabalho da Amazônia – GTA*) and the Cerrado Network in 1992; the National Coordination of Articulation of Black Rural Quilombola[2] Communities (*Coordenação Nacional de Articulação das Comunidades Negras Rurais Quilombolas – Conaq*) in 1996; the Brazilian Forum on Food and Nutritional Sovereignty and Security (*Fórum Brasileiro de Soberania e Segurança Alimentar e Nutricional – FBSSAN*) in 1998; the Articulation of the Semi-Arid (*Articulação do Semi-Árido – ASA*) in the northeast in 1999; and the National Articulation of Agroecology (*Articulação Nacional de Agrocologia – ANA*) in 2002. NGOs and pastoral groups of the Catholic Church, such as the Pastoral Land Commission (*Comissão Pastoral da Terra – CPT*), the Fisheries Pastoral Commission (*Comissão Pastoral da Pesca – CPP*) and the Indigenous Missionary Council (Conselho Indigenista Missionário – CIMI), linked to socio-environmental, indigenous, land, agro-ecological and social issues, participate and/or support these organizational processes. At the academic level, we can state the publication in 1986 of the book *Black Lands, Holy Lands, Indigenous Lands – common use and conflict* (*Terras de Preto, Terras de Santo, Terras de Índio – uso comum e conflito*) by Alfredo Wagner Berno de Almeida (Almeida 2008) and the works of Antonio Carlos Diegues (Diegues and Moreira 2001; Diegues 2011); and the New Amazon Social Cartography Project (*Projeto Nova Cartografia Social da Amazônia – PNCSA*), an instrument of knowledge also focused on strengthening social movements, initiated in 2005 and most recently extended to the whole country. It is also important to mention the work of numerous anthropologists and other experts that in the exercise of a citizen science have produced the necessary reports to recognize indigenous and *quilombola* territories.

In legal terms, the milestones are the 1988 Constitution and the International Labour Organization (ILO) Convention No.169 ratified by Brazil in 2002. It is also worth mentioning the creation of the National System of Nature Conservation Units (*Sistema Nacional de Unidades de Conservação da Natureza – SNUG*), established by Law 9.985 of 18 July 2000, which introduced the possibility of conservation areas with human presence and established the prevalence of socioenvironmentalism over classic preservationism (Santilli 2005, 112). A mark of the official recognition of the term 'territory' in relation to indigenous and traditional people is the Decree No. 6040 of 7 February 2007, which established the National Policy for the Sustainable Development of Traditional Peoples and Communities. Article 3 II states that:

> Traditional Territories: the spaces necessary for the cultural, social and economic reproduction of traditional peoples and communities, whether used in a permanent or temporary basis, observing, with regard to indigenous peoples and *quilombolas* respectively, the provisions of articles 231 of the Constitution and 68 of the Temporary Constitutional Provisions Act and other regulations.
>
> (BRASIL 2007)

This is preceded by the definition of traditional peoples and communities, demonstrating the unbreakable bond between peoples and territories:

> Traditional Peoples and Communities: culturally differentiated groups and that recognize themselves as such, which have their own forms of social organization, which occupy and use territories and natural resources as a condition for their cultural, social, religious, ancestral and economic reproduction, using knowledges, innovations and practices generated and transmitted by tradition.
>
> (Ibid.)

The Decree does not innovate; it only formalizes a definition of territory built by social actors through their historic struggles to exist and to be recognized. The diversity under this single concept is huge but the specific territoriality that configures the territory of each group, 'functions as a factor of identification, protection and strength' (Almeida 2008, 72). As such, the concept becomes a central category not only in analytical but also in political terms.

It is my understanding that along with the territories of traditional peoples and communities, we should also include the territories of peasant populations that, while not using common areas a priori, end up building something that can be identified as a result of their history and 'factor of identification' (see Isaguirre-Torres and Frigo 2013; Petersen 2014[3]; Ploeg 2014; Steinbock et al. 2013).

Considering the multiplicity of and differentiation among territories, at least four major components can be identified: indigenous territories, which strictly speaking compiles all the space that became the Brazilian territory with the Portuguese colonization but which today is confined to approved and disputed Indigenous Lands; common land brought to Brazil as a custom of European migrants and settlers; lands arising from occupation struggles carried out by reminiscent *quilombos* and traditional communities; and nuclei of the peasantry.

However, a careful observation of the constitution of these territories multiplies the possibilities and the forms they take. There are those that form a continuous space since the private and common areas are contiguous; those where private and common areas are not contiguous; cases where the Commons override properties that are not integrated into the Commons (Petersen 2014); cases where properties are an integral part of the Commons (Steinbock et al. 2013) but not managed according to the Commons; and cases where the private property is relativized for being part of the Commons' management strategy (Martins et al. 2014). There are also others where the Commons Territory overlaps with Integral Conservation Units (Favero and Zhouri 2013) or with private properties (Schmitz et al. 2006).

In this context, we are invited to adopt a deeper and renewed understanding of territories, a concept loaded with meaning: physical ecosystemic territories in line with their inhabitants; ancestral lands full of emotional and spiritual memory; laboratories of experiments and construction of knowledge; territories of struggle and the development of identities; territories constructed around family ties; territories where individuality is inseparable from the collective.

So how should these territories be qualified today? As a hypothesis to be debated, I present the idea that the Commons is analytical and politically the most appropriate concept to account for this field that unites nature and very diverse social groups. As such, I move towards an initial proposal in terms of definition. What makes up the Commons? A territory, a community or communities that occupy and generate the territory, their reproductive strategies and related skills, the mechanisms they use to maintain and consolidate their Commons, their culture. The Commons is made of material elements characterized and transformed over time by the 'fingerprints' of the human community occupying the territory; and immaterial because it produces a culture born out of the nature–community symbiosis.

We are talking about a minority of the population here; but considering the occupied area 'the natural resources under community control are not something residual' (Vianna Jr. 2013, 3), it assumes an eminently important role in the regional and global context of multiple environmental crisis. Looking at the sum of the officially recognized areas, i.e. 1,135,975 km^2 of Indigenous Lands (ISA 2015), 255,596 km^2 of sustainable use conservation areas such as extractive reserves and sustainable development reserves (BRASIL 2015c), 35,526 km^2 of quilombos (BRASIL 2015a), it appears that in early 2015, 1,427,097 km^2 were out of the market and under community control. Even taking into account the existence of double counting, we would have to add a further 801,000 km^2 of areas of family farming (BRASIL 2015b), where there are local/regional centres of agroecology, extractivists and *quilombos*. Considering that the size of Brazil is 8,515,767 km^2, we can argue that between 15 and 20 per cent of the Brazilian territory is officially outside the market, composed of areas managing the Commons, especially in the Amazon where Indigenous Lands occupy 22.25 per cent of its extension (ISA 2015). It is also noteworthy that territories occupied by indigenous peoples and traditional communities are much more extensive in reality but are not measurable simply because they lack land tenure or because they are fishing areas inadequate in terms of the institution of property.

If the amount of land being relatively preserved from environmental destruction should not be underestimated, nor can the meaning of this 'territorial turn' in relation to the institution of property be belittled. Professor Souza Filho, reflecting on the names given to Indigenous Territories, recalls that they have been identified as 'reserves', 'areas' and finally 'Indigenous Lands'; arriving at the heart of this evolution:

> The name territory was never used and instead it was intentionally denied. Of course there is a not very subtle difference between calling it land or territory. Land is the legal name given to individual property, whether public or private; territory is the legal name given to a jurisdictional space. Thus, territory is a collective space that belongs to a people. The same ideology that denies the existence of people... denies the use of the term territory.
>
> (Souza Filho 2003, 102)

However, the fact that the ownership regime does not respond to the state's need to attribute legal status to territorial conquests enshrined in the 1988 Constitution, has led to the application or invention of different forms of ownership, all of which still have property as a subliminal reference. Brazil has two categories of land registration and census, namely: establishments or units of exploitation, adopted by the agricultural census of the Brazilian Institute for Geography and Statistics (*Instituto Brasileiro de Geografia e Estatística – IBGE*) and rural property or domain unit, adopted by the registry of the National Institute of Colonization and Agrarian Reform (*Instituto Nacional de Colonização e Reforma Agrária – INCRA*) for tax purposes. The statistics that make up the agrarian structure in the country apply these, and only these, categories. Due to the tutelage regime, Indigenous Lands are registered in the Federal Heritage Department. The lands occupied by *quilombola* communities, also protected by the Constitution through Art. 68 of the Constitutional Addendum, should be converted into rural properties as final land title. Extractive Reserves continue to be a patrimony of the Union allocated to the extractivists as a Right of Use.

However, the clear identification of territories is one of the ways in which the Commons belonging to the peoples and communities here mentioned were imposed in the rights universe. This process has demonstrated how legal homogenization concerning the institution of property paralyses and impoverishes humanity and encloses survival strategies into a single model that no longer fits the unique moment of environmental crisis that we are experiencing and that demands innovative and differentiated solutions. The subjects of the Commons related to nature become unavoidable actors in the debate on tackling the climate crisis. But this is not all. Even if drawn in a context that is alien to them, the recognition of limits introduces these subjects into a universe where, like it or not, they have to struggle for their recognition and reproduction with the added interest that it can help them to better organize their strategies internally. According to an indigenous leader from Santarem: 'I think the mapping processes have been a basis for things, including mobilizations. Now the community is aware of what is around them' (Assis 2010, 181).

Not surprisingly, the territorialization movement that culminated in the occupation of significant areas of the country, is strengthened by the opposition it encounters but weakened by the violence of Brazilian elites and capital. The last decades combine the affirmation of new territorialities and the Commons of peoples and communities on one hand, with the relentless pursuit of broader economic and political interests to annihilate them, on the other. 'Non-market and non-capitalist-based modes of living ... are considered a barrier to the accumulation of capital and they therefore must be dissolved' (Harvey 2011, 65); and successive governments have adopted this logic in the desire to move from emerging to developed country.

The deconstruction of territories

The perverse dynamics affecting indigenous peoples, traditional communities and sectors of the peasantry, leads us to the idea of primitive accumulation identified by Marx (1980, 830) and the related movement for enclosure, which Hobsbawm (1989, 65) summarized as being accountable for 'the mass expropriation of peasants and the transformation of land into a commodity'. In contrast to Marx, Rosa Luxemburg examines colonial policy as the continuation of primitive accumulation and argues that 'force is the only solution open to capital; the *accumulation* of *capital*, seen as an *historical process*, employs force as a *permanent* weapon, *not only* at *its genesis,* but further on down to the present day' (Luxembourg apud. Composto and Navarro 2014, 39).

The old primitive accumulation characterized by Harvey (2011, 48) as 'accumulation by dispossession' and as the 'second enclosure' by the commons activist and scholar David Bollier, continues to be on the agenda. In an article on *quilombos* from the Marajó Island in Pará, Rosa Acevedo Marin (in Sabourin and Caron 1999) talks about how farms are advancing into *quilombola* territories and the limitations that the fences are imposing on the community's mode of existence.

It is also possible to imagine invisible fences; the communities or peoples that produce on an agroecological basis while surrounded by soya, corn or cotton and/or subjected to the aerial spraying of pesticides that corner them until they give up, as reported by Silvino, a family farmer from Santarém. In his testimony on video, in short he says that lured by offers, the residents end up selling their land; that due to the lack of movement, transportation becomes insufficient; that plagues fleeing from land clearing affect their land, and the poison used by agribusiness is killing bees and fruit trees... and that as a result, the only thing left to do is leave (Fase/Cepepo 2005). These symbolic fences are also strangling the indigenous peoples of *Volta Grande do Xingu*, fall line of the *Xingu* river in the Amazon, which are losing their territories in account of the diversion of the river's waters due to the construction of the Belo Monte hydroelectric complex.

Developments in the last decades have resulted in a new type of enclosure promoted on behalf of the environment. The United Nations Conference on Environment and Development – Rio-92 – initiated a sequence of Conference of the Parties to address the Convention on Climate Change and the Biodiversity Convention. In an economic and political environment dominated by neoliberalism, reduced UN influence and pragmatism of the main environmental organizations, it is no wonder that the solution to the environmental crisis has been attributed to the private sector. Large companies, in particular energy related multinationals, have been rewarded with the opportunity of offsetting their CO_2 emissions, partnering with traditional peoples and communities via the carbon market, *Reducing Emissions* from Deforestation and Forest Degradation (REDD+) initiatives and, to some extent, Payment for Environmental Services (PES) and the Brazilian federal programme *Bolsa Verde*[4] (Schlesinger 2014).

It's time to deploy a new 'Economics of Ecosystems and Biodiversity' (TEEB) because, says the director of Deutsche Bank and coordinator of TEEB, Pavan Sukhdev: 'We use nature because it's valuable, but we lose it because it's free' (Fatheuer 2014, 28). We return to Hardin's (2002) argument: that which belongs to everybody, belongs to nobody; that which is free is not well maintained. States don't take good care of public goods, because they have no value – a move from the public to the private.

As such, with the support of the Brazilian government, territories, peoples and communities are being recognized as long as they are placed at the service of these mechanisms: PES, REDD+ and carbon credits. In her book, which provides an in-depth analysis of the New Forest Code approved under Law 12.651 of October 2012, Larissa Parker (2015, 198) estimates that the Code (along with the numerous bills on PES and REDD+) coordinates a 'new legal engineering process… that meets the expansionary demands of primitive accumulation of capital… through new institutes or the modification of their legal nature'. According to Parker, this transfers the responsibility to protect the Commons from the public, as enshrined in the Constitution, to the market. The process represents a double advantage for capital: in addition to presenting itself as the solution to the environmental crisis obtaining, as such, recognition from broad sectors of society, it 'virtuously' opens new areas of expansion for businesses and profits and dismantles possible conflicts with peoples and communities now trapped in the tentacles of capital.

A further step is thus taken in the subtraction of the rights of peoples, traditional communities and the peasantry. If this is not accumulation by dispossession, then accumulation by submission could be a way of describing it. Indeed, NGOs, public sectors and companies (at best) present these market alternatives to the peoples and communities or harass them into accepting them. Considering the state of penury in which the majority of these communities live in, it is no wonder that unless they are politically mobilized, any kind of 'help' is welcomed, even when the final conclusion is that expectations were not met. Communities that are aware of what they want and already involved in the construction of their Commons may also consider the importance of receiving monetary support from the market in light of their economic strategy. The question will be: how to avoid being entrapped by this 'enclosure'? A survey restricted to the Amazon conducted in 2013, shows that the most frequent mechanism put in place refers to PES in the form of the *Bolsa Verde* Program (Schlesinger 2014). The impact of the financial aid is minimal. Movements and organizations of indigenous peoples and traditional communities are more concerned with the restrictions presupposed by these agreements in terms of the communities' autonomy and strategies of family and collective reproduction. As a result, a network of movements and organizations critical to this process was created: the '*Carta de Belém*[5]' group.

The expansion of agribusiness related to cattle ranching, monoculture of soybean, corn, sugarcane and cotton, eucalyptus, and, more recently palm oil (*elaeis guineensis*), the production of so-called 'industrial' fruits and fishing; the opening of roads, railways and ports, hydroelectric dams, mining, resorts on the coast… have constituted fighting fronts for the indigenous, traditional and peasantry populations. When the Map of Conflicts Involving Environmental Injustice and Health in Brazil was launched in 2009, 300 cases were presented. It noted that:

> Most of the cases raised in the map affect precisely populations and workers living in rural, forest and coastal areas (over 60%), in other words, in areas where disputes over natural resources related to Brazil's insertion in international trade affect traditional and agrarian communities.
>
> (Pacheco et al. 2013, 51)

> Most of the conflicts raised, concern indigenous peoples followed, in descending order, by family farmers, *quilombolas*, artisanal fishermen, riverine peoples, *caiçaras*[6] and extractivists in their various denominations.
>
> (Pacheco et al. 2013, 52)

Albeit very small, the universe of conflicts registered then was enough to point out certain trends. The reports published by CIMI and CPT for 2014 present numbers that confirm these trends. CIMI (2014) registered 135 suicides and 138 assassinations of indigenous peoples in 2014. In regards to the peasantry, in the same year, 12.188 families were evicted, 36 leaders assassinated and 56 suffered attempted murders (Medeiros 2015).

Less stressed than overt violence but perhaps more pernicious is the fact that peoples and communities are often victims of symbolic violence when economic or political forces impose their actions on them as legitimate, 'concealing the power relations that are the basis of its force' (Bourdieu 1972, 18). The analysis of the cases presented in the map mentioned previously led us to make the following comment:

> Legal decisions as well as information disseminated by major means of communications function as legitimisers of violence; legitimacy that forms the opinion not only of local society but also of the affected peoples themselves. As such, a huge effort is required on their part to respond and recognize this condition of peoples affected by this violence. We can imagine how many cases of injustice and violence do not publicly appear precisely because the victims internalize their unequal condition to such an extent that they don't even realize they have the right to react.
>
> (Leroy and Meireles 2013, 127)

Violence is more insidious than one might think. Therefore, 'what is taken as a spontaneous agreement is actually often a sort of coerced agreement, fruit of a decision taken under the spectre of past violence and the fear of repetition' (Alarcon and Torres 2014, 56), defined as 'participatory exclusion'.

In recent years, the so-called Rural Congressional Caucus, which brings together numerous representatives of agribusiness, undertook a systematic legislative offensive against traditional populations and the landless, in particular indigenous peoples, aimed at deconstructing the rights guaranteed in the 1988 Constitution. This attack is led by the Constitutional Amendment Bill n° 215, intended to transfer to Congress the authority to demarcate indigenous lands, create conservation units and provide land titles to *quilombos* currently in the hands of the executive, namely the National Indigenous Foundation (*Fundação Nacional do Indio – Funai*), the Brazilian Institute of Environment and Renewable Natural Resources (*Instituto Brasileiro do Meio Ambiente e dos Recursos Naturais Renováveis – Ibama*) and *Fundação Cultural Palmares* respectively. The foxes seek to seize the henhouse.

The federal government in turn, pursuing its goal of expanding the infrastructure required for its developmentalist project, seeks to reduce the legal obstacles to the occupation, by large enterprises, of lands legally held by traditional peoples or of those that are in the process of being recognized. In particular, the government appeals to successive simplifications in the environmental licensing process, ultimately rendering meaningless the legislative framework that protected, even if partially, the environment and traditional communities.

The resistance of the Commons

From the numerous statements resulting from meetings of organizations related to indigenous peoples, *quilombolas*, extractivists and agroextractivists, artisanal fishermen, peasants, women, agroecologists, etc., it is possible to affirm that they are all convinced that their future depends not only on guaranteeing permanent possession of their territories but also on their ability to find, within these territories, the necessary conditions to ensure their reproduction. In this context, the economic issue that we now propose to superficially address, becomes central.

In Elinor Ostrom (2010) we find eight 'Design Principles Illustrated by Long-Enduring Common Pool Resource (CPR) institutions' that allow us to reflect on the economic feasibility – or not – of the Commons addressed here. These principles relate to the adequacy of the natural resource base and those who use them, the institutions and rules, the participation of users in their formulation, forms of control, sanctions and conflict resolution required for the rules to be respected. Finally, Ostrom stresses the need for external governmental authorities to recognize these organizational systems. On the basis of these principles, it is possible to make a few observations.

Concerning this last principle, it is more than an issue of recognition. With regards to Brazil, the Commons are generally not consolidated and struggle to be recognized. It is worth noting the shortage of field studies reflecting the invisibility of the Commons populations and the lack of attention given to the issue. The cases presented by Ostrom demonstrate a relatively smooth integration of the Commons into the capitalist economy. This does not seem to be the case in Brazil where their very existence appears or is presented as a threat to development.

All of the Commons covered in this paper demand an expansion of the available resource base. In fact, the issue of recovery and expansion should not be considered as merely an afterthought since the Commons addressed here may not survive without a match between the natural resources base and their exploitation. Even in the case of indigenous peoples that dispose of what is still apparently extensive land, the shortage of hunting animals and fishery, for example, is severely felt. A rigorous cultural transformation – the transition from extraction to agroforestry and fisheries related management – becomes inevitable in the short or long term; something that not everyone will be able to assimilate. The developments and changes in the organization of the resource base are numerous, for example, when the function of lakes in the Amazon floodplains are redistributed or when space and its role is redefined by the intensified production of a forest area or by the installation of a fishery and irrigation project.

A mere subsistence economy through which families can ensure their survival from what the territory offers them, with little or no need to appeal to the market in order to sell and buy their products, is clearly not enough. The issue is to insert the Commons in the Commons-economy in order to guarantee the reproduction of families with a decent quality of life, access to essential goods and services and entertainment – demands expressed especially by women and the youth – and the insertion of increased and more diverse sets of products in the market, demonstrating that a Commons-economy can be more than marginal. Without the recovery and expansion of the natural resource base we will fall into the trap of over-exploitation and the abandonment of the territory. Apparently, albeit with some exceptions, this is not taking place. Increasingly and continuously the natural systems are being maintained.

On what social base are Commons-institutions built and forms of management triggered? Kinship appears as a fundamental element of cohesion facilitating rules of reciprocity, in particular in the case of slave descendent communities and indigenous peoples; it is also present through *compadrio*[7] relations in other types of social organizations. Reciprocity is horizontal but also vertical. Despite the inequality among community members, relations established between the more affluent and the poorest help maintain ties of solidarity. In larger communities, in peasant areas, for example, kinship ties are supplemented or replaced by relationships based on trust that preserve the 'exchange' of favours.

Territorialization processes put traditional institutions to a test and require their renewal and/or the creation of new institutions. Indeed, the 'territorial turn' demands a break from the status quo for which traditional leaders and institutions are not always prepared. The challenges are multiple and constant, both at the political as well as the technical level and require the creation of new institutional forms and/or the permanent renovation/adaptation of existing institutions. Thus, these new institutional forms should be twofold: political-organizational and technical-managerial or a combination of the two. Although some of the old institutions may be maintained, they will fall into disuse in the face of new demands. Other institutions may remain but under constant reform in order to take on new tasks and responsibilities.

The management of territories is the work of collectives. That said, we know that there are always leaders that show the way, encourage or direct the community or the peoples. Several studies refer to the 'elders' of the community, 'the older people'. In general, when approaching the reform or the creation of new institutions, the research consulted neglects to discuss the role of these leaders as if it was an obvious fact. The roles and responsibilities attributed to the new institutions demand a type of preparation from leaders and communities that will probably, in many cases, be beyond their capacity to actively participate in projects and other dynamics related to the construction and consolidation of the Commons. At the same time, the leaders would have to take on the role of maintaining the community's cohesiveness with their traditions; promoting internal capacity-building activities; defining, along with community members, the rules of management and conservation of natural resources and their economic exploitation; leading the struggle for territorial control if necessary; and raising and managing development projects and production, processing and marketing initiatives. In short, the leaders would have the overall responsibility to bridge the gap with the outside world! It would be no exaggeration to say that this transition represents an enormous effort for community leaders, almost completely excluded from the education system. This is especially true when the historical context characterizes them as mere testimonies of the past, at a time when economic forces and dominant policies attempt to extinguish them in every possible way and when agribusiness and the agro-industry have shaped tastes and conquered the food market.

The struggle for territorial recognition means access to citizenship or, more accurately, to the 'political city', its laws and power, decision-making and knowledge-related bodies to whom the Commons will have to address and from which they will receive information and techniques, exogenous standards, resources or techniques, prohibitions, restrictions, destruction. As such, communities and their leaders need to equip themselves with 'modern associations' (Sabourin and Caron 2009, 111) capable of acting as intermediaries between the community and the nation and to take on new arising tasks. The institutionalization of the Commons is not simply an internal process of construction. None of the articles consulted presented indications that the creation of the Commons takes place without the groups in question suffering previous external institutional influence. The influence of the Catholic Church, through its dioceses, pastoral and/or the Movement for Grassroots Education, for example, is remarkable. The influence goes so far as the statement that 'the Amazon community was literally invented by the Catholic Church. The two institutional axes in this process are the catechists ... and the community Councils, which take care of the problems of the place' (Castro et al. 2002, 277). This influence, which incidentally is not restricted to the Amazon, is seen as a positive movement, helping to form leaders, create or strengthen community organizations and raise awareness in communities on environmental issues.

Inasmuch as the 'territorial turn' takes consistency and that indigenous peoples, traditional communities and sectors of the peasantry fight for and/or conquer territories, other external actors come into play, be it to solve (or not) land tenure issues or to guarantee the implementation of projects and other initiatives aimed at consolidating the territories, especially in its relation to the economic sphere. These actors are the NGOs, pastoral sectors of the Catholic Church, popular organizations and social movements mentioned previously. Other actors include the government, state officials and public bodies such as the Chico Mendes Institute for Conservation and Biodiversity (*Instituto Chico Mendes de Conservação da Biodiversidade – ICMBio*) and IBAMA, Funai, the National Institute of Colonization and Agrarian Reform (*Instituto Nacional de Colonização e Reforma Agrária – INCRA*), the National Supply Company (*Companhia Nacional de Abastecimento – Conab*), the Public Prosecutor's Office, academia, etc.; private companies; external actors (international environmental and solidarity related NGOs, *Via Campesina*, the Slow Food Movement, international agencies of the UN system).

Enhancing the links between scientific research and local knowledge occupies a growing space in the Commons. It originates from the interest of researchers and institutions to contribute to the Commons populations but also from the communities' understanding that their territories are not frozen in the past and that their future depends in part on their ability to develop new knowledge that can contribute towards them seeing themselves and being seen as peoples that look to the future – their future and the future of humanity. This concept seems summarized in the Political Statement resulting from the IV ASA Conference (2003), held in the north-eastern municipality of Campina Grande:

[ASA] is convinced that in the diversity of experiences developed by farmers in the Brazilian semiarid, a type of knowledge is produced, which once inter-related with academically systematized knowledge will become knowledge capable of thrusting the sustainable development project for the semiarid.

Many Commons are included in the 'solidarity economy' through their presence in alternative local fairs, direct sales to consumers, in circuits that promote production exchange and/or the development of alternative certification mechanisms, as is the case of Cooperafloresta, an Association of agroecology farmers from the municipalities of *Barra do Turvo* in São Paulo and *Adrianópolis* and *Bocaiúva do Sul* in Paraná. This integration is more common in sectors of the agroecology-oriented peasantry briefly mentioned here. The obstacles are much greater in areas that are distant from urban centres, especially in the Amazon.

There is no lack of sectoral policies aimed at meeting the demands of these peoples but their effectiveness is not universal and suffers from discontinuity. Among them, the National School Food Programme (*Programa Nacional de Alimentação Escolar – PNAE*) the Food Acquisition Programme (*Programa de Aquisição de Alimentos – PAA*) and the National Family Farming Programme (*Programa Nacional da Agricultura Familiar – Pronaf*) stand out positively. However, claims are made that these credit, financing, production, storage, marketing, health standards and technical assistance policies need to better correspond to reality. Since they are not state policies, embodied on laws, sectorial policies are under threat by an overwhelming coalition of private interests nested in Congress. The Commons have barely appeared in the horizon and they are already interpreted, even if not clearly identified as the Commons, as an intolerable threat to the secular domain of the Brazilian elites. These elites, even if recycled, continue to irrigate their roots of slave masters.

Despite the Solidarity Economy, the Commons also depend on the market relations functioning in the capitalist economy. There are products of the extractive economy such as the *pirarucu* (*arapaima gigas*) and other valued fish, oils, nuts, *acai*,[8] shrimp and other goods of the peasant economy that have a guaranteed market, but which are facing difficulties when it comes to logistics: processing, industrialization, transportation. Present in local economies, the Commons also negotiate contracts with companies such as Natura (cosmetics) and Wickibold (bakery). Communities seek the market to certify their forest or agricultural products. There is no guide on how to behave in the face of companies. There is a risk that private sector initiatives related to the green economy, including those dedicated to the carbon or biodiversity market, end up suffocating the Commons, turning them into mere suppliers of commodities.

The responsibility of NGOs that support and facilitate this type of insertion in the market is central. Several of these organizations, the majority of which are the extension of large international NGOs with the mission of guaranteeing the conservation of natural resources, have chosen, by conviction or pragmatism, the path of partnerships with the private sector, including multinational corporations that promote the degradation of ecosystems and are primarily responsible for the climate crisis. It is worth raising the question of whether traditional peoples and communities are not considered by these NGOs only insofar as they serve their conservationist objectives. How else to interpret their silence in relation to those affected by the construction of the *Madeira*, *Xingu* and *Tapajos* hydroelectric dams? And their omission in regards to the genocide affecting the *Guarani Kaiowa* indigenous peoples in the state of Mato Grosso do Sul?

After all, the movement initiated by these social groups in the constitution of their own economy, distinct from the capitalist economy, which could be called oiko-economy or 'oikonomia', a result of the stubbornness of community members, its innovative character and the way it responds to the environmental crises, fulfils the necessary internal and external conditions (political and technical support) as a step towards its affirmation. The challenge of rendering these Commons irreversible, remains.

Conclusion

This chapter proposes a few reflections more aimed at opening discussions rather than concluding the issue. Local struggles have led to the creation of regional or national organizations and movements directed at strengthening the survivability of the Commons, serving as technical but above all political capacity-building spaces in order to gain sufficient collective force to stop processes that are destroying the Commons. Without them, 'collective action at the local level would hardly have been successful in the long run' (Diegues and Moreira 2001, 100).

The Commons of indigenous peoples, traditional communities and peasants are not open. Only these populations have the right to the immediate enjoyment of the resources available in their territory; but they are also common goods of humanity to the extent that societies, states and public international bodies support them. In the last two decades, since Rio-92, a system of rules aimed at preserving the environment has been established in Brazil. The aim is to seek a full encounter between general rules set by the state and rules already experienced by the inhabitants of the Commons. This combination of contributions is essential to the definition of territory as Commons. This is not just a secondary point; it introduces fundamental questions: are these Commons related to nature solely the Commons of its inhabitants, where access is limited to them or are they also society's Commons, as we inquired previously? Couldn't and wouldn't public officials also act as members of society, even in the exercise of their office, restoring the sense of public service: in service of society and not of the state/government? To a certain extent, isn't the state a co-participant of these Commons through the legislation?

The establishment of linkages between local Commons rooted in ecosystems and urban society in Brazil is urgent. If the water crisis is beginning to raise concerns, public authorities have not yet realized the importance of social sectors involved in the preservation of rural and forest environments. The same can be said in relation to the food contaminated by the production-model and to the climate. It is clear that the Commons discussed in this chapter have strong political weight, but little power. One of the main questions concerning the future of the Commons is related to the role of the State, or should we say, states and public international bodies when they become subordinate to the dictates of financial capital jeopardizing democracies. Isn't it time for global society to assume the common future?

Could the practices related to the Commons gain strength and stand as oikonomia, alternatives or at least parallel to the dominant economy if they were to be united? For this to happen we need to overcome the current phase of holding meetings among activists and intellectuals to discuss the subject and seek the formation of a social movement around each Commons and a global Commons. This assumes the creation of international networking processes, which is difficult in the current context of disintegration and dispersion among social movements as demonstrated by the World Social Forum held in Tunis in early 2015, according to several commentators such as Boaventura de Souza Santos and Candido Grzybowski. The lack of financial resources to promote encounters is also an obstacle worth adding, although virtual communications can remedy in part this failure.

Those that attend meetings of the peoples experiencing the Commons related to nature, frequently hear of *Buen Vivir*, roughly translated to 'good living' in English. The notion of *Buen Vivir* was brought to the public debate by different Andean peoples in the context of their confrontation against neo-extractivism. Each of these peoples expresses the notion with their own nuances, according to their traditions and world-views. The Andean Coordination of Indigenous Organizations (IOTC) presented the following summary:

> *Vivir Bien* is to live in community, in brotherhood and especially in complementarity. It is a communal, harmonious and self-sufficient life. *Vivir Bien* means to complement each other and share without competition, live in harmony between people and nature. It is the basis for the defense of nature, of life itself and all humanity.
>
> (CAOI 2010, 21–22)

The currently unpopular values of equality, solidarity, precaution and responsibility are contained in *Buen Vivir* and the Commons. Moreover, when claiming to be part of the natural world, the subjects addressed in this chapter are struggling against the dominant production and consumption model. It's worth listening to them.

Notes

1 It is important to mention that in Portuguese, the word Commons is not used. The concepts mainly used are common goods, common resources or resources of common use.
2 African-descendent rural communities, originated during the slavery period as a form of resistance, also known as maroon in English.
3 The reference consulted was a manuscript as listed in the references but a version of the article was also published in English as follows: Petersen P. (2015) 'Hidden treasures: reconnecting culture and nature in rural development dynamics', in *Constructing a New Framework for Rural Development* Research in Rural Sociology and Development, Volume 22, 157–194.
4 The 'Support for Conservation *Bolsa Verde* Programme' provides a quarterly benefit of R$ 300 to families in extreme poverty living in areas that are considered a priority in terms of conservation. The conditionality for receiving the financial aid is the conservation of ecosystems and the sustainable use of natural resources.
5 www.cartadebelem.org.br/site/about/
6 Traditional peoples of the Southeast and South of Brazil.
7 Non-biological form of kinship.
8 Berries, harvested from palm trees found around the Amazon River basin of South America.

References

Acselrad, H. (2010) 'Mapeamentos, identidades e territórios', in Acselrad, H. *Cartografia social e dinâmicas territoriais: marcos para o debate*, IPPUR/Universidade Federal do Rio de Janeiro – UFRJ, Rio de Janeiro 9–45.

Alarcon, D. F. and Torres, M. (2014) '*Não tem essa lei no mundo, rapaz!': a Estação Ecológica da Terra do Meio e a resistência dos beiradeiros do alto Rio Iriri*. ISA – Instituto Socioambiental, São Paulo, Amora – Associação dos Moradores da Reserva Extrativista Riozinho do Anfrísio, Altamira. E-book. (www.socioambiental.org/sites/blog.socioambiental.org/files/blog/pdfs/nao_tem_essa_lei_no_mundo_ebook.pdf) Accessed 10 November 2015.

Almeida, A. W. B. (2008) *Terra de quilombo, terras indígenas, 'babaçuais livre', 'castanhais do povo', faixinais e fundos de pasto: terras tradicionalmente ocupadas*, PGSCA-UFAM, Manaus.

Appel, K. O. (1987) *Sur le problème d'une fondation rationnelle de l'éthique à l'âge de la science*, Presses Universitaires de Lille, Lille.

ASA – Articulação no semi-árido brasileiro (2003) IV Encontro Nacional, Campina Grande (PB) 'Carta Política' (PDF).

Assis, W. F. T. (2010) 'Conflitos Territoriais e Disputas Cartográficas: Tramas sociopolíticas no ordenamento territorial do Oeste do Pará', in Acselrad, H. ed., *Cartografia social e dinâmicas territoriais: marcos para o debate*. Rio de Janeiro: IPPUR/Universidade Federal do Rio de Janeiro – UFRJ, 2010.

Bollier, D. (2008) 'Los bienes communes: un sector soslayado de la creación de riqueza', in Helfrich, S. ed., *Genes, bytes y emisiones: bienes comunes y ciudadanía*, Ediciones Böll, México.

Bourdieu, P. (1972) *Esquisse d'une théorie de la pratique*, Droz, Paris.

BRASIL. (2007) Presidência da República. Decreto N° 6040, do 7 de fevereiro de 2007. Institui a Política Nacional de desenvolvimento Sustentável dos Povos e

Comunidades Tradicionais. Art.3 I. (http://presrepublica.jusbrasil.com.br/legislacao/94949/decreto-6040-07). Accessed 4 September 2014.

——(2015a) INCRA – Instituto Nacional de Colonização e Reforma Agrária (www.incra.gov.br/search/node/Quilombolas) Accessed 20 June 2015.

——(2015b) Instituto Nacional de Colonização e Reforma Agrária. Reforma Agrária. (www.incra.gov.br/reforma_agraria). Accessed 30 June 2015.

——(2015c) MMA – Ministério do Meio Ambiente Coordenação nacional das Unidades de Conservação – CNUC, *Tabela consolidada das unidades de conservação.* Source: CNUC/MMA (www.mma.gov.br/cadastro_uc). Updated 17 February 2015. Accessed 10 July 2015.

Campos, N. J. (2011) *Terras de uso comum no Brasil. Abordagem histórico-espacial* Ed. Universidade Federal de Santa Catarina – FSC, Florianópolis.

CAOI – Coordinadora Andina de organizaciones Indígenas (2010) Buen vivir/Vivir bién, Filosofia, políticas, estratégias y experiencias regionals andinas (www.coordinadoracaoi.org/web/publicaciones/buen-vivir-vivir-bien/). Accessed February 2016.

Castro, F., McGrath, D. G. and Crossa, M. (2002) 'Adaptándo-se a los câmbios. La habilidade de las comunidades ribereñas en el manejo de los sistemas de lagos de la Amazônia brasileña', in Smith, R. C. and Pinedo, D. eds, *El cuidado de los bienes comunes. Gobierno y manejo de los lagos y bosques en la Amazonia*, Instituto del bien común; IEP – Instituto de Estudios Peruanos, Lima 277–302.

Cimi – Conselho Indigenista Missionário (2014). *Relatório Violência contra os Povos Indígenas no Brasil – Dados de 2014,* Coordenação da pesquisa: Lúcia Helena Rangel Cimi Brasília (www.cimi.org.br/File/Relatorio%20Violencia%20-%20dados%202014.pdf). Accessed 10 September 2015.

Composto, C. and Navarro, M. (2014) 'Claves de lectura para comprender el despojo y las luchas por los bienes comunes naturales en América Latina', in Composto, C. and Navarro, M. *Territorios en disputa. Despojo capitalista, luchas en defensa de los bienes comunes naturales y alternativas emancipatórias para América Latina*, Bajo Tierra Ediciones Mexico D.F. 33–75. (http://otrosmundoschiapas.org/docs/territorios_en_disputa_bienes_comunes.pdf). Accessed 7 September 2015.

Diegues, A. C. (2011) 'Repensando e recriando as formas de apropriação comum dos espaços e recursos naturais', in Diegues, A. C. and Moreira, A. C. eds, *Espaços e recursos naturais de uso comum*, Nupaub/Universidade de São Paulo – USP São Paulo 97–124 PDF version (http://nupaub.fflch.usp.br/sites/nupaub.fflch.usp.br/files/color/Espacos_UsoComum.pdf). Accessed 5 August 2015.

Diegues, A. C. and Moreira, A. C. (2001) *Espaços e recursos naturais de uso comum*, NUPAUB-USP, São Paulo.

Fase/Cepepo (2005) *O grão que cresceu demais. O caso da soja em Santarém e Belterra, Pará.* Video Fase/Cepepo Belém (www.youtube.com/watch?v=fC8AcAUwtQs). Accessed 12 September 2014.

Fatheuer, T. (2014) 'Nova Economia da Natureza. Uma introdução crítica', Vol. 35 da *Série Ecologia*. Fundação Heinrich Boell, Rio de Janeiro.

Fávero, C. and Zhouri, A. (2013) 'MG Parque das sempre-vivas: Expropriação territorial e violação dos direitos de quilombolas e comunidades tradicionais'. Blog Racismo ambiental. (http://passadicovirtual.blogspot.fr/2013/12/parque-das-sempre-vivas-expropriacao.html). Accessed 30 September 2014.

Hardin, G. (2002) 'La tragedia de los bienes comunes', in Smith, R. C. and Pinedo, D. Ed., *El cuidado de los bienes comunes. Gobierno y manejo de los lagos y bosques en la Amazonía*, Instituto de Estudios Peruanos, Instituto del Bien Comun, Lima 33–48.

Harvey, D. (2011) *O enigma do capital e as crises do capitalismo* Boitempo, São Paulo.

Helfrich, S. ed. (2008) *Genes, bytes y emisiones: bienes comunes y ciudadania*. Ediciones Böll, México.

Hobsbawm, E. J. (1989) *A Era das Revoluções – Europa 1789-1848* Paz e Terra, Rio de Janeiro.

IBGE – Instituto Brasileiro de Geografia e Estatística (2006) 'A Agricultura familiar em 2006. Notas técnicas', *Censo agropecuário 2006*. (www.ibge.gov.br/home/estatistica/economia/agropecuaria/censoagro/agri_familiar_2006_2/notas_tecnicas.pdf). Accessed 12 September 2014.

ISA – Instituto Socio-ambiental (2015) População indígena no Brasil. (http://pib. socioambiental.org/pt/c/terras-indigenas/demarcacoes/localizacao-e-extensao-das-tis). Accessed 12 September 2015.

Isaguirre-Torres, K. and Frigo, D. (2013) 'Desenvolvimento rural, meio ambiente e direitos dos agricultores, agricultoras, povos e comunidade tradicionais', *Série cadernos da agrobiodiversidade* Volume 2. Terra de Direito, Curitiba.

Leroy, J. P. and Meireles, J. (2013) 'Povos Indígenas e Comunidades Tradicionais: os visados territórios dos invisíveis' in Porto, M., Pacheco, T. and Leroy, J. P. eds, *Injustiça Ambiental e Saúde no Brasil. O mapa dos Conflitos*, Editora Fiocruz, Rio de Janeiro.

Lipietz, A. (2010) 'Questions sur les biens communs', in Petitjean, O. ed., *Les biens communs, modèle de gestion des ressources naturelles*, Passerelle No 02 05/2010, Ritimo 21 ter rue Voltaire, 75 011 Paris, 22–26 May (www.agter.asso.fr/IMG/pdf/coredem_dossier_biens-communs.pdf). Accessed 17 May 2015.

Mamani, F. H. (2010) Buen Vivir/Vivir Bien Filosofia, políticas, estrategias y experiencias regionales andinas, Coordinadora Andina de organizações indígenas – CAOI Lima www.mikandina.org (www.reflectiongroup.org/stuff/vivir-bien). Accessed 15 November 2015.

Martins, P., Porro, N. and Shiraishi Neto, J. (2014) 'O direito de propriedade ressignificado por quebradeiras de coco babaçu: a atualização da experiência no uso comum de recursos em uma comunidade tradicional' *Revista da Faculdade de Direito da UFG*, v. 38, n. 2. Jul./Dez.

Marx, K. (1980) *O Capital* Livro 1: O processo de produção do capital. Volume 2 Civilização Brasileira, Rio de Janeiro.

Medeiros, L. S. (2015) 'Conflitos fundiários e violência no campo', in CPT – Comissão Pastoral da Terra, *Conflitos no Campo – Brasil 2014*, CPT Nacional – Brasil, Goiânia (www.cptnacional.org.br/index.php/component/jdownloads/viewdownload/43-conflitos-no-campo-brasil-publicacao/2392-conflitos-no-campo-brasil-2014?Item id=23)Accessed 20 August 2015.

Ostrom, E. (2010) *Gouvernance des biens communs. Pour une nouvelle approche des ressources naturelles*, De Boeck, Bruxelles. Original: Ostrom, E. (1990) *Governing the Commons. The evolution of institutions for collective action*, Cambridge University Press, Cambridge.

Pacheco, T., Porto, M. F. and Rocha, D. (2013) 'Metodologia e Resultados do Mapa: uma síntese dos casos de injustiça ambiental e saúde no Brasil', in Porto, M., Pacheco, T. and Leroy, J. P. eds, *Injustiça Ambiental e Saúde no Brasil. O mapa dos Conflitos*, Editora Fiocruz, Rio de Janeiro.

Parker, L. A. (2015) *Novo Código Florestal e pagamentos por serviços ambientais. Regime Proprietário sobre os Bens Comuns*, Juruá, Curitiba.

Petersen, P. (2014) *Tesouros Escondidos. Reconectando Cultura e Natureza nas dinâmicas de desenvolvimento rural – um caso do semiárido brasileiro*, AS-PTA Rio de Janeiro Manuscript.

Ploeg, J. D. van der (2014) 'Dez qualidades da agricultura familiar', in *Revista Agriculturas: Experiências em agroecologia. Cadernos de debate* n. 1 Feb. 2014. AS-PTA Rio de Janeiro.

Sabourin, E. and Caron, P. (2009) 'Camponeses e fundos de pasto no nordeste da Bahia', in Godoi, E. P., Menezes, M. A. and Marin, R. A. eds, *Diversidade do campesinato: expressões e categorias*. v.2 'Estratégias de reprodução social' Editora Universidade Estadual de São Paulo – UNESP São Paulo, Núcleo de Desenvolvimento Agrário e Desenvolvimento Rural Brasília, DF 89–115.

Sabourin, E., Caron, P. and Silva, P .C. G. (1999) 'O manejo dos 'Fundos de Pasto' no nordeste baiano: um exemplo de reforma agrária sustentável' in *Raizes* n.20, Nov. 1999 Apud Campos, 2011.

Santilli, J. (2005) *Socioambientalismo e novos direitos. Proteção jurídica à diversidade biológica e cultural*, IEB, Peirópolis, São Paulo.

Schlesinger, S. (2014) 'Mapeamento de políticas e programas de pagamento de serviços ambientais na Amazônia brasileira' *Federação de Órgãos para Assistência Social e Educacional – Fase* Manuscript Rio de Janeiro.

Schmitz, H., Mota, D. M. and Silva Júnior, J. F. (2006) *Gestão Coletiva de Bens Comuns e Conflito Ambiental: o Caso das Catadoras de Mangaba*. III Encontro da Associação Nacional de Pós-graduação e Pesquisa em Ambiente e Sociedade – ANPPAS. 23–26 May 2006. Brasília-DF (http://ainfo.cnptia.embrapa.br/digital/bitstream/item/82578/1/Gestao.pdf) Accessed 15 November 2014.

SEPPIR – Secretaria de Políticas de Promoção da Igualdade Racial (www.seppir.gov.br/noticias/ultimas_noticias/2013/05/lancado-relatorio-de-gestao-2012-do-programa-brasil-quilombola) Accessed 10 September 2015.

Souza Filho, C. F. M. (2003) 'Multiculturalismo e direitos coletivos', in Santos, B. S. ed., *Reconhecer para libertar: os caminhos do cosmopolitismo multicultural*, Civilização Brasileira Rio de Janeiro 71–109.

Steinbock, W., Costa e Silva, L., Silva, R. O., Rodriguez, A. S., Perez-Cassarino, J. and Fonini, R. eds. (2013) *Agrofloresta, ecologia e sociedade*, Kairós, Curitiba.

Vianna, Jr. A. (2013) 'Destinação de terras públicas devolutas e terras comunitárias na Amazônia', in Lacerda, P. ed., *Mobilização social na Amazônia: a 'luta' por justiça e por educação*, E-Papers Rio de Janeiro, 109–118 (http://laced.etc.br/site/arquivos/Mobilizacao.pdf) Accessed 22 April 2015.

Part II

Controversy and disinformation

6 Planned disinformation

The example of the Belo Monte Dam as a source of greenhouse gases

Philip M. Fearnside

"Disinformation" (mendacious, deliberately incomplete or misleading information) is a hypothesis that haunts discussions of Amazonian development, particularly the Brazilian government's massive plans for building hydroelectric dams such as Belo Monte. Plans for hydroelectric dam construction in Brazilian Amazonia call for dozens of large dams and over a hundred smaller ones. Decision making in Brazil is critical to these developments, not only because of the large number of dams in Brazilian Amazonia but also because Brazil is financing and building many of the dams in neighboring countries (Fearnside 2014a). Impacts of dams include effects on indigenous peoples, such as loss of fish and other resources from the natural rivers. Resettlement impacts on both urban and rural people represent a concentration of the human cost of this form of development (Fearnside 1999). This is also true of impacts on the downstream residents who lose livelihoods based on fisheries and floodplain agriculture. Health impacts of reservoirs include proliferation of insects and methylation of mercury (transforming this metal into its poisonous form) (Leino and Lodenius 1995). Forest loss occurs not only from direct flooding but also from clearing by displaced residents, from road construction, from migrants and investments attracted to the area, and from agribusiness made viable by waterways associated with many of the dams (Alencar 2016; Alencar et al. 2015; Barreto et al. 2011; Fearnside 2001). Greenhouse-gas emissions from dams include carbon dioxide from decay of trees killed by flooding and emission of nitrous oxide and especially methane from the water in the reservoirs and from water passing through the turbines and spillways. Carbon credit for dams under the Kyoto Protocol's Clean Development Mechanism now represents a major additional source of impact on global warming because virtually all of the dams awarded credit would be built anyway even without this subsidy, meaning that the countries buying the credit can emit gases without there being a genuine offset to neutralize the impact of the emissions (Fearnside 2013a, b, 2015a).

Greenhouse-gas emissions by dams is an area where the hypothesis of disinformation arises as an explanation for official discourse. Hydropower continues to be portrayed as "clean" energy with zero or negligible emissions long after this has been known to be false. Although Belo Monte's Environmental

Impact Study (EIA) claims emissions would be negligible, this dam, along with at least one other that would be needed upstream to supply water for Belo Monte's turbines in the dry season, would have a negative impact on global warming for at least 41 years, with the magnitude of the impact exceeding that of greater São Paulo during the first ten years (Fearnside 2009a, b). This negative impact is based on comparison with the same energy generation with fossil fuels. Of course, the relative impact of the dams would be worse if compared with measures to increase the efficiency of electricity use or to generate with sources like wind and solar power. The option of simply not generating this electricity, part of which would be exported to other countries in the form of aluminum ingots, would give the best result (Fearnside 2016a). The idea that hydroelectric dams produce "clean energy," which is constantly repeated by the Brazilian government and by the hydroelectric and aluminum industries, is what dominates the view of the public.

Dams as the "only option"

"Disinformation," a euphemism for mendacious, deliberately incomplete or misleading information, is a term adopted by the US Central Intelligence Agency (CIA) (Agee 1975). This definition is essentially identical to that of the less-palatable term "lie," that is, a false statement presented as being true and intended to deceive.

One of the areas that best illustrates this is the promotion of hydroelectric dams in Brazilian Amazonia. The subject is almost always presented with the justification adopted by proponents of the construction projects, i.e. a choice between a dam and the development of the country, the only alternative being presented as blackouts and sacrificing the hopes of those who still live without electric lights. This is a case of disinformation that has gained widespread acceptance through constant repetition. Not mentioned is the underlying assumption that Brazil will continue exporting vast amounts of electricity in the form of aluminum and other electro-intensive metals. The first question has to be "what will be done with the electricity?" Only after addressing this question come the comparisons of impacts of each potential project, such as a hydroelectric dam. In the case of the Belo Monte Dam (Figure 6.1), proponents were successful in avoiding any discussion of the impacts of other dams planned upstream of Belo Monte. In all cases, the issue of greenhouse-gas emissions by dams has been absent, virtually always simply repeating the assertion that this is "clean" energy.

Almost every time the subject of dams arises, including the issue of greenhouse-gas emissions, the presumption is that we "need" more power, and therefore the decision is always a choice between the dam or another electricity source, generally fossil fuel. What is done with the energy is rarely questioned. However, this is the most basic question, and has to be answered before we can say what the net impact of the hydroelectric plant is. In the case of the Belo Monte Dam, for example, part of the energy is to make alumina and aluminum

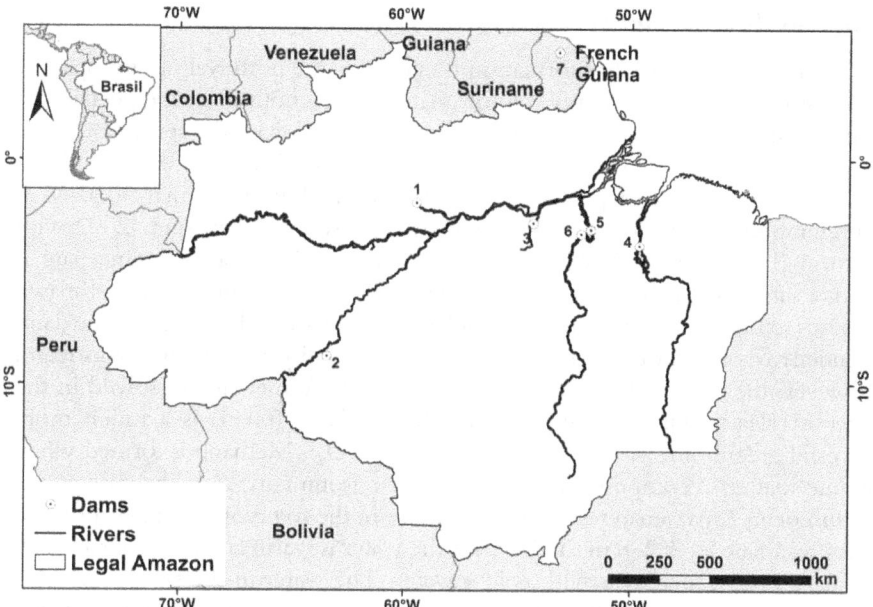

Figure 6.1 Map of dams mentioned in the text: 1) Balbina; 2) Samuel; 3) Curuá-Una;
4) Tucuruí; 5) Belo Monte; 6) Babaquara (Altamira); 7) Petit Saut

for export. Aluminum processing generates only 1.46 jobs per gigawatt-hour of electricity consumed (Fearnside 2016a), the only worse option in terms of creating employment in Brazil being iron alloys, which are also being exported (Bermann and Martins 2000, 90). Stopping export of aluminum and other electro-intensive commodities would be the first measure (Fearnside 2016a). Then there are many ways in which energy use could be more efficient (Baitelo et al. 2013; Bermann 2003; Moreira 2012). The most obvious item is the electric shower, which is an extremely inefficient way to obtain hot water for bathing. According to Brazil's National Climate Change Plan, 5 percent of all electricity consumption in Brazil is to heat water (Brazil, CIMC 2008). This is more than Belo Monte will generate. Much of Brazil's bath water could be heated with solar collectors, and what cannot be heated this way could be more effectively heated directly with gas rather than with electricity. Brazil is one of the only countries in the world that uses the electric shower. The lack of logic from the standpoint of the country is evident from the fact that a shower that costs approximately US$15 to the individual to install costs the country US$1000–1500 to build the generating capacity to supply electricity to the shower (Cidades Solares 2006). Only after advances in efficiency should we turn to other sources of power generation (solar, wind, etc.) and, finally, dams – with priority always given to options with the lowest impact.

Dams as "clean energy"

One of the areas where disinformation is evident is the portrayal of hydropower as emission-free "clean" energy. Unfortunately, this is not the case, especially for tropical dams. Dams emit gases that can be understood from the spillway design, for example in the Tucuruí Dam in Brazil's state of Pará (Fearnside 2004a). The water was taken at a depth of 20 m in Tucuruí-I (the original 4000-MW configuration inaugurated in 1984), which increased to 24 m since 2002 with Tucuruí-II (with 8000 MW of installed capacity). When a steel sluicegate is lifted, a slit opens at the bottom and water descends a "ski jump" and is thrown upwards to be sprayed as billions of droplets. This is part of the design of the dam, intended to oxygenate the water to decrease fish mortality in the river downstream. However, the other side of this coin is that all of the methane dissolved in the water is released into the air immediately. Methane (CH4) is a much more powerful greenhouse gas than carbon dioxide (CO_2). Methane is formed when organic matter decomposes in oxygen-free environments, as is the case at the bottom of an Amazonian reservoir. The water in the reservoir separates into two layers: a surface layer 2–8 m thick where the water is warm and is in contact with the air, and a deeper layer with colder water. The water in the two layers only rarely mixes, and the methane that forms remains trapped in the bottom layer. The intake for the spillways is below the thermocline (the division that separates the two layers), and the intake for the turbines is even deeper than the intake for the spillways. The concentration of methane measured in Tucuruí increases with depth, and reaches very high levels at the depths where water is drawn from the reservoir (Fearnside 2002, 2004b). This water emerges under high pressure, which is immediately reduced to one atmosphere at the outlet of the turbines. Solubility of gases in water is proportional to the pressure, and therefore most of the methane dissolved in water will come out in bubbles at the outlet of the turbines. This is what happens when one opens a bottle of Coca-Cola and CO_2 bubbles begin to emerge as soon as the pressure is released.

The organic matter that is converted into methane comes from sources in two groups. The first is the initial stocks, such as the leaves of the trees in the flooded area and the easily oxidized (labile) portion of the soil carbon. The second group is the renewable stocks, such as macrophytes (water weeds) growing in the water and terrestrial herbaceous vegetation that grows in the drawdown zone (the area exposed each year when the water level in the reservoir is lowered). The vegetation that grows in this area is soft, mainly composed of grasses, and quickly rots underwater (unlike wood, which decomposes very slowly). The vegetation in the drawdown zone is rooted in the bottom, where, when the water level rises, it decomposes in the anoxic zone and generates methane. When plants grow, photosynthesis removes carbon from the air in the form of CO_2, and when the plants later die when flooded this carbon is returned to the atmosphere in the form of CH4. Because this is an emission that is repeated every year in a sustainable way, the dam acts as a "methane factory" (Fearnside 2008).

ELETRONORTE (the government company that owns the Tucuruí Dam) reacted to my use of the phrase "methane factory" as follows in a text entitled "ELETRONORTE responds to the New York Times":

> Finally, ELETRONORTE no longer accepts, after 20 years of exhaustive and repeated explanations like this, that "scientists" continue claiming without any proof that "Tucuruí is virtually a methane factory." What is virtual has been these catastrophic predictions that only corroborate the opinion of those who, whether well informed or not, want nothing more than to speak ill of Brazil.
>
> (Brazil, ELETRONORTE 2004)

Another contribution of dams to global warming comes from the wood in the forest that is flooded when the reservoir is initially filled. This represents a substantial stock of carbon that leads to an emission of CO_2 by the decomposition of dead trees that are left projecting out of the water (Abril et al. 2013). This emission of CO_2 is added to the large pulse of methane production by underwater decomposition of the leaves that fall off the trees. The Balbina Dam is the worst example, with a large shallow reservoir that generates little energy. The reservoir has approximately 3,300 islands (Feitosa et al. 2007), increasing the dam's impact on the forest and also forming thousands of bays with standing water. Balbina has more global-warming impact than generating the same energy from fossil fuels (Fearnside 1995). Although a group in Canada had already identified hydropower plants in that country as sources of greenhouse gases two years earlier (Rudd et al. 1993), it was my 1995 publication that sparked a heated reaction from the hydroelectric industry worldwide, including in Brazil. The US Hydropower Association called the idea that dams have emissions "baloney" (IRN 2002). Other Amazonian dams also remain worse than fossil fuels for many years, as in the cases of Tucuruí, Samuel and Curuá-Una (Fearnside 2002, 2005a, b). The then "president" of ELETROBRÁS attacked me as subject to the "temptations" of the nuclear and thermal power lobbies (Rosa et al. 2004), and asserted that my estimates of emissions from dams were only "political claims" (Rosa et al. 2006; see replies: Fearnside 2004a, 2006a). Rosa et al. (2006) gave the following explanation of the phenomenon:

> Although he [Fearnside] selected Coca-Cola as an example, which is highly symbolic of his way of thinking, he could just as well have selected guaraná – a carbonated soft drink that is very popular in Brazil, flavored with Amazon berries. It is easier to see the bubbles as guaraná is transparent while Coca-Cola is dark. People in Brazil often sit around a table to chat as they drink it, with the bottles open and the glasses full for half an hour or more, without losing completely the bubbles. Instead of fast food, the Brazilian custom is a leisurely drink.
>
> (Rosa et al. 2006)

This is the origin of the term "fizzy science," with reference to the noise that the bubbles make when they emerge from a soft drink. "Fizzy Science" is the title of the publication by International Rivers (an environmental NGO) on the hydropower industry's conflict of interest in its research on emissions from dams (McCully 2006).

The head of the climate sector of Brazil's Ministry of Science and Technology (MCT), who was also responsible for the country's national inventory of greenhouse gases that was submitted to the Climate Convention in 2004, convened a meeting on emissions from dams, and subsequently put the transcript of the event on the MCT website. He stated explicitly that he had invited ELETROBRÁS to draw up this part of the report in order to avoid undesirable political consequences if large emissions from hydropower were accepted:

> We [the MCT climate sector] talked with Prof. Pinguelli [Rosa] and I asked the help of ELETROBRÁS [on the subject of greenhouse-gas emissions from dams]; actually, it was ELETROBRÁS that coordinated this work [i.e., the work reported in the hydroelectric portion of Brazil, MCT 2004] exactly because of this, because this subject was becoming political. It has a very great impact at the World level; we are going to suffer pressure from the developed countries because of this subject. And, this subject was little known. It is mistreated. It is mistreated and continues to be mistreated by Philip Fearnside himself, and we have to be very careful. The debate that is taking place now in the press shows this clearly; that is to say, you can take any one-sided statement to show that Brazil is not clean, that Brazil is very remiss, that Brazil, implicitly, will have to take on a commitment [to reduce emissions] in the future. This is a great political debate and we are preparing ourselves for it.
>
> (Brazil, MCT 2002)

In fact, the very small calculated emissions of dams in the national inventory completely omitted emissions from water that passes through the turbines and spillways (Brazil, MCT 2004, 152). The emission for the Tucuruí Dam in the National Inventory was only 0.56 million tons of carbon equivalent of CO_2 per year (for 1998–1999), a discrepancy of 1,437 percent compared with my value of 8.55 ± 1.55 million tons of carbon equivalent of CO_2 per year for 1990 (Fearnside 2002). For the Samuel Dam, the national inventory calculated 0.12 million tons of carbon equivalent of CO_2 per year (for 1998–1999), a discrepancy of 1,150 percent compared with my value of 1.5 million tons for 1990 or 146 percent compared with my value of 0.29 million tons for 2000 (Fearnside 2005a, b).

The same group persists in claiming that:

> Much controversy has come about recently from studies conducted in Amazonian reservoirs, especially from theoretical studies based on

extrapolations devoid of established scientific criteria. These studies have a strong bias against any type of hydroelectric development in the Amazon and put the viability of these enterprises in doubt with regard to greenhouse-gas emissions and were done for the Tucuruí, Samuel and Balbina hydroelectric dams (Fearnside 1995; Fearnside 1996; Kemenes et al. 2007).

(dos Santos et al. 2008)

Unfortunately, those who read the papers cited in the foregoing passage will find a different world. Kemenes et al. (2007) measured a large emission at Balbina and made calculations for other Amazonian dams indicating all of the other dams as being worse than fossil fuels (Kemenes et al. 2008). Additional errors in the calculations of the ELETROBRÁS group further worsen the conclusion for dams, approximately tripling the emission by bubbling and diffusion through reservoir surfaces (Fearnside and Pueyo 2012; Pueyo and Fearnside 2011). Projected emissions for planned storage dams show higher emissions than fossil fuels (de Faria et al. 2015; Fearnside 2016b). Brazil's president has announced a shift in priority from run-of-river to storage dams in Amazonia (Borges 2013).

Belo Monte and global warming

Belo Monte is a 11,233-MW hydroelectric dam that is under construction since 2011 on the Xingu River, in Brazil's state of Pará. The dam is expected to have severe environmental and social impacts, including impacts on indigenous peoples, and its licensing and implementation have involved multiple violations of international agreements and Brazilian constitutional protections and legislation (Fearnside 2012a; Magalhães and Hernandez 2009; Villas-Bôas et al. 2015).

Among the impacts of Belo Monte is emission of greenhouse gases (see: Fearnside 2011). The same group that prepared the hydroelectric dam portion of Brazil's 2004 national inventory was responsible for the section of Belo Monte's EIA on greenhouse-gas emissions. The estimation of methane emission from the future Belo Monte reservoir is described as follows:

> If the methane emission is similar to that of the Xingó reservoir, the projected area of the Belo Monte reservoir (400 km^2) will emit around 29 mg CH_4 m^{-2} d^{-1}. But, if it is similar to the Tucuruí reservoir it will emit 112 mg CH_4 m^{-2} d^{-1}. In the face of uncertainty we take the emission [of Belo Monte] to be the average of these two values, i.e. 70.7 mg CH_4 m^{-2} d^{-1}. Before flooding ... we arrive at a value of 48 mg CH_4 m^{-2} d^{-1} for the emission of the area to be transformed into the Belo Monte reservoir.
>
> (Brazil, ELETROBRÁS 2009, Vol. 8, p. 72)

Again, the imagined emission is minimal due to the omission of the main emission sources – turbines and spillways, in addition to dead trees rotting above the water (Fearnside 2009a). In the case of the Belo Monte Dam there is another major factor that raises real emissions to even higher levels. This is the effect of massive dams upriver to regulate the flow of the Xingu River at the site of the Belo Monte Dam. The EIA was all done under the assumption that these dams will not exist. The EIA document of nearly 20,000 pages would essentially become a work of fiction if other dams are later built. The impact is apparent from the technical data. The first would be the Babaquara Dam (now with its name changed to "Altamira"). The original plan calls for this dam to have a 6140-km^2 reservoir, or more than double the area of the notorious Balbina Dam. The vertical variation of water level in the reservoir would be 23 m, thus opening a 3580-km^2 mudflat every year as a drawdown zone. This would be a "methane factory" without parallel.

My calculations indicate a huge peak in concentrations of methane in the water of the Babaquara (Altamira) reservoir in the early years due to decay of the soft parts of the original vegetation and of the easily oxidized (labile) soil-carbon stock (Fearnside 2009b). These sources later decrease, but emissions at a lower level continue in subsequent years; the concentration of CH4 oscillates, with a peak every year when the drawdown zone is flooded. This represents an emission that would be sustained throughout the life of the dam. A form of validation of this result comes from measurements of methane in water in the Petit Saut reservoir in French Guiana, where a sustained oscillation of this type has been found since the reservoir was filled in 1994 (Abril et al. 2005).

The large initial emission, combined with a reasonable level of baseline emissions over the years, result in a time of 41 years for the Belo Monte hydroelectric complex with Babaquara (Altamira) beginning to have a net benefit in terms of emissions. Given climate threats in Amazonia and elsewhere, this is much too long a period to wait to begin mitigating global warming. Furthermore, the period of 41 years refers to a calculation without any value being given to time. If any discount rate greater than 1.5 percent per year is applied, the power plant remains worse than fossil fuel for more than a century. The time period considered is an essential factor. If only the first 10 years are considered, the average net emission totals 11.2 million tons of carbon equivalent to carbon CO_2 per year, or more than the emission of greater São Paulo (Fearnside 2009b). This assumes no discounting for the value of time, which would worsen the picture even more. This calculation considers the impact of each ton of methane as being only 21 times the impact of a ton of CO_2 (the conversion used by the 1997 Kyoto Protocol). The 2013 Fifth Assessment Report of the Intergovernmental Panel on Climate Change (IPCC) raises this conversion factor to a value of 34, or a 62 percent greater impact of methane from dams, if the same 100-year time horizon is used, or a value of 86 if a time horizon of 20 years is considered (Myhre et al. 2013, 714). The 20-year time horizon is what is relevant to maintaining global temperatures below the 2 degrees Celsius warming limit agreed in Copenhagen in 2009 as defining "dangerous" climate change or the 1.5 degrees Celsius aspiration endorsed in Paris in 2015 (Fearnside 2015b). The conversion value of 86 effectively quadruples the impact of methane from dams.

A key issue is the credibility of the official scenario of having Belo Monte as the only dam on the Xingu River. Disinformation is the most logical explanation for this scenario, which opponents of Belo Monte refer to as the "institutionalized lie" (Salm 2009). This scenario is based on the decision of the National Council on Energy Policy (CNPE) in July 2008 that only the Belo Monte Dam would be built on the Xingu River. However, there are strong indications that this official scenario does not correspond to the sequence of events that would be initiated with the construction of Belo Monte (de Sousa Júnior et al. 2006; Fearnside 2006b, 2012a). The CNPE is mainly composed of ministers, and these change with each presidential administration; this council is free to change its mind on previous decisions at any time. High authorities in the electrical sector have never accepted the CNPE resolution: the head of the National Electrical Energy Agency (ANEEL) called the decision "a typical case of giving up your rings to keep your fingers" (Pamplona 2008). At the highest levels of power there appears to be no intention of following the official scenario: when the then minister of environment Marina Silva proposed an extractive reserve in part of the area that would be flooded by dams upstream of Belo Monte, the then head of presidential staff (and current president of Brazil) Dilma Rousseff vetoed the proposal "because it could hinder the construction of dams additional to Belo Monte" (Angelo 2010). As president, Dilma Rousseff emphasized in a speech that Brazil "needs dams with large reservoirs" (Borges 2013), which appears to be an allusion to the Babaquara (Altamira) Dam.

Unfortunately, there is also a history of parallel cases of disinformation associated with dams that have already been built in the Amazon, where electrical authorities announced that they were not going to do something and instead did exactly what they had promised not to do. In the case of Balbina, a "public clarification" was released days before closing the dam, promising to fill the reservoir only to the level of 46 m above sea level, creating a 1580-km² impoundment (Brazil, ELETRONORTE 1987a). Filling to the 50 m level would only be done after years of study of the water quality. But what really ensued was filling the reservoir directly to the 50 m mark, and even a small amount (10 cm) above this mark. During the filling process itself this author obtained the plan being followed by ELETRONORTE engineers at the dam site, indicating the intention of raising the water level directly to the 50 m level (Brazil, ELETRONORTE 1987b). Today Balbina's reservoir covers 2995.5 km² according to our measurements from satellite images (Feitosa et al. 2007). The reservoir has been operated consistently above the 50 m level and in two years above the 51 m level (Feitosa et al. 2007).

The other case of documented disinformation providing a parallel for the Belo Monte/Babaquara complex is the Tucuruí-II project, which added 4000 MW of capacity to the Tucuruí power plant in the state of Pará. Under the law, any hydroelectric project with more than 10 MW needed an EIA (a limit subsequently raised to 30 MW), and ELETRONORTE was preparing to contract this report when Brazil's president simply flew to Pará and released the money. The rationalization was that the water level in the reservoir would not increase above the 70 m elevation of Tucuruí-I, and therefore would have no impact and did not need the study (Indriunas 1998). After the Tucuruí-II construction project was completed, the water level was simply raised, and Tucuruí has been operating at a 74 m water level since 2002 (Fearnside 2006b, c). Similarly, after the construction of Belo Monte it is likely that the construction of Babaquara (Altamira) would simply go ahead when the time for it arrives in the schedule. The timeline before launching the current official scenario foresaw this huge dam beginning operation seven years after Belo Monte (Brazil, ELETROBRÁS 1998, 145). The famous phrase of George Santayana (1905) that "Those who cannot remember the past are doomed to repeat it" has never been so relevant.

The logic of dams upstream of Belo Monte is apparent from the Xingu River hydrograph, i.e., the fact that during 3–4 months there would not be enough water to run a single turbine in the 11,000 MW main power house. An economic analysis, carried out by the Conservation Strategy Fund, in Minas Gerais, demonstrates the complete unviability of the Belo Monte Dam without water storage in major dams upstream (de Sousa Júnior et al. 2006). The financial temptation would be great to build Babaquara (Altamira) after a "planned crisis" when it is discovered that there is not enough water for the main power house at Belo Monte; adding Babaquara (Altamira) would increase the value of the energy generated per year at Belo Monte by approximately US$2.8–4.6 million (de Sousa Júnior et al. 2006, 76).

The Tapajós River provides a parallel case of disinformation concerning plans for building a particularly high-impact dam as part of a larger scheme that is apparently not to be publically revealed until after completion of a current project. In this case the São Luiz do Tapajós Dam plays the role of Belo Monte, and the Chacorão Dam the role of Babaquara (Altamira). Just as Babaquara (Altamira) would flood already-demarcated indigenous land, Chacorão would flood 11,700 ha of the Munduruku Indigenous Land. This dam is not included in the current ten-year plan for energy expansion (Brazil, MME 2015), nor in the "energy axis" of the Program for the Acceleration of Growth (PAC) (Brazil, PR 2011), but its locks are included in the PAC's "transportation" axis and represent a high priority to make the Tapajós River navigable as a waterway for transporting soybeans from Mato Grosso to ports on the Amazon River (Brazil, MT 2010). A sequence of events is currently playing out in the Tapajós River basin that repeats many of the worst features of the history of Belo Monte (Fearnside 2015c, d). Similarly, the multiple illegalities and injustices in the licensing and construction of Belo Monte repeat events that took place only a few years earlier in the case of the Madeira River dams (Fearnside 2013c, 2014b, c).

The backlash against criticism of Belo Monte has been unrelenting. Rogério César Cerqueira Leite (an influential member of the editorial board of the Folha de São Paulo newspaper) labled those who criticize the dam as "pseudo-intellectuals," "jugglers," "windbags," an "extemporaneous Brancaleone army" and by some new terms he contributed to the Portuguese language for the occasion: "ecopalermas," "ignocentes" and "verdolengos" (Leite 2010; see responses: Fearnside 2010; Medeiros 2010). Among other statements, Leite claims that the indigenous people should have no objection to the dams because they are "semi-nomadic" and can simply pick up and move to another part of the forest.

A dossier of pro-dam material was compiled by Bittencourt (2012), which culminates by implying that critics of Belo Monte are Marxists. Edification is provided by a long quotation from Lenin to the effect that the key to achieving true communism is to bring electric power to all of Russia in order to transform rural peasants into urban proletarians.

An example of the way information on Belo Monte was distorted is provided by a highly visible response to criticisms of the project that had been put forward in a video made by soap-opera stars from Brazil's Globo television network (Movimento Gota d'Água 2011). The TV stars had, indeed, made some factual errors in describing the Belo Monte project, but their basic criticisms of social and environmental impacts were correct. The video was responded to in a counter-video (Tempestade em Copo d'Água 2011) made by engineering students at Campinas State University (UNICAMP); the students each replied to a different statement in the TV stars' video. The counter-video culminates with the students' teacher (a former consultant to the Belo Monte consortium) declaring that Belo Monte is a great project for Brazil "in all aspects: economic, environmental and social." I recommend my debate with the teacher on the Terra internet television channel (Terra TV 2011). In response to a statement that indigenous peoples will be impacted, a student replied that he had done "research" and found that no indigenous areas will be flooded by the Belo Monte reservoir; clearly his research did not include impacts on the two downstream indigenous areas in the 100 km "reduced flow" stretch, let alone the implications of upstream complements to Belo Monte like the Babaquara (Altamira) Dam. Other students reply to a questioning of Belo Monte as producing "clean energy" by stating that the "very same water" that enters the reservoir will be released below the dam "just as clean as when it came in;" evidently, the students had missed out on studies showing high concentrations of methane and low concentrations of oxygen in water released by dams. The counter-video was then turned into a feature article and cover page of Veja, Brazil's largest news magazine (Eler and Diniz 2011). An image of each TV star and each student replying is shown with the statements in balloons in comic-book style, and each is accompanied by a drawing of a giant boxing glove "knocking out" the TV star. The article and cover-page were reprinted by the Belo Monte consortium and widely distributed in Altamira.

Unfortunately, the basic fact that the Belo Monte Dam would have a huge impact, far beyond what is officially admitted, still stands regardless of the discourse employed. Among these impacts is the emission of greenhouse gases. The best illustration of how these impacts have not yet managed to penetrate the curtain of discourse emerged at the fifteenth Conference of the Parties (COP) of the Climate Convention held in Copenhagen at the end of 2009.

A reporter from Amazonia.org.br (an environmental website run by Friends of the Earth-Brazilian Amazonia) interviewed Brazil's special ambassador for climate change, who was responsible for negotiating on the Brazilian side. The reporter asked: "But, isn't Belo Monte one of the hydroelectric projects that the government considers to be sources of renewable and clean energy?" The answer was: "Yes it is. But, what I'm saying is that I think it [Belo Monte] is not in Amazonia, right? So it is a different scheme" (Munhoz 2009). If key people in decisions regarding dams and climate change (such as negotiations on carbon credit for hydropower) do not even know that Belo Monte is located in the Amazon region, it is very difficult to imagine that they know the details of its impacts, including emissions of greenhouse gases.

Acknowledgements

This is translated, updated and adapted from Fearnside (2012b) "Desafios para midiatização da ciência na Amazônia: O exemplo da hidrelétrica de Belo Monte como fonte de gases de efeito estufa" In Fausto Neto A. (ed.) *A Midiatização da ciência: Cenários, desafios, possibilidades*, Editora da Universidade Estadual da Paraíba (EDUEPB), Campina Grande, Paraíba, Brazil. pp. 107–123. The author's research is financed by Conselho Nacional do Desenvolvimento Científico e Tecnológico (CNPq: Proc. 304020/2010-9; 573810/2008-7), Fundação de Amparo à Pesquisa do Estado do Amazonas (FAPEAM: Proc. 708565) and Instituto Nacional de Pesquisas da Amazônia (INPA: PRJ15.125).

References

Abril, G., Guérin, F., Richard, S., Delmas, R., Galy-Lacaux, C., Gosse, P., Tremblay, A., Varfalvy, L., dos Santos, M.A. and Matvienko, B. (2005) "Carbon dioxide and methane emissions and the carbon budget of a 10-years old tropical reservoir (Petit-Saut, French Guiana)," *Global Biogeochemical Cycles*, 19 GB 4007. doi:10.1029/2005GB002457.

Abril, G., Parize, M., Pérez, M.A.P. and Filizola, N. (2013) "Wood decomposition in Amazonian hydropower reservoirs: An additional source of greenhouse gases," *Journal of South American Earth Sciences*, 44 104–107. doi: 10.1016/j.jsames. 2012.11.007.

Agee, P. (1975) *Inside the Company: CIA diary* Penguin Books, New York, U.S.A. 640 pp.

Alencar, A., Piontekowski, V.J., Charity, S. and Maretti, C.C. (2015) "Deforestation scenarios in the area of influence of the Tapajós hydropower complex," Instituto de Pesquisa Ambiental da Amazônia (IPAM), Belém, Pará, Brazil. 3 pp. http://ipam.org.br/wp-content/uploads/2015/12/TapajosIPAM_2015.pdf

Alencar, A. (2016) "Cenários de perda da cobertura florestal na área de influência do complexo hidroelétrico do Tapajós," Instituto de Pesquisa Ambiental da Amazônia (IPAM), Belém, Pará, Brazil. 13 pp. http://ipam.org.br/bibliotecas/cenarios-de-perda-da-cobertura-florestal-na-area-de-influencia-do-complexo-hidroeletrico-do-tapajos/

Angelo, C. (2010) "PT tenta apagar fama 'antiverde' de Dilma," *Folha de São Paulo*, 10 October 2010, p. A-15. http://acervo.folha.com.br/fsp/2010/10/10/2

Baitelo, R., Yamaoka, M., Nitta, R. and Batista, R. (2013) [R]*evolução energética: A caminho do desenvolvimento* Greenpeace Brasil, São Paulo, SP, Brazil. 79 pp. www.greenpeace.org/brasil/pt/Documentos/Revolucao-Energetica-/

Barreto, P., Brandão, Jr. A., Martins, H., Silva, D., Sousa Jr., C., Sales, M. and Feitosa, T. (2011) Risco de Desmatamento Associado à Hidrelétrica de Belo Monte. Instituto do Homem e Meio Ambiente da Amazônia (IMAZON), Belém, Pará, Brazil. 98 pp. Available at: www.imazon.org.br/publicacoes/livros/risco-de-desmatamento-associado-a-hidreletrica-de-belo-monte/at_download/file

Bermann, C. (2003) *Energia no Brasil: Para quê? Para quem? Crise e alternativas para um país sustentável* 2nd edn. Editora Livraria da Física, São Paulo, SP and Federação dos Órgãos para Assistência Social e Educacional (FASE), Rio de Janeiro, RJ, Brazil. 139 pp.

Bermann, C. and Martins, O.S. (2000) Sustentabilidade energética no Brasil: Limites e possibilidades para uma estratégia energética sustentável e democrática. (Série Cadernos Temáticos No. 1) Projeto Brasil Sustentável e Democrático, Federação dos Órgãos para Assistência Social e Educacional (FASE), Rio de Janeiro, RJ, Brazil. 151 pp.

Bittencourt, F. (2012) "Um dossiê a favor de Belo Monte" Blog Luis Nassif, 29/11/2011 & 30/11/2012. http://jornalggn.com.br/blog/luisnassif/um-dossie-a-favor-de-belo-monte

Borges, A. (2013) "Dilma defende usinas hidrelétricas com grandes reservatórios" *Valor Econômico*, 6 June 2013. www.valor.com.br/imprimir/noticia_impresso/3151684

Brazil, CIMC (Comitê Interministerial sobre Mudança do Clima) (2008) Plano Nacional sobre Mudança do Clima – PNMC – Brasil. Ministério do Meio Ambiente, Brasília, DF, Brazil. 129 pp. Available at: www.mma.gov.br/estruturas/smcq_climaticas/_arquivos/plano_nacional_mudanca_clima.pdf

Brazil, ELETROBRÁS (Centrais Elétricas Brasileiras) (1998) Plano decenal 1999–2008. ELETROBRÁS, Rio de Janeiro, RJ, Brazil.

Brazil, ELETROBRÁS (Centrais Elétricas Brasileiras) (2009) Aproveitamento hidrelétrico Belo Monte: Estudo de impacto ambiental. Fevereiro de 2009. ELETROBRÁS, Rio de Janeiro, RJ, Brazil. 36 vols. Available at: http://philip.inpa.gov.br/publ_livres/Dossie/BM/DocsOf/EIA-09/EIA_%202009.htm

Brazil, ELETRONORTE (Centrais Elétricas do Norte do Brasil, S.A.) (1987a) "Esclarecimento Público: Usina Hidrelétrica Balbina. Modulo 1, Setembro 1987," ELETRONORTE, Brasília, DF, Brazil. 4 pp.

Brazil, ELETRONORTE (Centrais Elétricas do Norte do Brasil, S.A.) (1987b) "UHE Balbina: Enchimento do reservatório, considerações gerais" BAL-39-2735-RE. ELETRONORTE, Brasília, DF, Brazil. 12 pp + annexes.

Brazil, ELETRONORTE (Centrais Elétricas do Norte do Brasil, S.A) (2004) "Eletronorte responde The New York Times" Centrais Elétricas do Norte do Brasil S.A. ELETRONORTE, Brasília, DF, Brazil. (Posted at: www.eln.gov.br/ de 2004 until approximately 2007). Available at: http://philip.inpa.gov.br/publ_livres/Other%20side-outro%20lado/Hydroelectric%20emissions/Eletronorte%20em%20resposta%20ao%20artigo%20publicado%20na%20NY%20Times.pdf

Brazil, MCT (Ministério da Ciência e Tecnologia) (2002) "Degravação do workshop: Utilização de Sistemas Automáticos de Monitoramento e Medição de Emissões de Gases de Efeito Estufa da Qualidade da Água em Reservatórios de Hidrelétricas.

Centro de Gestão de Estudos Estratégicos do MCT, Brasília – DF, 06 de fevereiro de 2002" MCT, Brasília, DF, Brazil (Posted from 2002 to 2006 at: www.mct.gov. br/clima/brasil/doc/workad.doc). Available at: http://philip.inpa.gov.br/publ_ livres/Other%20side-outro%20lado/Hydroelectric%20emissions/Degravacao%20 de%20workshop-workad.pdf

Brazil, MCT (Ministério da Ciência e Tecnologia) (2004) Brazil's initial national communication to the United Nations Framework Convention on Climate Change. MCT, Brasília, DF, Brazil. 271 pp. Available at: www.mct.gov.br/upd_blob/ 0005/5142.pdf

Brazil, MME (Ministério das Minas e Energia) (2015) Plano decenal de expansão de energia 2024. MME, Empresa de Pesquisa Energética (EPE). Brasília, DF. 467 pp. Available at: www.epe.gov.br/PDEE/Relatório%20Final%20do%20PDE%202024. pdf

Brazil, MT (Ministério dos Transportes) (2010) "Diretrizes da política nacional de transporte hidroviário," MT, Secretaria de Política Nacional de Transportes, Brasília, DF, Brazil. 33 pp. Available at: www2.transportes.gov.br/Modal/Hidroviario/ PNHidroviario.pdf

Brazil, PR (Presidência da República) (2011) "PAC-2 Relatórios" PR, Brasília, DF, Brazil. Available at: www.brasil.gov.br

Cidades Solares (2006) *Boletim Informativo*, 1(4) September 2006. www.cidadessolares. org.br/conteudo_view_print.php?id=74

de Faria, F.A.M., Jaramillo, P., Sawakuchi, H.O., Richey, J.E. and Barros, N. (2015) "Estimating greenhouse gas emissions from future Amazonian hydroelectric reservoirs," *Environmental Research Letters*, 10(12) 124019. doi:10.1088/ 1748-9326/10/12/124019

de Sousa Júnior, W.C., Reid, J. and Leitão, N.C.S. (2006) Custos e benefícios do Complexo Hidrelétrico Belo Monte: Uma abordagem econômico-ambiental. Conservation Strategy Fund (CSF), Lagoa Santa, Minas Gerais, Brazil. 90 pp. Available at: http://conservation-strategy.org/sites/default/files/field-file/4_Belo_ Monte_Dam_Report_mar2006.pdf

dos Santos, M.A., Rosa, L.P., Matvienko, B., dos Santos, E.O., D'Almeida Rocha, C.H.E., Sikar, E., Silva, M.B. and Manuel, P.B. Junior A. (2008) "Emissões de gases de efeito estufa por reservatórios de hidrelétricas," *Oecologia Brasiliensis*, 12(1) 116–129.

Eler, A. and Diniz, L. (2011) "Nocauteados pela lógica" *Veja*, 44 (49) 140–146. (7 December 2011).

Fearnside, P.M. (1987) "Deforestation and International Economic Development Projects in Brazilian Amazonia," *Conservation Biology*, 1(3): 214–221. doi: 10.1111/ j.1523-1739.1987.tb00035.x

Fearnside, P.M. (1995) "Hydroelectric dams in the Brazilian Amazon as sources of 'greenhouse' gases," *Environmental Conservation*, 22(1) 7–19. doi: 10.1017/ S0376892900034020

Fearnside, P.M. (1996) "Hydroelectric dams in Brazilian Amazonia: Response to Rosa, Schaeffer & dos Santos," *Environmental Conservation*, 23(2), 105–108. doi:10.1017/ S0376892900038467

Fearnside, P.M. (1999) "Social impacts of Brazil's Tucuruí Dam," *Environmental Management*, 24(4), 483–495. doi: 10.1007/s002679900248

Fearnside, P.M. (2001) "Environmental impacts of Brazil's Tucuruí Dam: Unlearned lessons for hydroelectric development in Amazonia," *Environmental Management*, 27(3), 377–396. doi: 10.1007/s002670010156

Fearnside, P.M. (2002) "Greenhouse gas emissions from a hydroelectric reservoir (Brazil's Tucuruí Dam) and the energy policy implications," *Water, Air and Soil Pollution*, 133(1–4) 69-96. doi: 10.1023/A:1012971715668

Fearnside, P.M. (2004a) "Greenhouse gas emissions from hydroelectric dams: Controversies provide a springboard for rethinking a supposedly 'clean' energy source," *Climatic Change*, 66(1–2) 1–8. doi: 10.1023/B:CLIM.0000043174.02841. 23

Fearnside, P.M. (2004b) "Gases de efeito estufa em hidrelétricas da Amazônia," *Ciência Hoje*, 36(211) 41–44.

Fearnside, P.M. (2005a) "Brazil's Samuel Dam: Lessons for hydroelectric development policy and the environment in Amazonia," *Environmental Management*, 35(1) 1–19. doi: 10.1007/s00267-004-0100-3

Fearnside, P.M. (2005b) "Do hydroelectric dams mitigate global warming? The case of Brazil's Curuá-Una Dam," *Mitigation and Adaptation Strategies for Global Change*, 10(4) 675–691. doi: 10.1007/s11027-005-7303-7

Fearnside, P.M. (2006a) "Greenhouse gas emissions from hydroelectric dams: Reply to Rosa et al," *Climatic Change*, 75(1–2) 103–109. doi: 10.1007/s10584-005-9016-z

Fearnside, P.M. (2006b) "Dams in the Amazon: Belo Monte and Brazil's hydroelectric development of the Xingu River Basin," *Environmental Management*, 38(1), 16–27. doi: 10.1007/s00267-005-00113-6

Fearnside, P.M. (2006c) "A polêmica das hidrelétricas do rio Xingu," *Ciência Hoje*, 38(225), 60–63.

Fearnside, P.M. (2008) "Hidrelétricas como 'fábricas de metano': O papel dos reservatórios em áreas de floresta tropical na emissão de gases de efeito estufa," *Oecologia Brasiliensis*, 12(1), 100–115. doi: 10.4257/oeco.2008.1201.11 English translation available at: http://philip.inpa.gov.br/publ_livres/mss%20and%20in%20 press/Fearnside%20Hydro%20GHG%20framework.pdf

Fearnside, P.M. (2009a) "O Novo EIA-RIMA da Hidrelétrica de Belo Monte: Justificativas Goela Abaixo." In Santos S.M.S.B.M. and Hernandez F. del M. (eds.) *Painel de Especialistas: Análise Crítica do Estudo de Impacto Ambiental do Aproveitamento Hidrelétrico de Belo Monte. Painel de Especialistas sobre a Hidrelétrica de Belo Monte*, Belém, Pará, Brazil. pp. 108–117.

Fearnside, P.M. (2009b) "As hidrelétricas de Belo Monte e Altamira (Babaquara) como fontes de gases de efeito estufa," *Novos Cadernos NAEA*, 12(2), 5–56.

Fearnside, P.M. (2010) "Belo Monte: Resposta a Rogério Cezar de Cerqueira Leite," *Globoamazonia*, 7 June 2010. http://philip.inpa.gov.br/publ_livres/2010/Belo%20 Monte–GloboAmazonia-Resposta%20a%20Rogerio%20Cezar%20Cerqueira%20 Leite.pdf

Fearnside, P.M. (2011) "Gases de efeito estufa no EIA-RIMA da Hidrelétrica de Belo Monte," *Novos Cadernos NAEA*, 14 5–19.

Fearnside, P.M. (2012a) "Belo Monte Dam: A spearhead for Brazil's dam building attack on Amazonia?" GWF Discussion Paper 1210, Global Water Forum, Canberra, Australia. 5 pp. Available at: www.globalwaterforum.org/wp-content/ uploads/2012/04/Belo-Monte-Dam-A-spearhead-for-Brazils-dam-building-attack-on-Amazonia_-GWF-1210.pdf

Fearnside, P.M. (2012b) "Desafios para midiatização da ciência na Amazônia: O exemplo da hidrelétrica de Belo Monte como fonte de gases de efeito estufa." In Fausto Neto, A. (ed.) A Midiatização da ciência: Cenários, desafios, possibilidades, Editora da Universidade Estadual da Paraíba (EDUEPB), Campina Grande, Paraíba, Brazil. pp. 107–123.

Fearnside, P.M. (2013a) "Carbon credit for hydroelectric dams as a source of greenhouse-gas emissions: The example of Brazil's Teles Pires Dam," *Mitigation and Adaptation Strategies for Global Change*, 18(5) 691–699. doi: 10.1007/s11027-012-9382-6

Fearnside, P.M. (2013b) "Credit for climate mitigation by Amazonian dams: Loopholes and impacts illustrated by Brazil's Jirau hydroelectric project," *Carbon Management*, 4(6), 681–696. doi: 10.4155/CMT.13.57.

Fearnside, P.M. (2013c) "Decision-making on Amazon dams: Politics trumps uncertainty in the Madeira River sediments controversy," *Water Alternatives*, 6(2) 313–325. www.water-alternatives.org/index.php/alldoc/articles/vol6/v6issue2/218-a6-2-15/file

Fearnside, P.M. (2014a) Análisis de los principales proyectos hidro-energéticos en la región amazónica. Derecho, Ambiente y Recursos Naturales (DAR), Centro Latinoamericano de Ecología Social (CLAES) and Panel Internacional de Ambiente y Energia en la Amazonia, Lima, Peru, 55 pp. www.dar.org.pe/archivos/publicacion/147_Proyecto_hidro-energeticos.pdf

Fearnside, P.M. (2014b) "Impacts of Brazil's Madeira River dams: Unlearned lessons for hydroelectric development in Amazonia," *Environmental Science & Policy*, 38 164–172. doi: 10.1016/j.envsci.2013.11.004

Fearnside, P.M. (2014c) "Brazil's Madeira River dams: A setback for environmental policy in Amazonian development," *Water Alternatives*, 7(1), 156–169. www.water-alternatives.org/index.php/alldoc/articles/vol7/v7issue1/244-a7-1-15/file

Fearnside, P.M. (2015a) "Tropical hydropower in the Clean Development Mechanism: Brazil's Santo Antônio Dam as an example of the need for change," *Climatic Change*, 131(4), 575–589. doi: 10.1007/s10584-015-1393-3

Fearnside, P.M. (2015b) "Emissions from tropical hydropower and the IPCC," *Environmental Science & Policy*, 50, 225–239. doi: 10.1016/j.envsci.2015.03.002

Fearnside, P.M. (2015c) "Amazon dams and waterways: Brazil's Tapajós Basin plans," *Ambio*, 44(5), 426–439. doi: 10.1007/s13280-015-0642-z

Fearnside, P.M. (2015d) "Brazil's São Luiz do Tapajós Dam: The art of cosmetic environmental impact assessments," *Water Alternatives*, 8(3) 373–396. www.water-alternatives.org/index.php/alldoc/articles/vol8/v8issue3/297-a8-3-5/file

Fearnside, P.M. (2016a) "Environmental and social impacts of hydroelectric dams in Brazilian Amazonia: Implications for the aluminum industry," *World Development*, 77 48–65. doi: 10.1016/j.worlddev.2015.08.015

Fearnside, P.M. (2016b) "Greenhouse gas emissions from Brazil's Amazonian hydroelectric dams," *Environmental Research Letters*, 11(1) 011002 doi: 10.1088/1748-9326/11/1/011002

Fearnside, P.M. and Pueyo, S. (2012) "Underestimating greenhouse-gas emissions from tropical dams," *Nature Climate Change*, 2(6) 382–384. doi: 10.1038/nclimate1540

Feitosa, G.S., Graça, P.M.L.A. and Fearnside, P.M. (2007) "Estimativa da zona de deplecionamento da hidrelétrica de Balbina por técnica de sensoriamento remoto." In Epiphanio, J.C.N., Galvão, L.S. and Fonseca, L.M.G. (eds.) Anais XIII Simpósio

Brasileiro de Sensoriamento Remoto, Florianópolis, Brasil 21–26 April 2007. Instituto Nacional d e Pesquisas Espaciais (INPE), São José dos Campos, São Paulo, Brazil. pp. 6713–6720. http://marte.dpi.inpe.br/col/dpi.inpe.br/sbsr@80/2006/11.13.15.55/doc/6713-6720.pdf

Indriunas, L. (1998) "FHC inaugura obras em viagem ao Pará" Folha de São Paulo, 14 June 1998, pp. 1–17. www1.folha.uol.com.br/fsp/brasil/fc14069828.htm

IRN (International Rivers Network) (2002) "Flooding the land, warming the Earth: Greenhouse gas emissions from dams," IRN, Berkeley, California, USA. 18 pp. Available at: www.ircwash.org/sites/default/files/McCully-2002-Flooding.pdf

Kemenes, A., Forsberg, B.R. and Melack, J.M. (2007) "Methane release below a tropical hydroelectric dam," *Geophysical Research Letters*, 34 L12809. doi: 10.1029/2007GL029479.55

Kemenes, A., Forsberg, B.R. and Melack, J.M. (2008) "As hidrelétricas e o aquecimento global," *Ciência Hoje*, 41(145), 20–25.

Leino, T. and Lodenius, M. (1995) "Human hair mercury levels in Tucuruí area, state of Pará, Brazil," *The Science of the Total Environment*, 175 119–125.

Leite, R.C.C. (2010) "Belo Monte, a floresta e a árvore," *Folha de São Paulo*, 19 May 2010, p. A-3. http://acervo.folha.com.br/fsp/2010/05/19/2/

Magalhães, S.M.S.B. and Hernandez, F.M. (eds.) (2009) Painel de Especialistas: Análise crítica do estudo de impacto ambiental do aproveitamento hidrelétrico de Belo Monte. Painel de Especialistas sobre a Hidrelétrica de Belo Monte, Belém, Pará, Brazil. 230 pp. Available at: www.internationalrivers.org/files/Belo%20Monte%20pareceres%20IBAMA_online%20(3).pdf

McCully, P. (2006) "Fizzy science: Loosening the hydro industry's grip on greenhouse gas emissions research," International Rivers Network, Berkeley, California, USA. 24 pp. Available at: www.irn.org/pdf/greenhouse/FizzyScience2006.pdf

Medeiros, H.F. (2010) "Fatos sobre Belo Monte," *Folha de São Paulo*, 1 June 2010. www1.folha.uol.com.br/fsp/opiniao/fz0106201008.htm

Moreira, P.F. (ed.) (2012) Setor Elétrico Brasileiro e a Sustentabilidade no Século 21: Oportunidades e Desafios. 2a ed. Rios Internacionais, Brasília, DF, Brazil. 100 pp. Available at: www.internationalrivers.org/node/7525

Movimento Gota d'Água (2011) "Usina Hidrelétrica de Belo Monte – Movimento Gota D'água," YouTube www.youtube.com/watch?v=hzVIWvm99As

Munhoz, F. (2009) "'Só aceitamos a participação do REDD no mercado de carbono se ela for limitada', diz embaixador do Itamaraty" Amazonia.org.br, 7 December 2009. www.amazonia.org.br/noticias/noticia.cfm?id=337116

Myhre, G. and 37 others (2013) "Anthropogenic and natural radiative forcing." In Stocker, T.F., Qin, D., Plattner, G-K., Tignor, M., Allen, S.K., Boschung, J., Nauels, A., Xia, Y., Bex, V. and Midgley, P.M. (eds.) *Climate change 2013: The physical science basis. Working Group I contribution to the IPCC Fifth Assessment Report.* Cambridge University Press, Cambridge, UK, pp. 661–740. Available at: www.ipcc.ch/report/ar5/wg1/

Pamplona, N. (2008) "Aneel chama decisão de limitar usinas no Xingu de 'política'" Agência Estado, 22 July 2008. www.estadao.com.br/noticias/economia,aneel-chama-decisao-de-limitar-usinas-no-xingu-de-politica,209554,0.htm

Pueyo, S. and Fearnside, P.M. (2011) "Emissões de gases de efeito estufa dos reservatórios de hidrelétricas: Implicações de uma lei de potência," *Oecologia Australis*, 15(2) 114-127. doi: 10.4257/oeco.2011.1502. English translation available at: http://philip.

inpa.gov.br/publ_livres/mss%20and%20in%20press/Pueyo%20&%20Fearnside-GHGs%20FROM%20%20RESERVOIRS–engl.pdf

Rosa, L.P., dos Santos, M.A., Matvienko, B., dos Santos, E.O. and Sikar, E. (2004) "Greenhouse gases emissions by hydroelectric reservoirs in tropical regions," *Climatic Change*, 66(1–2) 9–21.

Rosa, L.P., dos Santos, M.A., Matvienko, B., Sikar, E. and dos Santos, E.O. (2006) "Scientific errors in the Fearnside comments on greenhouse gas emissions (GHG) from hydroelectric dams and response to his political claiming," *Climatic Change*, 75(1–2) 91–102.

Rudd, J.W., Harris, M., Kelly, C.A. and Hecky, R.E. (1993) "Are hydroelectric reservoirs significant sources of greenhouse gases?" *Ambio*, 22, 246–248.

Salm, R. (2009) "Belo Monte: Mentira institucionalizada," *Correio da Cidadania*, 4 December 2009. www.correiocidadania.com.br

Santayana, G. (1905) Reason in common sense. Vol. 1 in *The life of reason: The phases of human progress*. Dover Publications, Inc., New York, NY, USA, 5 vols.

Switkes, G. (2001) Leader of movement to stop Amazon dam murdered, *World Rivers Review*, 16(5): 13.

Tempestade em Copo d'Água (2011) "Alunos da Unicamp apoiam Belo Monte em paródia com vídeo de globais Estudantes rebatem argumentos do vídeo dos globais e defendem a hidrelétrica de Belo Monte." YouTube 26 November 2011. www.youtube.com/watch?v=gVC_Y9drhGo

Terra TV (2011) 6 December 2011. Belo Monte no Programa Sustentabilidade Debate busca esclarecer a grande polêmica do momento: A construção da hidrelétrica de Belo Monte. http://terratv.terra.com.br/videos/Noticias/Economia/Sustentabilidade/5180-393127/Sustentabilidade-Belo-Monte-06_12-Programa-completo.htm

Villas-Bôas, A., Garzón, B.R., Reis, C., Amorim, L. and Leite, L. (2015) Dossiê Belo Monte: Não há condições para a Licença de Operação. Instituto Socioambiental (ISA), Brasília, DF, Brazil. 55 pp. Available at: http://t.co/zjnVPhPecW

7 Biosafety regulations, practices and consequences in Brazil

Who wants to hide the problems?

Leonardo Melgarejo

After 10 years of the Biosafety Law approval, Brazil registers commercial releases of 77 transgenic events: 49 plants (6 soybeans, 29 maize, 12 cotton, 1 pinto bean, and 1 eucalyptus), 20 vaccines, 6 microorganisms, 1 mosquito and 1 enzyme.

The fact that the same set of rules has been used to evaluate and approve all such cases is significant. Developed to discuss biorisks associated with the cultivation of annual plants, such as soybean and maize, regulations that guide decisions on the requests for commercial release of transgenic events in Brazil, have already been used to authorize commercial trading of trees, such as eucalyptus and insects like the mosquito. Soon, following the same rules, Brazil will authorize commercial release of genetically modified fish and flies. The fragile assumption that the focus of interest should be centred on genetic modification and its derived proteins has been used to justify this reality. Giving poor attention to the body itself and practically ignoring associated technological packages and their socioeconomic implications, risk evaluators are guided by unilateral perspectives, over-rating a field of science (Genetics) regardless of others.

Furthermore, until now all commercial releases endorsed by the National Biosafety Technical Commission (CTNBio) were given, ignoring its own rules. All the information requested by Normative Resolution No. 5 (RN5), a guiding instrument for the evaluation of commercial releases requests, were never presented (Brasil 2008). The lack of long-term nutritional studies using genetically modified grains (nutritional evaluations conducted over two generations RN5, Annex III, A.4) stands out for its severity, as well as the lack of studies with pregnant animals, reviews of possible teratogenic effects (RN5, Annex III, A.6), or even the damage to non-target organisms important to the affected ecosystems listing the evaluated species, the reasons for their choice and used techniques (RN5, Annex IV, A.8). Even field trials in all ecosystems where transgenic events would subsequently be released (RN5, Annex IV, A. 3) have never been presented.

Justifications for the non-submission of these studies are negligible. Consider, for example, the fact that when RN5 (Annex III Risk Evaluation for Human and Animal Health (A.4, emphasis added) requests information regarding 'corresponding changes on animal performance when fed with genetically modified organisms or any of their parts, non-processed or after processing, even providing evaluation results of nutrition in experimental animals **for two generations,** indicating the species used in the tests, length of the test, physiological and morphological changes observed compared to control groups and nutritional quality changes, **if any**', most members of CTNBio seems to interpret that 'if there are no' studies, they are no longer relevant and it is not necessary to present them. Requests for postponement of the decision, until these referred studies were made available, were systematically rejected by majority vote.

The insistence on the adoption of short-term studies prevents the identification of subtle emergency impacts related to under-dosage. Consider, at this point, that the controversial study performed by the Séralini team (2012) outlined the development of tumours in rats, after 120 days of treatment, while studies performed by CTNBio are completed in only 90 days.

Discussions about this research (removal of publication from the journal, republication by another) and a publication of new studies confirming their findings should be considered (Oraby et al. 2015), in the observation of a singular fact: CTNBio refused by majority vote, to reopen discussion of the transgenic maize in question (NK603, from Monsanto), which remains available to Brazilian consumers in single or aggregate (combining transgenes such as for maize NK603xT25 or TC1507xMON810xMIR162xNK603 maize).

As if this was not enough to raise doubts as to the validity of decisions taken by CTNBio, plants modified to receive high doses of systemic herbicides, absorbed by them and metabolized in compounds, sometimes more toxic than the original active ingredients (Mañas et al. 2009), have been evaluated through tests that deny their real condition. In these tests, animals receive grains of maize and soybeans, tolerant to herbicide grown especially for the tests, in the absence of pesticides and therefore in opposite condition to that found in the million hectares designed for food and animal feed production. Even so, when reports reveal significant statistical differences, these are dismissed under the unjustified claim of biological irrelevance. Sometimes this occurs with the aggravating factor of denial by majority decision, of complementary studies, requested by board members of the minority group.

Studies indicating liver, kidney, testicles and mammary glands damages on animal experiment have been discarded by CTNBio, even when the recommendation for the rejection come from a minority of its members.

What arguments sustained these anomalies? What interests do they serve and what environmental implications[1] should be expected? This article reviews these questions.

Development

The robust scientific studies and mythology advantages versus transgenic lobby narratives

Risk assessments of genetically modified organisms seem to to studies performed by the companies that own technologies, focusing on genetic modification, resulting proteins and their technical functionality. These studies attribute scarce attention to possible effects of the product on socio-environmental and ecological relationships networks in environments where crops are carried out. Moreover, they are developed in laboratories and experimental sites that do not allow observing scale effects resulting from cultivation of successive crops, with continued use of pesticides related to genetic changes involved. The negligence of these and other aspects discussed below, weakens the government and its institutions revealing little justifiable tolerance from the Risk Evaluation Agency, to investigative methods of poor discriminatory power, of which conclusions systematically benefit private gains at the expense of collective interests.

The Evaluation Agency operates based on the assessment of requests from biotechnology companies. These requests, which exploit marketing elements, are justified based on studies made by the demanding organizations themselves. Performed in a few municipalities (which do not cover the country's climatic variability) and focusing on concerns on genetic modification and deriving proteins – in the ease justified by biased interpretation of the CTNBio rules – those documents do not take into account technological packages associated with genetic modifications nor ecological networks affected by them. It thereby decreases not only the effects of climate variability on the modified organisms, but also the residual impact of pesticides on soil and water communities, as well as on ecological functions and food webs.

As an aggravating factor, the non-scientific concept of 'substantial equivalence' (Millstone et al. 1999) is adopted on a general basis. Equating genetically modified plants (GMPs) to their parent/unmodified isogenic (NMPs), based on simplistic criteria have been used to block studies that contest its validity and also to feed the myth that there is international scientific consensus about safety of Genetically Modified Organisms (GMOs). After examining the body of scientific publications contained in specialized journals on this subject, Hilbeck et al. (2015) state: 'Claims of consensus on the safety of GMOs are not supported by an objective analysis of the refereed literature.'

It is almost embarrassing the fact that by adopting the 'equivalence', risk evaluators assume that despite the inexistence of differences (GMPs and NMPs are 'equivalent'), they also exist and are robust to the point of justifying patent rights and royalties charge. In fact, those scientists work with an analytical method that contradicts scientific methods, allowing them to conclude both by validity or denial of any assertion at the same time.

It is interesting to note that the lack of parameters that allows limiting the scope of 'acceptable' distinctions does not seem to bother them. Documented decisions are not uncommon, where to reaffirm the required 'equivalence', risk evaluators appointed to assemble CTNBio no longer considered significant statically differences exhibited in tests performed by the companies themselves. In these cases, when advisors of CTNBio request further studies, most of them accept the incoherent argument of 'biological irrelevance' of such differences, in order to reject them. Therefore, in the absence of levels for 'non-acceptance', 'equivalence' becomes unlimited.

Since most CTNBio members rarely identify doubts elements, it is presumed that in this risk assessment field, processes incorporate full information, eliminating uncertainties. Or, in a less generous interpretation, perhaps it may be considered that professional ethics intervention is needed or full transparency of the decisions taken, in order to have a rejection of any request for commercial release.

The obscurity surrounding the subject of GMOs, the lack of independent information, media campaigns involving opinion leaders and even government representatives help to consolidate this reality. This way, the spectrum of narratives that hide real problems inherent to genetically modified plants and their potential impacts on health and the environment are expanded.

This also reduces chances of aware and well-informed social actions that could minimize, through purchasing decisions, risks that technology imposes for health and the environment.

The narrative built by people interested in the adoption of transgenic plants in Brazil states, among other facts, that transgenic plants would be more productive (it would put an end to the hunger spectrum), safer (would be thoroughly studied based on strong regulations) and would protect the environment and health (by reducing the use of pesticides).

Table 7.1 Top 10 selling pesticides in Brazil: 2013

Active ingredient	Sales (ton. IA)	Ranking
Glyphosate salts	185,956.13	1
2,4-D	37,131.43	2
Atrazine	28,394.91	3
Mineral oils	28,347.06	4
Acephate	22,355.41	5
Vegetable oils	14,318.35	6
Chlorpyrifos	13,084.62	7
Methomyl	8,533.26	8
Mancozeb	8,419.01	9
Imidacloprid	7,940.82	10

Source: IBAMA (Brazilian Institute of Environment and Renewable Natural Resources)/Consolidation data provided by registrants companies of technical products and pesticides, as art. 41 of Decree No. 4,074/2002.

Sixty per cent of pesticides used in Brazil are applied on transgenic soybean, maize, and cotton crops. Only from glyphosate, from tie-in sales with those seeds, more than 180 million litres were sold in 2013/14 crops.

The massive use of transgenic seeds and associated poison has caused emergence and spread of resistant plants and insects, which demand growing use of increasingly toxic substances. Take, for example, the 'evolution' of RR technology (maize, cotton and soybeans tolerant to glyphosate, a carcinogenic herbicide classified under 'low toxicity') to ENLIST technology (corn, cotton and soybeans tolerant to 2,4-D, teratogenic herbicide classified as 'extremely toxic').

In Brazil about 45 million hectares of transgenic plants are grown. They are maize crops,[2] soybeans and cotton that receive herbicide applications, without being damaged (tolerance to herbicide technology – HT) or that produce insecticidal proteins (Bt technology), or both (combinations of Bt+HT technology). Therefore, there are no genetically modified plants with the purpose of increasing productivity. Given this, why would they be more productive?[3]

If they do not protect the environment or health, who might be interested in their expansion? According to the interpretation of the Minister of Science, Technology and Innovation (MCTI), communicated on 4 January 2015 to the members of CTNBio, the day before the date when these members have reacted regarding the request for commercial release of transgenic eucalyptus, 'it would be interesting to all, and only ignorant and uninformed people, contrary to national development might think the opposite'.

We should stress that, as warned by representatives of the Ministry of Agrarian Development (MDA), until that day, most studies approved by CTNBio, which should generate data to support that request on eucalyptus commercial release, were not yet completed.

The next day (transgenic eucalyptus was approved by CTNBio on 4 February 2015 (Fontes 2015), regardless of assessment insufficiency, regardless of the damage on the organic honey market, which involves about 350 thousand beekeepers, and still ignoring commitments made by the country to not deliberate on GM trees before the conclusion of all relevant studies, most of CTNBio members voted in favour of the biotechnology company's request.

Would the attitude of the Minister of Science, Technology and Innovation have been crucial to the fact that most members of CTNBio come to support that demand? Probably not. But the gesture was significant. It signalled the commitment of some part of the government and the fear of that minister, to an unlikely change in the routine of approvals accounted in this research field where, as evidenced by the repeated judgement of the majority of members of CTNBio, for whom the plaintiffs never fail to offer complete demonstrations of the absence of substantial risks related to their products on health and environment.

And this is the key point of this text: the assessment of possible environmental impacts on transgenic crops. How could these impacts be anticipated, caught and corrected?

Environmental impact

There are crucial moments for identifying environmental changes resulting from transgenic crops. They occur: (1) in previous studies to the approval of requests for commercial release and; (2) in monitoring activities, after commercial release.

In both cases, for obvious reasons, assessments should review actual circumstances associated with technology and their possible impacts on creatures and communities established in the involved ecosystems. For this purpose, relevant indicator organisms representative of the main ecosystems should be selected, considering aquatic and terrestrial environments, as well as the winged fauna, among others.

Furthermore, as previous studies developed as field trials and lab tests do not allow observing scale effects, it is relevant to establish monitoring procedures that allow tracking changes caused by a massive presence of transgenic plants, in environments where they will be grown. In this respect, post-commercial release monitoring emerges as a necessary condition for benchmarking validity of decisions taken in the absence of scale effects.

As an example of risks associated with the ineffectiveness of monitoring, when it does not issue alerts on the occasion of emergence of the first symptoms, consider the observed crisis in 2013, when it was decreed phytosanitary emergency in the state of Bahia (later extended to the whole of Brazil), for caterpillar control of *Helicoverpa armigera*, tolerant to toxic proteins present in Bt transgenic type maize and cotton crops. Damages have been more catastrophic after the caterpillar population explosion, at a later stage to emergence of small infestations that were not identified due to the absence or inefficiency of monitoring processes. The seriousness of the matter is illustrated by Tiburski (2013 p.5), which points risk of economic losses of up to five billion dollars to that particular case.

That crisis led the Ministry of Agriculture to allow import and use of neurotoxic insecticide (emamectin benzoate), banned in Brazil as recommended by the National Health Surveillance Agency, regulatory agency of the Ministry of Health, since 21 September 2007 (ANVISA, nd), in transgenic crops attacked by the caterpillar. It would be interesting to reflect at this point, about benefits resulting from the prevalence attributed to short-term economic interests, agribusiness, supported by the Ministry of Agriculture ('save the crop') over social interests of medium and long term, defended by the Ministry of Health ('protect the population in the crop areas').

Besides the issues related to deliberate on priorities, which assign even less relevance to the environment than to human health, we must consider that decision makers in charge of managing transgenic crops do not seem sufficiently informed about the implications of their practices. The continuing use of the same herbicides has reduced the effectiveness on herbicide-tolerant transgenic plants, encouraging more applications and justifying the 'entry' of new poisons on the market. Due to this, there are 32 plant species already known that are no longer controlled by main active ingredients. In the end, these plants became resistant to herbicide baths. We must call attention to *Coniza bonayriensis* (popularly known as 'Horseweed'), which after becoming a symbol of resistance to Monsanto starts to be included among the species whose presence justifies the determination of a phytosanitary emergency, allowing the Ministry of Agriculture to authorize import and application of unregistered toxic pesticides for use in Brazil (ordinance n5 from 08/21/2015, see www.sepaf.ms.gov.br/ministerio-da-agricultura-prioriza-controle-de-oito-pragas/). This way, the emamectin benzoate case is repeated, increasing evidences of fragilities on the monitoring of environmental impacts of transgenic crops as well as the political importance given to agribusiness sector interests at the expense of social aspects.

Risk assessments in analysis previous to commercial release decisions

The massive presence and then continued use of foreign substances to a certain environment will always lead to changes in such environment with ecosystem implications. Therefore, previous studies on the introduction of transgenic crops and associated pesticides must select and monitor indicator species that allows mapping changes in trophic chains. Selection of indicator species should take into account aspects of fragility and ecological importance for the major agricultural ecosystems involved. In addition, a separate analysis must be established for soil communities, aquatic environments, and winged species, as well as for the different biomes.

The Normative Resolution N.5 in its Annex IV (Risk Assessment to the Environment) is explicit in this matter requesting as a condition for commercial releases that previous studies with Genetically Modified Plants (items 3 and 8) generate information on (emphasis added):

> 3. Possible effects on **relevant indicator organisms** (symbionts, predators, pollinators, parasites or competitors of the GMO) **in ecosystems where it is intended to be cultivated,** compared to the parental organism of GMO in a conventional production system; and 8. Negative and positive impacts on target and non-target organisms that may occur with the release of GMOs, **listing the assessed species, reasons for choice and techniques used to demonstrate the impacts.**

Unfortunately, as explained below, these requirements have never been met. As a result, there is no adequate basis for interpretation of the environmental impacts caused by transgenic crops and associated pesticides in different affected ecosystems or even in large biomes in which they are contained.

Besides, previous studies should also have considered that transgenic events correspond to creatures that did not exist until then, and who will present biological behaviour mediated by environmental conditions. Like all living beings, they will be affected by the weather and under adverse conditions will reveal unusual behaviour. Therefore, they should be previously studied under biotic and abiotic stress conditions. However, this does not occur. CTNBio evaluators only access summary data of studies carried out under optimized laboratory tests and controlled field conditions. Several studies show the inadequacy of this approach, asking for more comprehensive risk analysis, structured in a way to search for abnormalities that occur in nature, when GMPs are faced with natural stresses caused by extreme temperatures, water oscillations, disease attacks, salt concentrations and other factors prevalent in the real world (Chen et al. 2005; Matthews et al. 2005; Then and Lorch 2008; Traavik 2008).

We will summarize the discussion of this issue by stating that these conditions are not met by the risk assessment analysis performed in Brazil. There are no field tests in relevant biomes such as the Amazon region, the Semi-Arid and *Caatinga*. Also, there are no impacts tests in biotic and abiotic stress conditions. The responses of new transgenic maize varieties when grown by family farmers in the Northeast and Northern Brazil and then offered as food to their families or feed to animals, will be under unprecedented conditions never evaluated with wide open risks implications, and until then completely neglected. These are real conditions tests, where all beings associated with those crops will take part as associated test subjects.

Obviously the same reasoning applies to all living communities in the soil and aquatic environments throughout all the national territory.

It is not only about the shortage or poor structuring of studies that assess impacts of transgenic and associated technological packages in the environment. Also, information generated in other environments is presented (and accepted), examining 'indicator' species of dubious quality for ecosystem assessments, because they do not exist in the Southern hemisphere. How could they attest environmental safety of these technologies in the environments? Who could justify their choice or accept justifications in this sense taking into account the complexity and sophistication of such a diverse country like ours?

Yet, analysis of processes approved by CTNBio to certify safety of GM crops waste on the aquatic environments reveal acceptance of information regarding Northern hemisphere fishes, which clearly do not cover Brazil specificities. The entire Brazilian winged fauna, with its bulky mega biodiversity – with so many species at risk of extinction already mapped by IBAMA – has been represented in many of the studies approved by the CTNBio by a single US, Mexico and the Caribbean native bird (*Colinus virginianus*), which does not even exist in Brazil territory.

That same routine applies to insects when it is observed that for most of the commercial releases authorized in Brazil, with favourable vote by the majority of the members of CTNBio, studies involving *Chrysoperla carnea* are being accepted. This insect species is from Palearctic regions (north Europe) and does not exist in Brazil.

On the other hand *Doru luteipes* (earwig) probably the most important and increasingly rare natural predator of caterpillars in maize crops has not been studied.

Evidently this information, even though worthy of audits, does not sensitize scientific community representatives or even environmental departments of the

Ministry of Environment. They were and are being recorded and denounced in CTNBio meetings, government agencies and civil society seminars without any resonance. Similarly, warnings about the disappearance of bees, genetic erosion and depletion of the soil seed banks of agricultural areas, or even the preference of animals for non-transgenic crops, fall on deaf ears. The lack of studies on the effect of transgenic in local environment is misinterpreted as absence of the problem. This understanding disowns ICMBio to summon up companies to provide information, a norm followed in case ICMBio admits a problem, for which information is unavailable. CTNBio could demand, or at least not prevent (by using majority vote in favour), that evaluators interested in further information forwarded relevant demands to biotechnology companies, even if only as a safeguard for favourable decisions on requests for commercial release.

Unfortunately, damage in the microscopic world of bacteria and fungi (which ensure life and fertility to soils) may be occurring in a more noticeable way with a minimum possibility of recovery.

In a very brief way, and considering the inadequacy of the adopted indicator organisms, negligence to the endangered ecological functions and the fact that as of today even species listed by IBAMA as endangered were associated with transgenic crops and their pesticides, it only remains to assume that previous studies to commercial release are sparsely informative and that environmental impacts of transgenic crops will be known through post-commercial release monitoring activities.

Risk assessment in post-commercial releases monitoring activities

Individual development is expressed by phenotypic characteristics that translate its genetic possibilities, conditioned by the environment. Once the genome indicates possibilities of expression, which will be guided by environmental pressures, a modification introduced therein will open unusual possibilities. Any predictions about the direction of these possibilities will be unrealistic. The fact is that genes operate in interactive networks. Thus, the incorporation of a transgene can affect the equilibrium between that genetic information and the rest of the genome, triggering unpredictable effects, in man or other living being. Another risk is because humans have many genes similar to those of other animals. There are, for example, about 480 regions identical to those of mice, which are kept in men and mice from the time of species differentiation. They exert very relevant functions that the current scientific knowledge does not explain explains even the natural selection theory. Additionally, as we are on accelerated climate change, any instability in the genome implies unusual possibilities. Even if nothing is verified in the first generation, the risk extends to the future.

It is well known that massive presence of herbicides and toxic proteins related to large-scale cultivation of these crops exert selective pressures to the living beings and communities established in those environments. As a result, changes in plant, insects, fungi and bacteria communities are observed, as well as the emergence of herbicides-resistant plants and insects that are no longer affected by insecticidal proteins. In response to these cases, farmers increase the number of applications on their crops and migrate to pesticides that are even riskier, entering a vicious circle that seriously threatens human health (Carneiro et al. 2015; Ferment et al. 2015).

Therefore, one of the obvious indicators to be sought in monitoring processes concerns the emergence of changes both in transgenic plants as the resistant insects population and tolerant plants. While revealing a decrease in effectiveness of the technology – with socio-economic implications – these cases also point to the expansion in the risks to health and the environment. Less obvious indicators correspond to non-target organisms population, indirectly affected (ACRE 2007; Wolfenbarguer et al. 2008; Trevisan et al. 2012), consisting of insects that act as natural predators of the species addressed by transgenic proteins, to scavenger insects, herbivores, birds, rodents and many other living beings who depend on seed banks (ACRE 2007), as well as the entire community of fungi and soil bacteria (Tarafdar et al. 2012), and those beings that inhabit aquatic environments (Rosi-Marshall et al. 2007). Endophytic microbial communities that fix nitrogen from air in both leguminous (King et al. 2001) and non-leguminous plants, such as maize, cotton and sugarcane (Stuart 2006), may have their performance affected by changes in the transgenic plant's cellular environment.

Among the indirect effects, there are changes in microbial communities from the rhizosphere (Kremer and Means 2009), with emphasis on the reduction of populations of Rhizobium (King et al. 2001), which are responsible for atmospheric nitrogen fixation in association with leguminous plants' roots. The critical point is loss of fertility and its long-term implications, notwithstanding the relevance of these populations for revenue and profitability of crops as well as national accounts and the agenda of nitrogenous fertilizers imports.

In the case of insecticidal proteins, which are expressed in different ways at levels that vary quite significantly for different tissues and plant organs, target insects end up facing doses that stimulate the emergence of tolerances complexifying their control. Also, non-target organisms are subjected to toxic loads that may exceed 2,500 kg per hectare.

Note, in the table opposite, that maize DAS 59122-7 has up to 88 nanograms of Cry34Ab1 protein, per dry matter milligram (whole plant). In one hectare accounting for about 30 tons of dry matter 2,640 kg of toxic protein could be released.

And maize MON 863-5, which contains up to 54 nanograms of Cry3Bb1 protein per dry matter milligram, could add 1,620 kg of toxic protein to the environment. In such quantities it is clear that environmental impacts over a wide network of non-target organisms tend to be huge and are completely unknown.

Reviews of studies available on international scientific literature published in specialized journals with peer review allow to affirm that 'Claims of consensus on the safety of GMOs are not supported by an objective analysis of the refereed literature' (Hilbeck et al. 2015).

Table 7.2 An example of variation in expression levels of δ-endotoxin in different maize constructs expressing five different δ-endotoxins

Active ingredient/ OECD unique ID	Leaf	Root	Pollen	Seed	Whole plant
Cry1Ab SYN-BT11-1	3.3 ng/mg	2.2–37.0 ng/mg protein	< 90 ng Cry1Ab/g dry wt. pollen	1.4 ng/mg (kernel)	–
Cry1Ab MON-00810-6	10.34 ng/mg	–	< 90 ng Cry1Ab/g dry wt. pollen	0.19–0.39 ng/mg (grain)	4.65 ng/mg
Cry1F DAS-01507-1	56.6–148.9 ng/mg total protein	–	113.4–168.2 ng/mg total protein or 31 to 33 ng/mg pollen	71.2–114.8 ng/mg total protein	803.2– 1572.7 ng/mg total protein
Cry3Bb1 MON-00863-5	30–93 ng/mg	3.2–66 ng/mg	30–93 ng/mg	–	13–54 ng/mg
Cry34Ab1 DAS-59122-7	5–302 ng/mg dry weight	24–102 ng/mg dry weight	63–88 ng/mg dry weight	29–85 ng/mg dry weight	9–88 ng/mg dry weight
Cry35Ab1 DAS-59122-7	2–113 ng/mg dry weight	1–16 ng/mg dry weight	0–0.2 ng/mg dry weight	1–2 ng/mg dry weight	1–16 ng/mg dry weight

Source: EPA – US Environmental Protection Agency. Consensus document on safety information on transgenic plants expressing *Bacillus thuringiensis* – derived insect control proteins. Organization for Economic Co-operation and Development ENV/JM/MONO (2007) 14. p.28.

We also highlight over 750 studies discarded by Risk Assessment Agencies (Ferment et al. 2015) as well as references by The Nature Institute (2008) for Unintended Effects of Genetic Manipulation.

In the face of so many evidences it should be expected that monitoring processes approved by CTNBio would be structured to identify or not anticipated adverse effects by previous assessments as specified in the Normative Resolution 9 (Brasil 2011).

However they are being approved without answering questions as basic as: 'what' will be monitored, 'how' monitoring will be conducted, 'where' and 'when' it will be held or even 'who' will perform 'what kind' of analytical observations. Also, they do not tell 'how and under what circumstances' (methodology, protocols and criteria) the monitoring will be held.

Plainly, without specifying how information that allows mapping adverse effects will be obtained, monitoring proposals approved by CTNBio will also not meet their goals. In fact, they only validate inoperative systems that conceal impacts and obstruct the perception of damages, contributing to the consolidation of the narrative that sustains their absence.

Conclusion

Who benefits from misleading narrative about GMOs?

The advance of transgenic crops in Brazil has been supported by assertions that do not correspond to reality. These include statements that contradict statistics, evidences and common sense. As examples, we can list the alleged strength of biosafety studies, the excellence of legal regulations, or the positive association between the expansion of transgenic crops and productivity gains, reduction in the use of pesticides, the expansion of agricultural income and even gains in environmental preservation.

In fact, in all cases we see the opposite. The genetic modifications that take the national territory in addition to not contributing to the expansion in productivity, delay the release of more productive varieties. Nor do they reduce the use of pesticides but instead increase it. They stimulate the emergence of resistant plants and insects and complexify their control. Increasingly *Bt* genes are being introduced for the same pests leading to the emergence of multiple tolerances. The same occurs in the control of undesirable plants where herbicide combinations can be seen applied to annual crops that alternate, without intervals for environmental recovery. This brings enormous selective pressure, resulting in the explosion of plants/insects populations that develop tolerance/resistance to those pesticides.

The evolution of transgenic varieties happens in the opposite direction of society's wishes, amplifying the problems that in theory justified their presence. They are also bringing back more dangerous pesticides to the market – old formulas that should be abandoned will have their use increased in Brazil. The switch from RR (Monsanto) technology (glyphosate tolerant plants) to ENLIST (DOW) technology (plants tolerant to 2,4D) corresponds to the substitution of a poison classified by ANVISA as having low toxicity to another one classified as extremely toxic. To this case, add up the repercussions of the approval of maize and soybeans varieties tolerant to *Dicamba* and herbicides from the imidazolinones group. What evolution is this and whose interests does it serve?

These technologies are still far from being analysed in a properly, competent and robust way. Even worse, particularities of the ecosystems are neglected and despite tolerant evaluators and decisions made in a disguised and non-transparent way, legal regulations are not being respected. The environmental impacts are not being adequately examined and absence of information is being confounded with absence of problems. The scientific community contributes to the concealment of risks and society does not receive guidance that allows conscious and well-informed purchasing decisions. Government agents help by operating as supporting elements to marketing campaigns developed in the interest of companies which already have the connivance of the authorities responsible for the construction of instruments that could act on limiting damage.

The intellectual property rights and the oligopoly of the transgenic seed market, as well as the associated technological packages expand the profit of those companies and at the same time also increase in a disproportionate way production costs. Facing reduction in profit per unit of area, farmers are compelled to expand the size of crops fields. The growth of economically viable minimum size is exclusionary for most small farmers, forcing them to leave the countryside. This weakens social fabric, small business and rural schools, prevents milk collection lines, emptying small towns and causing a kind of reverse agrarian reform which reduces food supply and increases labour costs, generating inflationary pressures.

Political and environmental impacts are also dramatic. Changes in legislation designed to serve interests of the same companies allow forgiveness of environmental crimes in a reverse pedagogy that encourages expansion of monocultures on protected areas, wetlands and springs with long-term implications still unpredictable. The concentration of power in the hands of a few big transnational companies also threatens food security and national sovereignty.

On one hand, the balance of payments depends on maize and soybean crops. On the other, a few big biotechnology companies (Monsanto, Bayer, DuPont, Syngenta, BASF, Dow) are the owners of transgenic seeds that Brazil needs to plant every year in order to keep the economy standing. A simple shortage threat would be sufficient to characterize farmer's limited autonomy in this field as well as to justify the occupation of public posts by defenders of interests that go against the objectives of a sovereign nation.

In times of climate changes the alternative to ensure food security for people demand concerns and investments that enable non-transgenic seed banks controlled by farmers and dispersed in different regions of the country. However, movements observed in the Congress indicate the opposite direction. Bills that aim to end the labelling of products made from transgenic crops are being developed, which would prevent consumers from signalling the government showing annoyance with the channelling of public resources for the expansion of genetically modified crops. Another bill also being developed aims to end with the moratorium on reproductive restriction technology (GURT or Terminator), which in the future will prevent farmers controlling their own crops, definitively ending any possibility of self-supply.

All these implications meet outside interests, facilitated by people and groups who benefit from them in some way, at the expense of the needs of Brazilian society, threatening the biodiversity of this territory so significant for the future of the planet.

Overcoming this scenario requires much more than simple change in the way the legal regulations associated with GMOs have been conducted in Brazil. It is significant the fact that risk assessment procedures disregard basic topics in the area and neglect the warnings issued by the international bibliography, thus expanding the risks to the national economy, people's health and environmental stability. These facts should not remain merely in restricted grounds; the Brazilian population has to have access to this type of information, once the future of the country is being committed. Disregarding public opinion can lead people to disbelieve in their institutions, threatening Brazilian democracy and the sovereign nation foundation.

The alternative, designed to overcome this situation, looks fragile and concerns the multiplication of reports that stimulate participation and nourish the collective awareness of what is at stake and who is interested in maintaining or changing trends in progress.

Particularly, I trust that this apparent fragility will allow enormous changes. Just as the hope overcame fear, indignation will overcome apathy.

Notes

1   Take into consideration at this point, the publication entitled Lavouras Transgênicas – Riscos e Incertezas. Mais de 750 estudos desprezados pelos órgãos reguladores de OGMs (GM Crops – risks and uncertainties: More than 750 studies neglected by risk assessment bodies) (Ferment et al. 2015).
2   GM eucalyptus and beans approved by CTNBio are still not commercially grown.
3   It should be stressed that there are still traditional researches in productivity increase, and that these researches are successful. However, as biotechnology companies dominate these markets and only GM varieties can be protected under patents that ensure collection of royalties, the most productive varieties are not marketed after being identified. They only enter the market when they become transgenic. So, they will not be more productive for being GM, but will be GM for being more productive. As the transgenes incorporation process is time consuming, national economy loses productivity gains, because farmers are unable to access the most productive varieties when they are obtained, but only when companies can genetically transform them.

References

ACRE – UK Advisory Committee on Releases to the Environment (2007) 'Managing the Footprint of Agriculture: Towards the Comparative Assessment of Risks and Benefits for Novel Agricultural Systems,' *Report of the ACRE Sub-group on Wider Issues raised by the Farm-Scale Evaluations of GM Herbicide Tolerant Crops revised after public consultations*, 3 May.

ANVISA – National Agency for Sanitary Surveillance. Analysis of rejection of technical product based on the active ingredient Benzoate of Emamectin (based on the technical note summary) (http://portal.anvisa.gov.br/wps/wcm/connect/880a100 047457e298a06de3fbc4c6735/parecer_indeferimento_ativo_benzoato_emamectin. pdf?MOD=AJPERES) Accessed 30 August 2015.

Bohn, T., Primicerio, R., Hessen, D. O. and Traavik, T. (2008) 'Reduced fitness of Daphnia magna fed Bt-transgenic variety Mayse', *Archives of Environmental Contamination and Toxicology*, 55, 584–92.

Brasil (2008) Ministério da Ciência, Tecnologia e Inovação – Comissão Técnica Nacional de Biossegurança – Resolução Normativa Nº 05 de 12 de março de 2008

– Brasília (www.ctnbio.gov.br/index.php/content/view/11444.html) Accessed 1 November 2015.

Brasil (2011) Ministério da Ciência, Tecnologia e Inovação – Comissão Técnica Nacional de Biossegurança – Resolução Normativa N° 09, 2 December 2011 (www.ctnbio.gov.br/index.php/content/view/16781.html) Accessed 4 May 2015.

Brasil (2016) Ministério da Ciência, Tecnologia e Inovação – Comissão Técnica Nacional de Biossegurança – Resumo Geral de Plantas Geneticamente modificadas aprovadas para Comercialização (www.ctnbio.gov.br/index.php/content/view/20559.html) Accessed 2 April 2016.

Buyer, J. S. and Blackwood, C. B. (2008) 'The effects of Bt corn on soil and rhizosphere microbial communities', United States Dept. of Agroc. Research Service.

Carneiro, F. F., Pignati, W., Rigotto, R. M., Augusto, L. G. S., Rizollo, A., Muller, N. M., Alexandre, V. P., Friedrich, K. and Mello, M. S. C., (eds.) (2012) Dossiê ABRASCO – Um Alerta Sobre os Impactos dos Agrotóxicos na Saúde. Associação Brasileira de Saúde Coletiva, Rio de Janeiro 625p (www.abrasco.org.br/UserFiles/Image/Dossieing.pdf) (in English). Accessed 2 January 2016.

Carneiro, F. F., Pignati, W., Rigotto, R. M., Augusto, L. G. S., Rizollo, A., Muller, N. M., Alexandre, V. P., Friedrich, K. and Mello, M. S. C., (eds.) (2015) Dossiê ABRASCO – Um Alerta sobre os Impactos dos Agrotóxicos na Saúde. Associação Brasileira de Saúde Coletiva, Rio de Janeiro 625p (www.abrasco.org.br/site/2015/03/dossie-abrasco-um-alerta-sobre-os-impactos-dos-agrotoxicos-na-saude/) In Portuguese: Accessed 15 March 2016 (http://www.abrasco.org.br/UserFiles/Image/Dossieing.pdf) In English: Accessed 2 January 2016

Chen, D., Ye, G., Yang, C., Chen, Y. and Wu, Y. (2005) 'The effect of high temperature on insecticidal properties of Bt Cotton,' *Environmental and Experimental Botany*, 53: 333–342.

EPA – US Environmental Protection Agency. Consensus document on safety information on transgenic plants expressing Bacillus thuringiensis – derived insect control proteins. Organization for Economic Co-operation and Development. ENV/JM/MONO (2007) 14.

Ferment, G., Melgarejo, L., Fernandes, Gabriel, B., and Ferraz, J. M. (2015) 'Lavouras transgênicas – riscos e incertezas: Mais de 750 estudos desprezados pelos órgãos de avaliação de risco' (GM Crops – risks and uncertainties: More than 750 studies neglected by risk assessment bodies) *ADEN/MDA* Brasilia, 555p. (www.mda.gov.br/sitemda/pagina/nead talk) Accessed 1 December 2015.

Fontes, S. Brasil (2015) é o 1° 'país a aprovar uso comercial de eucalipto transgênico', Valor, São Paulo 9 April 2015 (www.valor.com.br/empresas/3999630/brasil-e-o-1) Accessed 1 December 2015.

Hilbeck, A., Binimelis, R., Defarge, N., Steinbrecher, R., Székács, A., Wickson, F., Antoniou, M., Bereano, P. L., Clark, E. A., Hansen, M., Novotny, E., Heinemann, J., Meyer, H., Shiva, V. and Wynne, B. (2015) 'In the scientific consensus on GMO safety', *Environmental Sciences Europe*, 27:4 (www.enveurope.com/content/27/1/4/abstract) Accessed 30 August 2015.

King, C. A., Purcell, L. C. and Vories, E. (2001) 'Plant growth and nitrogenase activity of glyphosate-tolerant soybean in response to foliar glyphosate applications', *Agronomy Journal*, 93 (1), 179–186 (www.agronomy.org/publications/aj/abstracts/93/1/179) Accessed 3 November 2015.

Kremer, R. J. and Means, N. E. (2009) 'Glyphosate and glyphosate-resistant crop rhizosphere interactions with microorganisms', *European Journal of Agronomy*, 31 (3): 153–161.

Mañas, F., Peralta, L., Raviolo, J., Garci, O. H., Weyers, A., Ugnia, L. Gonzalez, C. M., Larripa, I. and Gorla, N. (2009) 'Genotoxicity of AMPA, the environmental metabolite of glyphosate, evaluated by the Comet assay and cytogenetic tests', *Ecotoxicology and Environmental Safety*, 72 (www.ncbi.nlm.nih.gov/pubmed/ 19013644) Accessed 2 November 2015.

Matthews, D., Jones, H., Gans, P., Coates, S. M. J. and Smith, L. M. J. (2005) 'Toxic secondary metabolite production in genetically modified potatoes in response to stress', *Journal of Agricultural and Food Chemistry*, 53 (20), pp. 7766–7776.

Millstone, E., Brunner, E. and Mayer, S. (1999) 'Beyond "substantial equivalence"', *Nature*, 401, 525–526.

Oraby, H., Kandil, M., Shaffie, N. and Ghaly, I. (2015) 'Biological impact of feeding rats with a genetically modified-based diet', *Turkish Journal of Biology*, 39 (http:// journals.tubitak.gov.tr/havuz /biy-1406-61.pdf) Accessed 2 November 2015.

Rosi-Marshall, E. J., Tank, J. L., Royer, T. V., Whiles, M. R., Pokelsek, J. and Stephen, M. L. (2007) 'Toxins in transgenic crop by products may affect headwater stream ecosystems', Proceedings of the National Academy of Science USA 104: 16204–16208.

Séralini, G. E., Clair, E., Mesnage, R., Gress, S., Defarge, N., Malatesta, M., Hennequin, D. and Vendômois, J. S. (2012) 'Long term toxicity of a Roundup herbicide and a Roundup-tolerant genetically modified maize', Food and Chemical Toxicology (http://dx.doi.org/10.1016/j.fct.2012.08.005) Accessed 2 November 2015.

Stuart, R. M. R. (2006) Comunidade de fungos endofiticos associada à cana-de-açúcar convencional e geneticamente modificada. Dissertação (Mestrado em Agronomia) (Community of endophytic fungi associated with sugarcane conventional and genetically modified. MS thesis ESALQ-USP Piracicaba).

Tarafdar, J. C., Rathore, I., Shiva, V., and Jagadish, C. (2012) 'Effect of Transgenic Bt-cotton on Biological Soil Health', *Applied Biological Research*, 14 (1).

The Nature Institute (2008) Unintended Effects of Genetic Manipulation at (http:// natureinstitute.org/txt/ch/nontarget.php.) Accessed 30 August 2015.

Then, C. and Lorch, A. (2008) 'A simple question in a complex environment: How much Bt toxin genetically engineered MON810 maize plants actually produce', in Breckling, B., Reuter, H. and Verhoeven, R. (eds) *Implications of GM-crop Cultivation at Large Spatial Scales*, Theorie in der Ökologie 14 Frankfurt Peter Lang.

Tiburski, L. (2013) Manejo integrado da Helicoverpa armigera na cultura do milho (Zea mays) na região de Curitibanos – SC (Integrated Management of *Helicoverpa armigera* in maize (Zea mays) Curitibanos Region – SC) Federal Univ of SC, (https:// repositorio.ufsc.br/bitstream/handle/123456789/117342/PROJETO%20DE%20 CIENCIAS%20RURAIS%20-%20MIP.pdf?sequence=1) Accessed 30 August 2015.

Traavik, T. (2008) 'GMOs and their unmodified counterparts: Substantially equivalent or different?' in Breckling, B., Reuter, H. and Verhoeven, R. (eds) *Implications of GM-crop Cultivation at Large Spatial Scales*, Theorie in der Ökologie, 14 Peter Lang Frankfurt.

Trevisan, H., Carvalho, A. G., Aguiar, A., Wood, B. and Abreu, I. (2012) 'Polen do milho transgênico – Possível efeito ecológico nas colônias de abelhas' (Pollen from GMO maize – Possible ecological effects on bee colonies) *Agrotec*, n1 pp. 80–81.

Wolfenbarguer, L. L., Naranjo, S. E., Lundgren, J. G., Bitzer, R. J. and Watrud, L. S. (2008) 'Bt Crops Effects on Functional Guilds of Non-target arthropods: A meta-analysis', *PLoS ONE*, 3(5): e 2128.

8 Tax incentive for pesticides

A debate on its (un)constitutionality from the environmental rule of law and the environmental public order

João Alfredo Telles Melo and Geovana de Oliveira Patrício Marques

> Ser capaz, como um rio
> que leva sozinho
> a canoa que se cansa,
> de servir de caminho
> para a esperança
>
> (Mello 2001, 170)[1]

This chapter aims to discuss the (un)constitutionality of normative provisions that grant a series of tax benefits aimed at the production, import and marketing of pesticides in the country, from the understanding that if Brazil is the world's largest consumer of these agrochemicals, this happens due to a number of tax, financial and technical incentives oriented to the use of these substances in agriculture.

The data from this poisonous industry are staggering: since 2008 Brazil consumes around 1 million tons per year of these agents, representing a consumption of about 5 liters per inhabitant/year (Londres 2011, 19). In Brazil, in 2007, around 1,400 pesticides, 430 active ingredients, and 750 technical products were registered (Carneiro 2015, 53). A business that in 2011 has generated 8.5 billion dollars (Zafalon 2012), and whose consequences have brought, as will be seen below, a number of serious impacts and damages to human health and environmental soundness in the country, which occupies the sad first place in this lurid championship of poison consumption.

The presentation of our subject begins with the contextualization of the widespread use of these pesticides in a global scenario of social and environmental crisis, where pollution, in its various aspects, is a sign – but not the only one – of this civilizational 'crossroads' in which humanity finds itself.

Leaning on the crisis, Morin and Kern (2005), still in the '90s of the last century, when analyzing the 'planetary agony', conceptualize the state of the art of the 'Homeland Earth' and the 'Humanity-community of destiny' as a 'polycrisis', an interrelation of development, modernity and society crisis; a civilizing crisis therefore. It is also a crisis of the capitalist civilization, and of the merchandise production system.

Then this work will debate the Constitution of 1988 – which, in the words of Marlmenstein, created 'the basis for the flowering of a new democratic order' (Marmelstein 1988, 15), its consistent and expressive evaluative aspects, that has set up a true 'environmental public order', established by the political formula of the 'Environmental Rule of Law' in the conception of Benjamin (2008) and other scholars. Some principles of the Environmental Law will be presented, which will serve as a parameter for the debate in the last part of this chapter. Therefore, the analysis of the constitutionality of these rules will be made from these above-mentioned parameters.

The fact: the pesticide contamination in the context of a global social–environmental crisis

> The natural world outside our farms and cities is not there as decoration but serves to regulate the chemistry and climate of the Earth, and the ecosystems are the organs of Gaia that enable her to maintain our planet habitable.
>
> (Lovelock 2010, 27)

Today, it can be said that there is almost a consensus that the planet is immersed in a social and environmental crisis of global nature and whose proportions have never before been experienced by the human society. Its most severe and most visible face, but not the unique, is the overheating of the earth and the climate change. Although it has already been submitted to the Fifth Climate Change Assessment Report of the IPCC (Fifth Assessment Report), it was the release, in February 2007, of the previous report that caused a great impact, given its very serious conclusions, observing, on the changes in climate and its effects, that the warming of the climate system is unequivocal and that its causes, linked to the emission of greenhouse gases (GHGs), are anthropogenic and unnatural, and its impacts on nature and society are already being felt (Fifth Assessment Report).

On a daily basis occurrences are observed everywhere in the world of some extreme weather and environmental phenomena: droughts, typhoons, floods, etc., phenomena that have been increasingly intense and recurrent to the point that a term of war vocabulary has been adapted to the ecological repertoire: the 'climate refugee' or 'environmental refugee', who already account for millions on the planet. The International Red Cross, which published in 2001 the 'World Disasters Report', projects the existence of 200 million climate refugees by 2050 (Hood 2007).

However, as mentioned before, global warming and climate change are just the most visible face of a major crisis, which is related to the current configuration of the capitalist mode of production, with its model of development based on a fossil-fuel matrix and a productivist vision that sustains a way of life of global economic elites based on wasteful consumption, which is, at the same time, environmentally unsustainable and socially unjust; not only on a regional or national scale, but on a planetary level.

Foster, author of the classic *Marx's Ecology: Materialism and Nature* (Foster 2005a), in a provocative 2005 article entitled 'Organizing Ecological Revolution' lists the warning signs of the global environmental crisis, demonstrating the unsustainability of the path of humanity in current times, among which he highlights the most significant and serious, yet not so often mentioned, aspect of the crisis: the breakdown of the regenerative capacity of the planet, its balance. In his words:

> According to a study published by the National Academy of Sciences in 2002, the world economy exceeded the earth's regenerative capacity in 1980 and by 1999 had gone beyond it by as much as 20 percent. This means, according to the study's authors, that 'it would require 1.2 earths, or one earth for 1.2 years, to regenerate what humanity used in 1999'.
>
> (Foster 2005b)

Air pollution is among the consequences of these activities. The World Health Organization states that only in 2012, 7 million people died from exposure to air pollution, which has become the greatest environmental risk factor in the world. Cancer and cardiovascular and pulmonary diseases are death causes of this dire statistic (OMS 2014).

Certainly, a part – even if it cannot be properly measured – of this criminal pollution is caused by spraying (via tractors, or manually) of pesticides (although there are other forms of pollution caused by agricultural poisons). The difficulty of having this measure is due to its underreporting. The researchers Augusto et al. state that 'for every recorded case of poisoning by pesticides another 50 occur without notice or erroneous notification'; they also state that 70 per cent of poisonings occur in poor or developing countries (Augusto et al. 2011, 259), which clearly constitute another case of environmental injustice.[2]

Nevertheless, contamination data is very significant in our country. Rosa, Pessoa and Rigotto, in a study on the impact of pesticides on human health, reported that, according to data from the National Poison Information System (SINITOX, in Portuguese), between 1989 and 2004, there were more than 1 million cases of human poisoning by the chemical poisons, of which 6,632 died (Rosa, Pessoa and Rigotto 2011, 240).

Among these deaths, one particularly draws attention, that of Vanderlei Matos da Silva, a worker of Del Monte Fresh Produce Brasil Ltda, in Limoeiro do Norte, Ceará, on 30 November 2008, given the fact that labour courts in both first and second degree recognized that his death occurred due to chronic liver disease resulting from contact with toxic substances (Júnior 2014; Silva 2014).

In its turn, Nodari writes that 'hundreds of independent researcher's studies (report) potential chronic or acute toxicological effects of many products' and exemplifies the effects such as 'reproductive, teratogenic, mutagenic, carcinogenic and nervous' (Nodari 2012, 119). He also warns to the fact that such effects can result from both occupational activity and the intake of pesticide residues.

Over the years, it has been brought to light the fact that these chemicals eliminate insects necessary for the balance and reproduction of the plants, and pollute the land, air and groundwater. Thus, these chemicals pollute[3] and cause incalculable damage to the environment. In its application, they end up dispersing in the air and hence are carried by rain into the rivers, contaminating the soil and groundwater. It is impossible not to relate the increase in the use of pesticides – and the pollution they cause – with the expansion of agribusiness in the country, a model of which, in addition to the use of pesticides, leads to other major environmental impacts such as deforestation, mono-cultivation in large extensions, changings in the soil microfauna, etc.

According to Braga, the damage to human health due to handling in industrial activities, environmental contamination, including water, and eating contaminated products are enormous: 'the pesticides in the soil are transferred partly to the plant's cell tissue, with different levels of concentrations that depend, among other factors, on the existing concentration in the soil and to the type of plant' (Braga 2014, 143).

The intake of food contaminated with residues of chemical poisons by the population is very serious in our country; as it can be seen in Tables 8.1 and 8.2, the assessment of some foods by the Pesticide Residue Analysis in Food Program (FOR, in Portuguese) of ANVISA has shown a high and unsatisfactory level of contamination (Relatório complementar relativo 2012):

The assessment carried out in these foods is based on a level of safety for human consumption. However, these tests do not analyse all the possible active ingredients present in the food, which does not allow sustaining the absence of other kinds of ingredients.

The costs for the treatment of diseases caused by the use of pesticides are taken over by society, as Codonho says, while the companies that use them in their production chain just reap the profits generated by this practice. Also, according to the researcher, 'a study was conducted in the state of Paraná, where it was estimated that for every dollar spent on the purchase of pesticides in the state, about $1.28 could be generated in external costs for poisoning' (Codonho 2014, 40–41).

This work cannot give an exhaustive list of the quantitative and qualitative data regarding the gravity of the impacts of pesticides on the environment, and on people's life and health. The literature is vast, as well as the news (Castilho 2013). The fact is that one can already state, supported by several researchers, that the safe use of pesticides in our country is impossible for a number of factors. Rosa, Pessoa and Rigotto, in the important article mentioned above, list the following factors: the magnitude of pesticide use in the country; the

Table 8.1 Analyzed samples according to the presence/absence of pesticide residues

Unsatisfactory samples	25%
Satisfactory samples	75%
with detected residues	42%
without detected residues	33%

Table 8.2 Analyzed unsatisfactory samples according to the presence/absence of unauthorized or authorized but above limits pesticide residues

Product	Number of analyzed samples	NA (1)		>MRL (2)		>MRL + NA (3)		Total unsatisfactory (1+2+3)	
		N°	%	N°	%	N°	%	N°	%
Zucchini	229	104	45	5	2.2	1	0.4	110	48
Lettuce	240	93	39	2	0.8	12	5.0	107	45
Bean	245	10	4.1	4	1.6	4	1.6	18	7.3
Corn meal	208	2	1.0	4	1.9	0	0.0	6	2.9
Tomato	246	28	11.4	6	2.4	5	2.0	39	16
Grape	229	57	25	5	2.2	5	2.2	67	29
Total	1397	294	21	26	1.9	27	1.0	347	25

(1) Samples containing only unauthorized active ingredients (NA)

(2) Samples containing only authorized active ingredients but above the authorized limits (>MRL)

(3) Samples containing the two irregularities (NA and >MRL); (1+2+3) Sum of all kinds of irregularities

extent of the territory that would have to be monitored and controlled; the lack of institutional and structural conditions for the control organisms; the inadequate education and training of farmers etc. (Rosa, Pessoa and Rigotto 2011, 242). The same conclusion was reached by Abreu, researcher at the Public Health Post-Graduate Program of the State University of Campinas (Unicamp) (Sales 2015). The fact that the Brazilian population consumes 22 types of pesticides that are banned in the European Union only reinforces this insecurity (Carneiro 2015, 53).

Given all these facts, one question arises: if the use of pesticides is not safe, and if they have caused such severe damages to the environmental soundness and to human health, what does justify the incentive to its production, importation, comercialization and use? Would the norms that provide them with tax benefits be constitutional? Before answering these questions, it is important to analyse the evaluative content of our Constitution, especially regarding its environmental aspect.

The value: the principles of the Constitution, the Environmental Public Order and the State Environmental Law

> The greening of the Constitution brings a certain heretical flavor, away from the foregoing formulas, by proposing the solidarist recipe – temporal and materially expanded (and, therefore, a prisoner of utopian traits) – of the we-all-in-favor-of-the-planet.
>
> (Benjamin 2008, 58–59)

The Environmental Law – understood as a systematization of standards, principles, doctrine and jurisprudence that attempts to discipline the relationship of human society with its natural environment – is ultimately the result of the current environmental crisis (as mentioned above) that endangers not only the survival of our species, but of life as a whole, on our little planetary spacecraft. So much so that the emergence of this new legal branch, on an international scale, can be traced back to the documents produced during the First Conference on the Human Environment, held in Stockholm in 1972, convened to discuss the environmental problems that have already reached a global dimension (Milaré 2014, 251).

The impact of this crisis on the law takes place in the phenomenon described by Benjamin (2008) as the 'Constitutionalization of the Environment and Ecologization of Law', which leads the doctrine to find in our Constitution (and others that resemble by the same political formula) elements of what might be called, with Canotilho (2008), 'Democratic and Environmental Rule of Law' or, in Sarlet and Fensterseifer (2010), 'Socio-ecological Rule of Law', or even in Leite and Ayala (2011, 53), 'Environmental State of Law', for whom this form of organization should seek 'an environmental situation capable of providing harmony between ecosystems and thus ensure full satisfaction of dignity beyond the human being'.

The idea that our Constitution of 1988 political formula is a State of Environmental Law (or, in a more detailed definition, Socio-environmental Democratic State of Law), derives from the 'post-positivist' dialectical synthesis which surpasses, in the words of Belchior (2011, 90), with a new concept, the jusnaturalism versus positivism antinomy (Belchior 2011, 90), in which the principles are recognized as the legal norm, and are thus self-executing (and not mere bills of rights).

Marlmenstein, referring to what he defines as the 'triumph of Constitutionalism' with the renewal of the members and thoughts of the Supreme Court, notes, by analyzing the production of the Court, that 'today is pacified in the Supreme Court's jurisprudence the understanding that, by virtue of the maximum effectiveness of the constitution, *it is possible to extract from the constitutional principles direct commands to the legislator*' (emphasis added) (Marlmenstein 2013, 29).

To know the principles of Environmental Law enshrined in our Bill of Rights one should start by Article 225, transcribed below:

Art. 225 – All have the right to an ecologically balanced environment, which is an asset of common use and essential to a healthy quality of life, and both the Government and the community shall have the duty to defend and preserve it for present and future generations.

This chapter does not aim to make a full presentation of the principles of the Environmental Law, but merely of those that focus on the analysis of the constitutionality of regulations that prescribe tax incentives for pesticides.

After these considerations, this analysis must start by the principles contained in this legal measure by the Fundamental Right to an Ecologically Balanced Environment. It is a fundamental right of third generation – in the view of Bonavides (2015), which includes a cumulative and qualitative process of political victories – or third dimension – according to Sarlet (2014), who also maintains the unity and indivisibility of all generations and dimensions of fundamental rights arising from the principle of human dignity, recognized in in article I, section III, of the Federal Constitution.

It is the very dictum of art. 225 which relates, as essential to a healthy quality of life, the environmental quality, its solidity, its balance. It is not possible, therefore, to consider a dignified life in an environment – whether in its natural, artificial or cultural dimension – that is not balanced, healthy, sustainable. Although there are different principles in the doctrine, there is almost a consensus that the Fundamental Right to an Ecologically Balanced Environment constitutes a sort of principle of principles.

Machado articulates the principle of the fundamental right to an ecologically balanced environment with the principle of the right to a healthy quality of life. In his words:

> The human beings health is not defined by the existence or not of a diagnosed disease. It must be taken into account the state of the elements of nature – water, soil, air, flora, fauna and landscape – to assess whether these elements are in a state of sanity and if its use provides health or causes disease or discomfort to humans.
>
> (Machado 2014, 66)

This definition is in conformity with the concept of health from the World Health Organization: 'a state of complete physical, mental and social well-being and not merely the absence of disease or infirmity' (Conceito de Saúde 2015). It is this conception that is present in art. 196, section VIII, where health is defined as one of the fundamental social rights; and, more than that, when stating, in its art. 200, it is for the Unified Health System (SUS in Portuguese) to 'collaborate in protecting the environment, including in this scope the work environment'.

Another major innovation introduced by the Environmental Law is that the fundamental environmental right goes along with a fundamental duty, surpassing both the liberal view of individual rights – as it requires the abstention of the state for the individual enjoyment – as well as the notion of social rights, whose classic formulation is 'rights of all, and duty of the state'. Not here. The fundamental right belongs to everyone and so it is everyone's (the state and the community, in the constitutional diction) duty to defend it and to preserve it for present and future generations.

We are dealing with one of the three paradigmatic fractures as identified by Benjamin: 'the effacing of rigid formal positions between creditors and debtors (both have the right to an ecologically balanced environment and the duty to protect it)' (2008, 59).

When referring to the fundamental duty of the State, art. 225 also emanates another correlative principle, with different terminologies in the doctrine such as the 'Principle of the Obligation of Government Intervention' in Machado (2014, 134), or, in a broader, comprehensive view, the 'Principle of Public Nature of Environmental Protection' according to Milaré (2014, 262). The manifestation of this principle in the Constitution can be found in the tasks set out in paragraph 1, of the same art. 225.

In addition to art. 225, the principle of fundamental duty of the State or obligation of government intervention can also be found in art. 23, which deals with the material competences of the federal entities, which, in its paragraphs VI and VII establishes on the environment: VI – to protect the environment and fight pollution in any of its forms; VII – to preserve the forests, fauna and flora.

The wording that the duty to protect the environment must turn not only for the present generation but also for the future, englobes two principles at the same time. The first is focused on the 'human family' as Milaré (2014, 261) puts it, which is the so-called Intergenerational Solidarity (or Equity) that, given the finite nature of natural resources, understands that future generations also have the right to an ecologically balanced environment.

This vision comes from the very understanding of what the concept – which is also a principle (the second, therefore, of the statement above) of the Environmental Law – of 'Sustainable Development' means, a concept whose origin is traced back to the document entitled 'Our Common Future', published in 1987, also known as the Brundtland Report, named after the Norwegian Prime Minister of that time who coordinated the work under the World Commission on Environment and Development. The expression and the goal of the so-called sustainable development have their origins in this text: '(the sustainable development) meets the needs of current generations without compromising the needs of future generations.' Nevertheless, this goal is highly questionable in the context of the capitalist social organization based on the unlimited economic growth, the transformation of natural resources into commodities and the unbridled pursuit of profit.

Regarding other important principles of the debate about pesticides, one could not help but mention what Leite and Ayala (2011, 32) designate as the preventive and precautionary principle, which is based on Principle 15 of the Rio Declaration on Environment and Development, of which Brazil is a signatory, namely:

In order to protect the environment, the precautionary approach shall be widely applied by States according to their capabilities. Where there are threats of serious or irreversible damage, lack of full scientific certainty shall not be used as a reason for postponing cost-effective measures to prevent environmental degradation.

(Declaração do Rio)

Leite joins the precautionary principle with the preventive action in a very didactic manner; he adds that just like the precautionary principle, the preventive action 'seeks anticipatory measures against environmental damage, i.e., to create conditions to prevent any situations of environmental degradation' (Leite and Ayala 2011, 56).

Last but not least, this interpretive effort must include the concept of environmental public policy as developed by Benjamin in his classic text already quoted in this chapter. This important concept – that is embodied in Article 225 as well as in several other constitutional provisions – presents the notion that one is not 'dealing with a simple rereading of the private nature [...] but of a private order submitted to a higher public order under the rule of constitutionally fixed commandments and previous limits' (Benjamin 2008, 123).

This advanced conception – inserted in a contemporary interpretation that recognizes the normative force of the Constitution, its evaluative load, expressed in the principles underpinning the rule of environmental law – follows the understanding that the Fundamental Right to a Balanced Environment is unappropriable, inalienable and imprescriptible; therefore, one must not recognize (by the way, it is abominable according to the author's strong expression) any acquired right to pollute, 'since there is no law against the law, much less against the Constitution itself' (Benjamin 2008, 125).

Knowing and interpreting the constitutional requirements applicable to the issue that we intend to discuss, we might move on to the presentation of the standards that benefit, with tax incentives, the production, import and sale of pesticides in our country.

The norms: tax incentives for pesticides

'Show me the coin used for paying the tax.' They brought him a denarius, and he asked them, 'Whose image is this? And whose inscription?' 'Caesar's,' they replied.

Then he said to them, 'So give back to Caesar what is Caesar's, and to God what is God's.'

(Mathew, 22, 19–32)

In Brazil, the incentive policy to the use of agrochemicals started in 1975 with the National Agricultural Development Plan, which, as Rosa, Pessoa and Rigotto (2011, 218) note, 'encouraged and required the use of pesticides, offering great investments for its use, increasing, also, the number of industries producing pesticides in the country, from 14 plants in 1974 to 73 in 1985'.

Nevertheless, it is through extra-fiscality – when tax rules have its purpose established by the legislature beyond the simple collection, using it as a means of influencing human behaviour – that the current public policy incentives to agricultural poisons takes place.

Extrafiscality, however, has a positive side when it's applied, for example, to cigarettes, in which the increase of the tax causes the increase of the product price by internalization of its cost. The goal is not to punish the practice or to compensate the public treasury, since this is not the taxation's function, but to discourage consumption (by the impacts that it causes to human health and life and by the expenses it causes to the public health system) by increasing prices, and thereby fulfilling the tax collection function.

Extrafiscality also has a negative side (incentives, tax exemption, subsidy), having by some tax experts a concern with decreasing the tax collection by applying these norms (Cavalcante 2014, 143–58). However, it should be noted that the adoption of healthy environmental practices involves reduction in public spending. Encouraging protective practices prevents pollution and environmental degradation, which cause damage to the healthy quality of life of an entire society, generating costs for the state. The idea is that by decreasing the tax collection at a certain period, the public expenses resulting from harmful activities would also decrease in the long term.

Such is not the case for norms that grant tax benefits to the production and marketing of pesticides, as listed below:

1   In the CONFAZ[4] agreement No. 100/97, ratified by the act of the Permanent Technical Committee (COTEPE in Portuguese) No. 17/97 established a reduction of the tax basis of pesticide products by 60%:
    First term – The ICMS[5] tax on interstate exits are reduced by 60% (sixty per cent) from the following products: I – insecticides, fungicides, formicides, herbicides, parasiticides, germicides, miticides, nematicides, rodenticides, defoliant, desiccant, spreader, adhesives, growth stimulators and inhibitors (regulators), vaccines, serums and medicines, produced for use in agriculture and livestock, including inoculants, forbidding their application when given to the product diverse destination;

2   The Federal Decree 5630 of December 22, 2005, which provides the reduction to zero of contribution rates for PIS[6] and COFINS[7] in the import and domestic marketing of fertilizers, agricultural pesticides and other products;

3   The Federal Decree 7660, from 23.12.2011, exempting pesticides from Tax on Industrialized Products, the IPI.

One should add that some states still grant a further reduction, as is the case of Ceará: under the State Decree 24.569 of July 31, 1997, the tax base is reduced by 100 per cent, configuring a total exemption of ICMS.

Hence there is a sort of 'backwards extrafiscality' from both an ethical and environmental point of view, where substances responsible for serious damage to the environment and to the health and life of humans – to the point of causing thousands of deaths across the country – have similar incentives as the agricultural products of the basic food basket,[8] when the most fair and rational measure would be to overtax pesticides, just like tobacco and alcohol in order to inhibit rather than encourage its use.

One cannot help but to share the same 'strangeness' expressed by Pacobahyba: 'the use of an award sanction as developed by Bobbio in order to stimulate a clearly harmful trade to the environment' (Pacobahyba 2011, 272).

The question, of course, is not restricted to the debate in the tax context, but should also turn to the legal, constitutional and environmental hermeneutics, which is this chapter's object and that will be faced in the following conclusions.

Conclusions

A abelha por Deus foi amestrada
Sem haver um processo bioquímico
Até hoje não houve nenhum químico
Pra fazer a ciência dizer nada...

(Autor da Natureza, Zé Ramalho)[9]

Brazil is the world's largest consumer of pesticides and the reason for this sad title is, of course, the public policy of encouragement of the use of agricultural poisons, since the National Agricultural Development Plan in the 1970s, until the current policy that grants tax benefits to the use, marketing, production and import of pesticides.

The use of pesticides in our country is responsible for the poisoning of the natural environment – water, air, soil, plants and animals – and human populations, from those who work in factories and farms, to the consumer, given its high degree of contamination. The safe use of pesticides is highly questionable.

The Constitution of 1988 instituted a new public policy in Brazil, the environmental public order, based on the broader political formula of the Environmental Rule of Law – based on principles that ensure the right and the fundamental duty of the environment, the right to health, to intergenerational solidarity, and to preventive action.

The norms that grant tax benefits to the utilization, marketing, production and import of pesticides, based on a sort of 'backwards extrafiscality', backwards because unfair, unethical and unsustainable, hurts the constitutional principles, the public order and the environmental rule of law, which embody the values elected by the Political Charter. One can therefore state that the above norms may be considered unconstitutional.

The end of tax exemptions to pesticides is one of five key areas (which also include the end of aerial spraying, the immediate ban of pesticides that are already banned in other countries, the protection of water resources, and the creation of pesticides and genetic modified organism free areas) of the Permanent Campaign Against Pesticides and for Life, which brings together more than 100 national civil society entities and aims to sensitize the population to the risk of these poisons to health and the environment, aiming also the discussion of a new food production model based on the principles of agroecology (Campanha Permanente Contra 2011).

In this process of sensitization and mobilization, two tools have shown to be of fundamental importance: the dossier of the Brazilian Association for Community Health (ABRASCO in Portuguese) on the impacts of pesticides on health (Impactos dos agrotóxicos na saúde 2015), as well as the documentaries 'The Poison is on The Table' 1 and 2, written by filmmaker Silvio Tendler,[10] that addresses both the complaints of the damages caused by pesticides, and of agroecological alternatives.

It is a fact that the decision on the constitutionality of law or regulatory act, under Brazilian law, is the responsibility of our Constitutional Court, the Supreme Court, in the light of the provisions of article 102, I, 'a' of our Political Charter. But in this process – which involves social struggles, academic research, and legal clashes – the decision made by the Ceará's Labor Court to hold a company accountable for the death of a worker who handled pesticides (as mentioned above), and actions such as the Public Prosecutor's Office of São Paulo proposition to ban carcinogenic poisons (MPF/SP 2015), are the most important initiatives to strengthen the understanding that, in defence of health and the environment, we must fight for the unconstitutionality of any tax benefit to production, consumption, import and use of pesticides in our country.

Notes

1   'Be able, like a river / leading alone / the canoe which tires, / of serving as a path / to hope'.
2   Environmental injustice: 'mechanism by which unequal societies, from an economic and social point of view, direct the greatest burden of the damage of environmental development to low-income populations, discriminated racial groups, traditional ethnic communities, working class neighborhoods, and marginalized and vulnerable populations' (Acselrad et al. 2009).
3   'Pollution consists, in general lines, on the contamination of the environment. Can also be understood as any change in the physical, chemical or biological component of the environment that could constitute directly or indirectly damage to the fauna and flora, the health, well-being and development of human populations' (Pinheiro 2010).
4   The National Council for Financial Policy – Conselho Nacional de Política Fazendária (CONFAZ) – gathers the finance secretaries of the member states. According to its bylaws, its creation was intended to develop and harmonize procedures and standards inherent in the tax jurisdiction of the States and the Federal District. Under Decree No. 7050, of December 23, 2009, the CONFAZ is a collegiate body that integrates the structure of the Ministry of Finance.

5 Imposto sobre Circulação de Mercadorias e Serviços – Tax on Circulation of Goods and Services.
6 Programa de Integração Social – Social Integration Program.
7 Contribuição para o Financiamento da Seguridade Social – Contribution to the Financing of Social Security.
8 Basic basket refers to the Brazilian economic measure that considers the value of a set of basic food products for the subsistence of a family. In the State of Ceará, the basic food basket products realize a reduction of only 58.82% in the ICMS (Tax on trading of goods and services) (Law no 12.670, of December 30 of 1996).
9 'The bee by God has been tamed/ Without there being a biochemical process/ Until today there have been no chemical/ To make science say anything …' (Author of Nature, Zé Ramalho, Brazilian singer and composer).
10 Available at www.youtube.com/watch?v=a6Lawf6CTek

References

Acselrad, H., Mello, C.C.A. and Bezerra, G.N. (2009) *O que é Justiça Ambiental*, Garamond, Rio de Janeiro.

Augusto, L.G.S., Gurgel, A.M., Bedor, C.N.G., Gurgel, I.D.G., Friedrich, K., Mello, M.S.C. and Siqueira, M.T. (2011) 'O contexto de vulnerabilidade e de nocividade do uso dos agrotóxicos para o meio ambiente e a importância para a saúde humana', in Rigotto, R. ed., *Agrotóxicos, trabalho e saúde: vulnerabilidade e resistência no contexto da modernização agrícola no Baixo Jaguaribe/CE* Edições UFC, Fortaleza.

Belchior, G.P.N. (2011) *Hermenêutica Jurídica Ambiental* Saraiva, São Paulo.

Benjamin, A.H. (2008) 'Constitucionalização do ambiente e ecologização da Constituição Brasileira', in Canotilho, J.J.G. and Leite, J.R.M. eds, *Direito Constitucional Ambiental Brasileiro*, Saraiva, São Paulo.

Bonavides, P. (2015) *Curso de Direito Constitucional* Malheiros, São Paulo.

Braga, B. (2014) *Introdução à Engenharia Ambiental*, Pearson, São Paulo.

Campanha permanente contra os agrotóxicos e pela vida. (2011) Contra os agrotóxicos. Available at: www.contraosagrotoxicos.org/

Canotilho, J.J.G. (2008) 'Direito Constitucional Ambiental Português e da União Europeia', in Canotilho, J.J.G. and Leite, J.R.M. eds, *Direito Constitucional Ambiental Brasileiro*, Saraiva, São Paulo.

Carneiro, F.F. (2015) *Dossiê ABRASCO: um alerta sobre os impactos dos agrotóxicos na saúde*, EPSJV, Rio de Janeiro; Expressão Popular, São Paulo.

Castilho, I. (2013) O Caso do envenenamento das abelhas é certamente um dos mais trágicos, ainda que não seja o único *Outras Palavras*, 10 September. Available at: http://outraspalavras.net/posts/o-envenenamento-das-abelhas/

Cavalcante, D. (2014) 'Instrumentos fiscais na efetivação da política nacional de resíduos sólidos: do poluidor-pagador ao protetor-recebedor', in Cavalcante, D. *Tributação ambiental: reflexos na Política Nacional de Resíduos Sólidos*, CRV, Curitiba.

Codonho, M.L.P.C.F. (2014) *Desafios para a concretização da agricultura sustentável no Brasil: uma contribuição do direito para a regulação do uso dos agrotóxicos*, O Direito por Um Planeta Verde, São Paulo.

Conceito de Saúde segundo a OMS (2015) *Medicina Tropical*. Available at: www.alternativamedicina.com/medicina-tropical/conceito-saude

Declaração do Rio sobre Meio Ambiente e Desenvolvimento (1992). Available at: www.mma.gov.br/port/sdi/ea/documentos/convs/decl_rio92.pdf

Fifth Assessment Report (AR5) IPCC 2014, Available at: www.ipcc.ch/report/ar5/index.shtml

Foster, J.B. (2005a) *A Ecologia de Marx: materialismo e natureza*, Civilização Brasileira, Rio de Janeiro.

Foster, J.B. (2005b) Organizing Ecological Revolution, *Monthly Review*. Available at: http://monthlyreview.org/2005/10/01/organizing-ecological-revolution/

Hood, M. (2007) *Mudança climática: aumento dos perigos para futuros refugiados*. UOL Noticias (http://noticias.uol.com.br/ultnot/afp/2007/01/31/ult1806u5420.jhtm) Accessed 14 March 2016.

Impactos dos agrotóxicos na saúde (2015) *Dossiê Abrasco*. Available at: http://abrasco.org.br/dossieagrotoxicos/

Júnior, M. (2014) TRT mantém decisão que condena multinacional *Diário do Nordeste*, 14 November. Available at: http://diariodonordeste.verdesmares.com.br/cadernos/cidade/trt-mantem-decisao-que-condena-multinacional-1.1150525

Leite, J.R.M. and Ayala, P. (2011) *Dano Ambiental: do individual ao coletivo extrapatrimonial*, Editora Revista dos Tribunais, São Paulo.

Londres, F. (2011) *Agrotóxicos no Brasil: um guia para a ação em defesa da vida*. AS-PTA – Assessoria e Serviços a Projetos em Agricultura Alternativa, Rio de Janeiro.

Lovelock, J. (2010) *Gaia: alerta final* Intrínseca, Rio de Janeiro.

Machado, P.A.L. (2014) *Direito Ambiental Brasileiro*, Malheiros, São Paulo.

Marlowe, H. (2007) Mudança climática: aumento dos perigos para futuros refugiados, *UOL*, 31 January. Available at: http://noticias.uol.com.br/ultnot/afp/2007/01/31/ult1806u5420.jhtm

Marmelstein, G.L. (2013) '25 anos da Constituição de 1988: presente passado e futuro', in Carvalho, P.R.M. and Rocha, M.V. eds, *25 anos da Constituição de 1988: os direitos fundamentais em perspectiva*, Expressão Gráfica, Fortaleza.

Mello, T. (2001) *Poemas preferidos pelo autor e seus leitores: edição comemorativa dos 75 anos do autor*, Bertrand Brasil, Rio de Janeiro.

Milaré, E. (2014) *Direito do Ambiente*, Editora Revista dos Tribunais, São Paulo.

Morin, E. and Kern, A-N. (2005) *Terra-Pátria*, Sulina, Porto Alegre.

MPF/SP quer que agrotóxicos contendo ingrediente cancerígeno sejam proibidos no país (2015) *Ministério Público Federal*. Available at: http://noticias.pgr.mpf.mp.br/noticias/noticias-do-site/copy_of_meio-ambiente-e-patrimonio-cultural/mpf-sp-quer-que-agrotoxicos-contendo-ingrediente-cancerigeno-sejam-proibidos-no-pais

Nodari, R.O. (2012) 'Risco à saúde dos seres vivos adindo dos agrotóxicos – ênfase nos herbicidas', in Aragão, A, Leite, J.R.M., Ferreira, J.S. and Ferreira, M.L.P. eds, *Agrotóxicos; a nossa saúde e o meio ambiente em questão: aspectos técnicos, jurídicos e éticos*, FUNJAB, Florianópolis.

OMS: 7 milhões de mortes em 2012 foram associadas à poluição (2014) Agência Brasil Available at: http://agenciabrasil.ebc.com.br/internacional/noticia/2014-03/oms-7-milhoes-de-mortes-em-2012-foram-associadas-poluicao

Pacobahyba, F.M.O.M.C. (2011) *A concessão de benefícios fiscais pelos Estados na contramão da democracia brasileira*, Meritum, Belo Horizonte.

Pinheiro, C. (2010) *Direito Ambiental*, Saraiva, São Paulo.

Relatório complementar relativo à segunda etapa das análises de amostras coletadas em 2012 (2014) *Programa de Análises de Resíduos Agrotóxicos em Alimentos*. Available at: http://portal.anvisa.gov.br/wps/wcm/connect/d67107004634368583a5bfec1b28f937/Relat%C3%B3rio+PARA+2012+2%C2%AA+Etapa+-+17_10_14-Final.pdf?MOD=AJPERES

Rosa, I.F., Pessoa, V.M. and Rigotto, R.M. (2011) 'Introdução: Agrotóxicos, saúde humana e os caminhos do estudo epidemiológico', in Rigotto, R. ed., *Agrotóxicos, trabalho e saúde: vulnerabilidade e resistência no contexto da modernização agrícola no Baixo Jaguaribe/CE*, Edições UFC, Fortaleza.

Sales, C. (2015) Dissertação de mestrado derruba mito do uso seguro de agrotóxicos *JUSBRASIL*. Available at: http://carollinasalle.jusbrasil.com.br/noticias/1465059 23/dissertacao-de-mestrado-derruba-mito-do-uso-seguro-de-agrotoxicos?utm_ campaign=newsletter-daily_20141021_215&utm_medium=email&utm_source= newsletter

Sarlet, I.W. (2014) *A eficácia dos Direitos Fundamentais* Livraria do Advogado Editora, Porto Alegre.

Sarlet, I.W. and Fensterseifer, T. (2010) 'Estado socioambiental e mínimo existencial (ecológico?): algumas aproximações', in Sarlet, I.W. and Fensterseifer, T. eds, *Estado socioambiental e direitos fundamentais*, Livraria do Advogado Editora, Porto Alegre.

Silva, C. (2014) Trabalho, agrotóxicos e morte no campo: uma longa espera, *Ecodebate*, 28 March. Available at: www.ecodebate.com.br/2014/03/28/trabalho-agrotoxicos-e-morte-no-campo-uma-longa-espera-por-justica-por-claudio-silva/

Zafalon, M. (2012) Vendas de defensivos agrícolas são recordes e vão a US$ 8,5 bi em 2011 *Folha de S. Paulo*, 20 April. Available at: http://bit.do/sindag2256

Part III

Facing the consequences of climate change

9 From co-leader to loner

Brazilian wavering positions in climate change negotiations

Larissa Basso and Eduardo Viola

From 1992 to 2015, commitments to mitigate climate change have advanced in the climate regime, but very slowly and lacking the ambitiousness that science indicates is needed to successfully tackle the issue. Different countries have different positions: some, a minority, push for more ambitious pledges, therefore aiming to reform the current regime, while others offer much lower standings, which are too conservative to implement real global climate change mitigation.

While having the potential to become a leader in the climate regime, given its position in the global carbon cycle and the technological and human resources available to decarbonization, Brazil has remained a conservative power. Brazilian position has been wavering: conservative from 1992 to 2005, moderate conservative between 2006 and 2010, conservative between 2011 and 2015, and moderate conservative in 2015, after presenting the Brazilian Intended Nationally Determined Contribution (INDC). INDCs provide detailed information regarding the commitment each country is willing to undertake in new climate deal, the Paris Agreement of 2015, including quantifiable information on the reference point and base year, time frames and periods for implementation, scope and coverage, planning processes, assumptions and methodological approaches for estimating and accounting GHG emissions and removals. This chapter argues that in order to understand Brazilian wavering positions in the climate regime and why the country is not – and probably will not become, in the near future – a climate leader, it is necessary to analyse the trajectory of Brazilian domestic policies and politics in the major Greenhouse Gases (GHG) emitting sectors.

The objective of the chapter is to demonstrate this proposition. In order to do so, this chapter is divided into four parts. The first analyses the evolution of commitments to reduce GHG emissions in the climate regime. The second focuses Brazilian power assets and its positions in the regime. The third, divided into two smaller parts, presents the profile and evolution of Brazilian GHG emissions, as well as the evolution of policy and politics in three main emitting sectors: Land Use, Land Use Change and Forestry (LULUCF), energy and agriculture. In the conclusion, it is explained why the Brazilian INDC is a reflection of current struggles to advance climate change mitigation in Brazil and why the current economic and political situation in the country indicates that Brazil is hardly going to become a climate leader in the near future.

Evolution of commitments to reduce GHG emissions in the climate regime, 1992–2015

Between the end of the 1980s and the beginning of the 1990s, climate change entered the international policy agenda. In 1988, the Intergovernmental Panel on Climate Change (IPCC) was established and, in 1990, it called for a global treaty on the issue. The United Nations Framework Convention on Climate Change (UNFCCC) was discussed and negotiated for two years and was opened for signature during the Earth Summit, in June 1992. By signing it, members agreed to pursue development paths that would decouple economic growth and GHG emissions. The UNFCCC entered into force in March 1994.

In 1997, the Kyoto Protocol was adopted. It is the first agreement on targets to reduce GHG emissions. Developed countries (Annex I of the Protocol) agreed on limited compulsory GHG reduction emission targets (7 per cent reduction in 20 years). Emerging economies rejected any commitment to reduce their curve of emissions growth (moving out from business as usual scenarios), so non-Annex I countries would propose voluntary reductions only. For this reason, and fearing what it considered an unfair competition in global markets, the United States withdrew from the Protocol in 2001 (Viola 2002). This was the first setback for the protocol: having 1990 GHG emissions as baseline, in 1997 mandatory commitments to cut emissions bound countries that were responsible for 65 per cent of the global GHG emissions; the amount was reduced to around 45 per cent when the United States left; in 2001 this share was further reduced to around 40 per cent (because of the dramatic increase of Chinese, Brazilian and Indian emissions); when the Protocol entered into force, in 2005, the share was around 30 per cent only.

In 2007, the Bali Road Map, the result of decisions to reach a secure climate future, was enacted. It includes the Bali Action Plan, a new negotiation process to address climate change in five tracks: shared vision regarding climate change, mitigation, adaptation, technology, and financing. These negotiations were to be conducted by the Ad Hoc Group on Long-term Cooperative Action under the Convention, and were supposed to be completed by 2009, when it was expected that UNFCCC members would reach an agreement on a New Global Climate Deal, to replace the Kyoto Protocol after 2012 – according to Annex I countries – and, a second period of the Kyoto Protocol – according to non-Annex I countries.

Conference of Parties (COP) 15, in Copenhagen, 2009, was initiated amidst a relatively optimistic atmosphere. For the first time since 1992, a great number of heads of state were present in a COP, although there were major divergences among countries on many issues, including the type of outcome (new treaty, second period of Kyoto) and the nature of commitments (according to developed countries mandatory for all, according to developing countries mandatory for developed countries and voluntary for themselves). The Copenhagen Accords, negotiated between the BASIC (Brazil, South Africa, India and China) countries and the United States, were last-minute agreements that members took note of – it was not officially adopted due to broad disagreements between UNFCCC members and opposition from some (Bolivia, Sudan and Venezuela), preventing consensus. Some goals, including long-term goal of limiting to 2°C from pre-industrial levels the maximum average temperature rise, were agreed; but the accords contain no measure to implement them. The same was true for measurement, reporting and verification of developing country actions. On the same occasion, industrialized countries pledged up to USD 30 billion in fast-start finance for the period 2010–2012.

In 2010 (COP 16, Cancun), expectations were lower but results were a little more significant. The Cancun Agreements are a set of decisions to address climate change comprehensively. Their main objectives cover climate change mitigation, transparency of actions, technology, financing, adaptation and forests. An extent package to help developing countries in dealing with the issue was also agreed. In the occasion, the mandate of the Ad Hoc Group on Long-term Cooperative Action was extended as well. Yet, no agreement regarding the text that would replace the Kyoto Protocol after 2012 was reached. Neither was it at COP 17, in Durban, in 2011. In the occasion, it was agreed that another legal instrument or outcome with legal force under the Convention, applicable to all UNFCCC members and to be implemented from 2020 would be completed by 2015. The outcome of Durban removed any expectation regarding a new treaty starting in 2013. In December 2011, immediately after Durban, Canada withdrew from the Protocol having accumulated emissions well beyond its 1997 commitments, without receiving any (even rhetorical) threat of sanctions – a clear evidence of Kyoto Protocol's toothlessness.

In 2012, the Kyoto Protocol faced its final setback: the period in which the protocol was to be in force came to an end, and Japan and Russia decided not to sign the Doha Amendment, which created a second commitment period, 2013 to 2019, for the Protocol. For this reason, countries – European Union, Switzerland, Norway and New Zealand – currently bound to reduce GHG emissions accounted for only 13 per cent of 2013's global emissions and 9 per cent of 2019's projected amount. In 2013, in Warsaw, UNFCCC members agreed that each member should present their INDC by 1 October 2015. INDCs should provide detailed information regarding the commitment each country would be willing to undertake in the New Global Climate Deal, including quantifiable information on the reference point and base year, time frames and periods for implementation, scope and coverage, planning processes, assumptions and methodological approaches for estimating and accounting GHG emissions and removals.

The Paris Agreement of December 2015 represents progress in general normative goals in the regime, but it is deficient in the correspondent means of implementation: countries' INDCs are voluntary; even if all pledges are successfully implemented, they are far away from reaching a 2 degrees Celsius increase in average Earth temperature, and even more if the target of 1.5 degrees Celsius is considered; quantitative global goals for 2050 and global decarbonization were eliminated from the Agreement; and nothing was stated about the elimination of subsidies to fossil fuels, which correspond to around 5 per cent of global GDP. The Paris Accord promotes some gradual decarbonization but it is far from the urgently needed deep de-carbonization. A brief analysis of key INDCs already indicated that results would not meet reduction in carbon emissions needed to avoid dangerous climate change. Only the European Union's pledge to reduce emissions is in tandem with what science defines as needed to avoid serious climate change. The United States, Canada and Japan pledges do not mean substantial reduction when translated into the 1990 baseline: the US and Canada chose 2005 as baseline, and Japan indicated 2013, a substantial retrogression from its Kyoto and Copenhagen commitment to 1990 as a baseline. China committed to peak its emissions by 2030, but Chinese emissions are expected to reach 35–40 per cent of total global emissions by then. India did not commit to peak emissions. Mexico promises reductions regarding a *business as usual* scenario. Russian commitments are poor considering that the baseline is 1990 and after that Russian emissions were reduced by more than 50 per cent due to economic collapse. Brazil pledged to cut 37 per cent of its GHG emissions until 2025, compared to 2005 levels, and 43 per cent until 2030, and a detailed analysis of its content is provided below.

Brazilian climate power assets and evolution of Brazilian positions in the climate regime

Climate change cannot be mitigated if climate powers do no engage with the cause. From all UNFCCC members, relatively few answer, together, for the largest share of GHG emissions, as well as for the technological and human capital needed to mitigate climate change – they are climate powers (Viola et al. 2012). Climate powers could be divided into three broad categories, according to their ability to influence the international climate regime: climate superpowers – China, United States and European Union; climate great powers – India, Russia, Japan, South Korea and Brazil; and climate medium powers – among these, the most relevant are Indonesia, Canada, Mexico, Australia, Iran, Saudi Arabia, South Africa and Turkey (Viola et al. 2012).

Climate powers standings in the climate regime can be classified as reformist or conservative, according to the forces that prevail in influencing their position in the global carbon cycle and mindset regarding climate change. When the country is mostly in favour of measures that really push climate change mitigation forward, thus reforming the current regime, it will be classified as reformist; when it opposes such measures, it will be classified as conservative. Of course this is a broad classification, which takes into account the prevailing standings of a country regarding climate change. Countries are complex entities composed of governments, businesses, scientific communities and civil society, which might not share the prevailing standings.

Brazil is a big country: it has a territory of 8,514,215 km², the fifth greatest in the world, and 205 million inhabitants. It is among the ten greatest world economies, but is now facing a severe economic crisis, instigated by economic policy decisions from 2009 to 2014 (artificial inflation control, major mistakes in allocation of public spending, corruption in Petrobras and in other state-owned corporations): economic growth was average 3.6 per cent a year between 2003 and 2012, compared to 2.8 per cent during the previous decade (World Bank 2014); economic stagnation was the picture in 2014; an estimated 3.8 per cent contraction is expected in 2015, and an estimated 3 per cent to 4 per cent contraction in 2016. Public debt has grown from 52 per cent of GDP in 2013 to 65 per cent in 2015, and it is estimated to reach 75 per cent in 2017. It is expected that Brazilian per capita income of 2013 will be recovered only around 2022. Brazil has considerably reduced inequality in the period 2003–2013, but, since 2014, inequality has been rising again. Brazil has limited military power, despite its increased role in peace-keeping missions, but has important soft power assets (Nye 2011): it is a pacific, multi-ethnic, multi-cultural and multi-religious country, with magnificent nature and extraordinary tourism potential.

The country is very vulnerable to climate change, and extreme weather events have intensified in recent years: severe droughts became common, especially during the winter in central regions; drier weather patterns have affected local river flows, with important effects on the generation of electricity by hydropower; rainfalls have become much more intense in the south of the country, and has dislocated an important share of the population. It is important to remember that Brazil is among the greatest producers and exporters of agriculture commodities (FAO 2013), so changes in climate patterns that affect the performance of crops will have some impact in international food markets.

Brazil is a great climate power. It is among the top 10 GHG emitters (see Table 9.1). It is currently the eleventh greatest energy producer and seventh greatest energy consumer (World Bank 2014). The country is a great producer of hydroelectricity and has increased its production of oil, but imports of diesel and gasoline have also increased, with impacts to Brazil's balance of payments and several setbacks to the ethanol production chain (see below). Brazil also imports natural gas from Bolivia. Brazil has an important role to play in energy technology: it developed worldly known technology for exploring deep offshore oil reserves, and it is also world leader in hydropower technology (especially in designing reservoirs) and in producing electricity and fuel from sugar cane. Regarding other renewable energy technology,[1] it occupies the fifth place among the producers (WWF and Roland Berger 2012; The Pew Charitable Trusts 2012).

Brazil is an important player in the climate change regime, and Brazilian international positions in climate negotiations reflect the advances and setbacks in domestic policies.

Until 2005, Brazil remained very conservative: (i) it strongly opposed commitments for developing countries to reduce their carbon emissions growth from business as usual scenarios; (ii) it promoted an extremely distorted interpretation of the principle of common and differentiated responsibilities – almost everything was differentiated (Viola 2002); (iii) it strongly opposed the introduction of avoided deforestation in the Clean Development Mechanism;[2] and (iv) it formed alliances with highly carbonized energy matrix countries. Brazil's inability to deter deforestation of the Amazon at the time and the fear of being internationally charged for it were the main reasons for the conservative position.

From 2006 to 2010, Brazil moved gradually towards reformism. This move was largely driven by successful implementation of domestic policies to tackle deforestation, especially in the Amazon region. In the international climate regime, Brazil accepted measures it had previously opposed (such as international forest regulation), and, in 2009, at COP 15, the country pledged a voluntary 36.1 per cent to 38.9 per cent reduction of the curve of emissions growth in 2020 emissions. The 2009 pledge was an advance compared to previous reluctance to accept targets, but a poor commitment considering that the baseline was an inflated business as usual emissions scenario. In the period 2009–2010, it also started to abandon the discourse that Brazil was a poor country that needed international assistance to tackle climate change.

In 2011, due to changes in domestic policy and politics regarding especially the energy sector, but also LULUCF and agriculture, Brazil went back to the traditional conservative standing. It joined other countries in requesting developed countries to commit to a second period of compulsory emissions reduction targets before emerging economies accepted binding targets. In 2012, Brazil set the tone of the Earth Summit (Rio+20) by lobbying to exclude climate change from the negotiations and promoting a diffuse definition of green economy, against a more consistent one defended by the European Union. In 2013, at COP 19, Brazil reinstated the doctrine of historical emissions – in order to define carbon rights, emissions should be measured since 1850. This doctrine had never been accepted by developed countries and has been strongly criticized by most scientists and analysts; it had been abandoned by Brazil in 2009. In 2014, at COP 20, it presented a proposal of concentric differentiation of countries, in which emerging economies depend, to adopt binding emissions reduction targets, exclusively on domestic political decisions.

In late September 2015, Brazil went back to a moderate conservative standing. In its INDC, Brazil pledged to reduce GHG emissions by 37 per cent until 2025 and 43 per cent until 2030, compared to 2005 levels (Brazil 2015). It is an important advance, as, for the first time, Brazil presented an absolute GHG reduction target compared to a base year, instead of business as usual scenarios. The Brazilian pledge is also more ambitious than pledges of almost all major GHG emitters (US, Japan, South Korea, China, India, Russia, Canada, Mexico, Australia, Indonesia, South Africa and Turkey). Yet, the INDC is not enough to be considered a step towards reformism.

Compared to the Brazilian pledge in 2009, the INDC is an advance because Brazil is now committed with an absolute goal, setting 2005 emissions as baseline and abandoning the deviation from the business as usual curve. However, 2005 is a poor baseline because Brazilian GHG emissions skyrocketed in 2005, due to peak deforestation (LULUCF answered for 70 per cent of total Brazilian GHG emissions in 2005) – in fact, Brazilian per capita emissions in 2005 were 12 tCO$_2$e, higher than European, Russian and Japanese per capita averages and two to four times higher than Chinese or Indian.

In addition, pledges for some specific sectors fall short of ambitiousness. Regarding the energy sector: (i) increasing the share of sustainable biofuels to approximately 18 per cent by 2030 is not challenging when bioenergy already answered for 17.6 per cent in the transport sector in 2014 (EPE 2015a, 25); (ii) achieving 45 per cent of renewables in the energy mix by 2030 is also not challenging, as their share has been average 43.64 per cent between 2004 and 2014 and average 45.32 per cent between 2004 and 2009 (EPE 2015b, 24), having decreased due to policy issues reported in section 3.2, below; (iii) increasing renewable energy sources other than hydro in the total energy mix between 28 per cent and 33 per cent by 2030 when they already accounted for 27.9 per cent in 2014 (EPE 2015a, 20; EPE 2015b, 24) is also conservative. Regarding LULUCF, pledging to reach zero deforestation by 2030 is a surprisingly negative target, as it signals that an activity that is mostly illegal and finds no support in Brazilian society will only be fully enforced in 15 years. The latter is, by far, the greatest setback in Brazilian discourse regarding climate change mitigation to date.

Finally, an analysis of policies that have been implemented by the federal government, as will be detailed below, show that they do not point towards decarbonization: subsidies to fossil fuels and large investments in oil production; substantial increase of the number and the production of fossil fuel thermal power plants; little incentives to biofuels; substantial credit to conventional agriculture and little amounts to low carbon agriculture. Observatorio do Clima, a think tank, had published another proposal that could have been based Brazilian INDC to make it truly reformist: Brazil could reduce its emissions in 25 per cent by 2030 compared to 1990 levels, and the reduction would be pushed by consistent promotion of biofuels, solar and wind electricity, smart grid and low carbon agriculture (Observatorio do Clima 2015a).

Brazilian climate policies and politics: advances and setbacks

Evolution of Brazilian GHG emissions

Brazil is among the top 10 GHG world emitters. Between 1990 and 2012, it has descended from fifth to seventh position not due to the decrease of its absolute emissions, but due to the increase in world and other countries shares, which changed Brazil's relative position among the top 10. In fact, between 1990 and 2012 Brazilian emissions have increased 13.48 per cent considering all emissions and 79.18 per cent if LULUCF emissions are excluded.[3] In 1990, the difference between Brazilian emissions including and excluding LULUCF

Table 9.1 Top 10 GHG emitters in 2012 and their emissions in 1990 (MtCO$_2$e)

	1990 including LULUCF	1990 excluding LULUCF	2012 including LULUCF	2012 excluding LULUCF	δ 1990–2012 including LULUCF	δ 1990–2012 excluding LULUCF
China	3,218.45	3,320.97	10,684.29	10,975.50	231.97%	230.49%
United States	5,743.98	5,936.93	5,822.87	6,235.10	1.37%	5.02%
European Union–28	5,138.52	5,235.35	4,122.64	4,399.15	-19.77%	-15.97%
India	1,212.02	1,239.06	2,887.08	3,013.77	140.18%	143.23%
Russia	2,776.78	2,776.78	2,254.47	2,322.22	-18.81%	-16.37%
Indonesia	1,025.74	392.22	1,981.00	760.81	93.12%	93.97%
Brazil	1,606.59	565.09	1,823.15	1,012.55	13.48%	79.18%
Japan	1,116.38	1,190.22	1,207.30	1,344.58	8.14%	12.96%
Canada	681.93	568.79	856.28	714.12	25.56%	25.55%
Mexico	477.07	435.27	748.91	723.85	56.98%	66.30%
World	33,937.21	30,423.75	47,598.55	44,815.54	40.25%	47.30%

Source: World Resources Institute, *CAIT – Climate Data Explorer* (database). Available at <http://cait.wri.org/>, accessed 15 September 2015.

was 184.30 per cent, the largest among the top 10 emitters; in 2012, it was 80.05 per cent, the second largest after Indonesia's (160.38 per cent).[4]

Brazilian emissions have traditionally been driven by LULUCF, more specifically deforestation. In 1990, their share in total Brazilian CO_2 emissions was almost 68 per cent (see Table 9.3), and 77 per cent in 1995, when deforestation in the Amazon region peaked. This share has decreased from 2004 to 2012; recently – 2013 and 2014, and probably 2015 as well – deforestation has been in an upward trend compared to 2012 numbers (see Table 9.5). Compared to 1990 levels, however, Brazil has managed to decrease its CO_2 emissions from LULUCF by 56.49 per cent in 2013 (see Table 9.4). CO_2 emissions from LULUCF could restore its downward trend if some policy adjustments were made (see below); they would, then, be overtaken by emissions from energy and agriculture, which would become the main drivers of Brazilian emissions.

CO_2 emissions from energy have been consistently rising in Brazil. Between 1990 and 2013, they more than doubled (see Table 9.4). After 2005, when deforestation has started to be successfully tackled, the share of energy in Brazilian CO_2 emissions rose to around a quarter of the total (Table 9.3), and recently it has approached 30 per cent (Table 9.3). Between 2010 and 2013, CO_2 emissions from energy have risen 22.58 per cent, an amount previously compared only to the increase that took place between 1995 and 2000 (23 per cent).[5] When net CO_2 emissions are considered (accounting carbon capture and sequestration from reforestation and afforestation), energy answered for the largest share in total Brazilian CO_2 emissions in 2013, 39 per cent (Observatorio do Clima 2015b, 09).

Agriculture and cattle grazing emissions are rising as well. Between 1990 and 2013, their total amount increased 45.67 per cent (Table 9.4). Agriculture and cattle grazing have traditionally occupied the second position among Brazilian CO_2 emissions drivers – although its shares have been much lower than LULUCF's. In 2013, they were surpassed by energy, though: agriculture and cattle grazing generated 26.84 per cent of total Brazilian CO_2 emissions (Table 9.3) and 36 per cent of total net emissions (Observatorio do Clima 2015b, 09).

In 2013, industrial processes and waste had, jointly, less than 10 per cent of total Brazilian CO_2 emissions (Table 9.3). Before deforestation has started to be successfully tackled, their share was around 5 per cent or less. Nevertheless, emissions from both sectors are increasing: between 1990 and 2013, they almost doubled in the case of industrial processes and rose almost 70 per cent in the case of waste (Table 9.4). And they tend to keep increasing.

Brazilian policies and politics, 1990–2015

LULUCF: tackling deforestation

Deforestation, especially in the Amazon forest and Cerrado Savannah, has been the largest source of Brazilian GHG emissions. Deforestation has driven the

Table 9.2 Brazilian CO_2 emissions 1990-2013, total and per end sector (t of CO_2)

	1990	1995	2000	2005	2010	2013
LULUCF	1,246,826,158	2,203,818,031	1,457,940,462	1,506,174,138	598,966,415	542,467,957
Energy	220,842,015	241,128,263	296,746,885	312,086,341	366,514,702	449,278,127
Agriculture and cattle grazing	286,975,574	316,514,619	327,973,366	392,044,726	406,454,587	418,040,777
Industrial processes	51,496,756	65,203,752	76,412,495	83,309,599	94,629,265	99,282,988
Waste	28,951,842	33,421,205	38,209,466	41,229,018	48,770,448	48,738,582
Total	1,835,092,345	2,860,085,870	2,197,282,674	2,334,843,822	1,515,335,417	1,557,808,431

Source: Own elaboration, based on data from *Observatorio do Clima, Sistema de Estimativa de Emissão de Gases de Efeito Estufa*, available at <www.seeg.eco.br/tabela-geral-de-emissoes/>, accessed 14 September 2015.

Table 9.3 Shares of end sectors CO_2 emissions: 1990–2013 (%)

End-use sector	1990	1995	2000	2005	2010	2013
LULUCF	67.94	77.05	66.35	64.51	39.53	34.82
Energy	12.03	8.43	13.51	13.37	24.19	28.84
Agriculture and cattle grazing	15.64	11.07	14.93	16.79	26.82	26.84
Industrial processes	2.81	2.28	3.48	3.57	6.24	6.37
Waste	1.58	1.17	1.74	1.77	3.22	3.13
Total	100	100	100	100	100	100

Source: Own elaboration, based on data from *Observatorio do Clima, Sistema de Estimativa de Emissão de Gases de Efeito Estufa*, available at <www.seeg.eco.br/tabela-geral-de-emissoes/>, accessed 14 September 2015.

Table 9.4 Variation in Brazilian CO_2 emissions, total and per end sector (%)

	δ 1990–2013	δ 1995–2013	δ 2000–2013	δ 2005–2013	δ 2010–2013
LULUCF	-56.49	-75.38	-62.79	-63.98	-9.43
Energy	103.44	86.32	51.40	43.96	22.58
Agriculture and cattle grazing	45.67	32.07	27.46	6.63	2.85
Industrial processes	92.79	52.26	29.93	19.17	4.92
Waste	68.34	45.83	27.56	18.21	-0.07
Total	-15.11	-45.53	-29.10	-33.28	2.80

Source: Own elaboration, based on data from *Observatorio do Clima, Sistema de Estimativa de Emissão de Gases de Efeito Estufa*, available at <www.seeg.eco.br/tabela-geral-de-emissoes/>, accessed 14 September 2015.

colonization around the Amazon forest: during several decades, the exploitation of tropical timber was a profitable activity. Until the middle of the twentieth century, when settlements were mostly spontaneous or a strategy to defend the country's borders from foreign invasion, deforestation was not a concern. The situation changed in the 1970s when the military governments, besides promoting the occupation of the region in order keep the territory securely under national sovereignty, encouraged both migration to the area, so to avoid land reform in other highly populated parts of the country, as well as the use of the land to produce commodities, in order to improve Brazilian balance of payments. Official policy included the construction of highways (Arima et al. 2015), and fiscal and financial incentives to clear the forest for crops and livestock, and other subsidies (Carvalho 2012). In the second half of the 1980s, when debates on sustainable development came onto the international agenda and Amazon deforestation was under the spotlight, the first national measures to tackle it were debated in government but very poorly implemented.

Poor countries have typically answered for GHG emissions from deforestation. Between 1992 and 2005, however, Brazil, a middle income economy, led them: in fact, Brazilian emissions from deforestation answered for yearly 2 per cent to 5 per cent of total CO_2 increases in the atmosphere. In addition, timber exploitation and cattle grazing were the main drivers of those emissions, meaning that they were helping aggravate climate change while not contributing to Brazilian development – emissions from China, e.g., driven mainly by the increase in the numbers and production of thermal power plants, were also helping aggravate climate change, but at least were generating energy to produce goods and jobs for the Chinese population.

It was only in 2006 that deforestation started to be reduced. Deforestation in the Amazon decreased from annual 19,000 km^2 in 2005 to annual 4,500 km^2 in 2012 (INPE 2015). The cutback was due to legal and institutional changes: political priority was given to the issue; law enforcement and institutional capacity was enhanced; coalitions by multi-stakeholders against the consumption of soy beans and beef produced in deforested areas were formed; the influence of NGOs and the scientific community on the media increased; new and extensive national parks and conservation units were created; and cooperation between state and national governments was boosted (Viola and Franchini 2013). Deforestation rates were the lowest in 2009, the year in which outcomes from specific policies were coupled with the effects of the financial crisis, which decreased agribusiness commodities prices, reducing the incentives to deforest. 2009's numbers would be repeated again only in 2012.

2009 brought other changes. Following the greater attention given to climate change by the media and the public due to expectations of good results at COP 15, the federal government was pressured by the Amazon-region state governments to change its international position regarding forests: they demanded that the country accept the inclusion of REDD+[6] into the Clean Development Mechanism or any other market mechanism. This and other requests from corporate coalitions – to accept market mechanisms for avoided deforestation, to decelerate emissions, and to commit to emissions reduction by 2020 – helped to shape the Brazilian pledge made in Copenhagen, i.e., a voluntary commitment to reduce the expected 2020 (according to a business-as-usual scenario) GHG emissions by 36–39 per cent. The pledge was a substantial change from the previous unwillingness of the Brazilian government to accept emissions targets. It was incorporated into the National Climate Change Policy. In the same legal piece, the Plan of Action to Prevent and Control Deforestation of the Legal Amazon pushes further deforestation policies. Despite remaining enforcement issues, the results of the plan have been significant, especially when compared to the measures envisaged for other sectors/regions, which have been much less successfully implemented.

Deforestation rates have increased since 2013, though. According to official data, from January/December 2013, deforestation rates were 28.87 per cent higher than from January/December 2012; from January/December 2014, deforestation rates decreased 15 per cent compared to January/December 2013, but were still 9.65 per cent higher than from January/December 2012; from January/December 2015, rates were 16.34 per cent higher than from January/December 2014, but 27.56 per cent higher than January/December 2012 (Table 9.5).

These numbers were driven by deforestation in private areas or areas under many stages of ownership – deforestation to increase the size of properties or accelerate land ownership: 55 per cent of the deforestation between August 2012 and July 2013, 57 per cent between August 2013 and July 2014 and 63 per cent between August 2014 and July 2015 were due to this cause (Martins et al. 2013; Victor et al. 2014; Imazon 2015). The new Forest Code, enacted in 2012, which grants amnesty to deforestation instead of pushing for the restoration and use of degraded areas in agribusiness (responding to short-term private interests to enlarge the size of grazing and cropping areas) has definitely played a role in the increase.

In order to tackle this new wave of deforestation, several measures could be undertaken: developing a better, more complete and transparent registration of rural properties in forest areas; fighting deforestation related to land speculation; improving monitoring; correctly collecting taxes over rural properties; creating economic incentives to forest conservation; improving agreements with the private sector against deforestation; and removing barriers to forest investment (Observatorio do Clima 2015c, 19–22).

Energy: less hydroelectricity and ethanol, and negligible energy efficiency

In comparison to the world average, Brazil has a lower carbon energy matrix. While fossil fuels – oil, coal and natural gas – answered for 81 per cent of world TPES (Total Primary Energy Supply: the total amount of energy available for use in the country, including domestic production and imports) in 2003 and for 81.7 per cent in 2012 (IEA 2005; IEA 2014), in Brazil they answered for 54.27 per cent of TPES in 2003 and 57.52 per cent in 2014

Table 9.5 Annual deforestation in Brazilian Legal Amazon (km²/year)

State	1990	1995	2000	2005	2010	2012	2013	2014	2015*	Σ 1990–2015
Acre	550	1,208	547	592	259	305	221	309	279	12,173
Amazonas	520	2,114	612	775	595	523	583	500	769	19,730
Amapá	250	9	–	33	53	27	23	31	13	1,316
Maranhão	1,100	1,745	1,065	922	712	269	403	257	217	20,542
Mato Grosso	4,020	10,391	6,369	7,145	871	757	1,139	1,075	1,508	128,724
Pará	4,890	7,845	6,671	5,899	3,770	1,741	2,346	1,887	1,881	127,122
Rondônia	1,670	4,730	2,465	3,244	435	773	932	684	963	52,648
Roraima	150	220	253	133	256	124	170	219	148	6,250
Tocantins	580	797	244	271	49	52	74	50	53	6,180
Total	13,730	29,059	18,226	19,014	7,000	4,571	5,891	5,012	5,831	374,686

(*) Estimated.

Source: INPE, PRODES Project (database). Available at <http://www.obt.inpe.br/prodes/prodes_1988_2015.htm>, accessed 11 December 2015.

(Table 9.6). Yet, their share in the Brazilian energy matrix, compared with low carbon energy sources share, is increasing. From 2003 to 2014, Brazilian energy production increased 48.37 per cent and TPES increased 43.29 per cent; however, during the same period, energy production and TPES from non-renewable sources increased, respectively, 56.04 per cent and 48.52 per cent, and the ones from renewable sources increased, respectively, 39.71 per cent and 36.61 per cent. This data shows that non-renewable energy sources answered for a greater share of the energy production and supply in Brazil during the 2003–2014 decade.

Low carbon energy sources, led by hydropower, are the main source of Brazilian electricity, but their share has also been declining. In January 2009, 96.01 per cent of total Brazilian electricity in the Interconnected National System (in Portuguese, SIN) was generated by low carbon energy sources (hydro, nuclear, solar, wind and biomass) (MME 2009a, 20); in July 2015, however, the amount was 78.83 per cent (MME 2015, 15). Brazil has some of the world's greatest hydropower potentials, but 70 per cent of its remaining potential is located on Amazonian basins (Eletrobras 2012), a region with complex ecosystems and great importance for maintaining the continent's climate and biodiversity. For this reason, run-of-the-river hydropower plants have been constructed in the region – Jirau and Santo Antonio in the Madeira River, and Belo Monte in the Xingu River. On the one hand, run-of-the-river hydropower plants produce electricity according to the natural river flows, allowing smaller dams to be built and reducing environmental impacts. On the other hand, because electricity demand is constant, run-of-the-river hydropower plants are less efficient from an economic point of view and must have back-up systems – so their total environmental impact depends on the chosen back-up system. Sadly, in recent years, this role has been played by fossil fuel thermal power plants.

Electricity generation by fossil fuels has been increasing. When two periods are compared – January 2006 until December 2010 as period A, January 2011 until August 2015 as period B – the amount of GWh generated at SIN by conventional thermic power plants increased 192.41 per cent – from total 129,091.14 GWh in Period A to 377,479.23 GWh in Period B (ONS 2015). The number of fossil fuel power plants as well as their share in total power plants operating in Brazil have also been increasing: in July 2009, from a total of 2,080 power plants in operation in Brazil, 914 generated electricity using either coal, oil or natural gas (43.94 per cent of total); in July 2015, they were 2,200 from a total of 4,250 power plants (51.76 per cent of total) (MME 2009b; MME 2015). These numbers represent a 140.70 per cent increase in the number of fossil fuel power plants in the period, and a 17.79 per cent increase of their share in total power plants in operation in the country in the period (MME 2009b; MME 2015).

New renewable energy sources – wind, biomass and solar – are in the Brazilian energy matrix. In 2002, the federal government enacted the Alternative Energy Sources Incentive Program (in Portuguese, PROINFA).

Table 9.6: Domestic energy production and TPES, by source: 2003–2014 (10^3 tep (toe))

Source	Energy production 2003	% from total	Energy production 2014	% from total	δ 2003–2014	TPES 2003	% from total	TPES 2014	% from total	δ 2003–2014
Oil	77,225	42.03%	116,705	42.80%	51.12%	80,688	40.16%	108,012	37.52%	33.86%
Gas	15,681	8.53%	31,661	11.61%	101.90%	15,512	7.72%	41,343	14.36%	166.52%
Coal	1,823	0.99%	3,059	1.12%	67.80%	12,848	6.39%	16,236	5.64%	26.37%
Uranium	2,745	1.49%	681	0.25%	-75.19%	3,621	1.80%	1,747	0.60%	-51.75%
Total	97,474	53.05%	152,106	55.79%	56.04%	112,669	56.09%	167,338	58.13%	48.52%
Hydro	26,283	14.30%	32,116	11.78%	22.19%	29,477	14.67%	32,116	11.15%	8.95%
Wood	25,965	14.13%	24,728	9.70%	-4.76%	25,973	12.93%	24,728	8.59%	-4.79%
Sugarcane	28,357	15.43%	49,232	18.05%	73.61%	27,093	13.48%	49,232	17.10%	81.71%
Other	5,663	3.08%	14,451	5.30%	155.18%	5,663	2.82%	14,427	5.01%	154.75%
Total	86,268	46.95%	120,527	44.20%	39.71%	88,206	43.91%	120,503	41.86%	36.61%
Total	183,742		272,633		48.37%	200,875		287,841		43.29%

Energy production = energy production in Brazil; TPES = total primary energy supply (domestic energy production + energy imports − energy exports +/− energy stock changes).

Source: Own elaboration, using data from: *Balanço Energético Nacional Séries Completas 1970–2013*, available at <https://ben.epe.gov.br/BENSeriesCompletas.aspx> e *Matriz energética nacional 2014*, available at <https://ben.epe.gov.br/BENRelatorioSintese.aspx?anoColeta=2015&anoFimColeta=2014>, both accessed 07 July 2015.

PROINFA mandated that 10 per cent of Brazil's 2020 energy supply should be produced by new renewable sources, meaning small hydropower plants (with installed capacity between 1 and 30 MW and reservoirs smaller than 3 km²), wind and biomass – solar was not contemplated at this point due to concerns over prices.[7] After 2004, these energy sources took place in federal government's auctions for electricity: between 2004 and 2011, small hydropower plants sold around 1,800 MW; wind sold 5,399.5 MW; and biomass was responsible for around 2,500 MW (Nogueira and Costa 2012). Despite their increased participation in the Brazilian electricity market, new renewable electricity production still faces many challenges: (i) their share in the electricity production is small; (ii) they face difficulties to compete with traditional sources due to the minimum cost criteria employed in the auctions; (iii) they are opposed by an alliance of large hydropower technicians, Eletrobras (the state-owned corporation for operating hydropower plants and electricity transmission), bureaucratic circles and large private corporations who construct dams; and (iv) there is no official policy that establishes long-term incentives, quotas and minimum prices for alternative energy sources, or legal obligation to auction alternative energy production periodically (Viola and Basso 2015).

Nuclear is an important low carbon energy source for a large part of the world, and could most successfully applied as a transitional source: since it is highly effective in generating electricity, it can substitute fossil fuels until technology for renewable sources is developed (Carvalho 2012). Brazil has two nuclear power plants and a third under construction, but public opinion does not favour nuclear energy: nuclear has the strongest opposition in the country – even most of the academic community is against it. The lobby in favour of nuclear power, headed by nuclear engineers, military personnel, and some diplomats, is losing power because most of them are retiring. Corruption scandals around the construction of Angra 3, involving the president of Electronuclear (state-owned corporation that run the nuclear plants), a retired Admiral from the Navy who was imprisoned, have made matters worse. Nuclear will never be an important source of energy in Brazil, although it will remain in place due to technological know-how, security reasons and the fact that the country has huge uranium reserves.

Oil remains the main source of fuel in Brazil. Biodiesel production is still in its infancy. The National Biodiesel Programme, created in 2005, established a minimum of 8–10 per cent biodiesel in the diesel mix. Civil society, led by entities of the health sector, has been pressing to increase this percentage, especially in metropolitan areas, in order to reduce health issues related to air pollution. They are supported by the Brazilian Union for Biodiesel and Biokerosene (in Portuguese, UBRABIO), but opposed by the oil sector. Brazil is a traditional producer of ethanol from sugarcane. This production was propelled by both scarcity and high prices of oil during the crises of the 1970s. In the 1990s, a severe crisis of supply reduced the use of ethanol. It came back to the spotlight in 2003, after the development of the flexible fuel technology for automobiles.

Competition between oil/derivatives and ethanol has not been fair. Until 2006, the domestic prices of oil/derivatives in Brazil followed international ones. After that, however, due to the discovery of the deep offshore reserves – and the illusion that Brazil would rapidly become a great producer and exporter of oil – the federal government started subsidizing oil prices to maintain high economic growth rates, changing the relative prices of gasoline/ethanol and undermining the competitiveness of the latter. In addition, tax exemptions were given to the automobile industry, and led to dramatic increase in fuel demand while ethanol prices were still not competitive. These heterodox policies enhanced short-term economic growth but increased long-term macroeconomic imbalance, and penalised both Petrobras – which faced huge financial losses – and the ethanol production chain (Viola and Basso 2015).

Expectations about the deep offshore oil reserves were high until 2010. Most environmentalists were silent about 'popular pre-salt', fearing they could become isolated: there was no strong environmental movement to defend low carbon energy sources instead of exploiting the new oil reserves. Nevertheless, it was a victory without a battle: while it was expected that reserves exploitation would start soon after the discovery, a dispute among Brazilian state governors concerning the future share of pre-salt exploration revenue, ended up postponing the reserves auction schedule by a couple of years. In this period, the economic output of the technology (dramatically improved in the second half of the 2000s) to exploit shale gas was achieved in the US – supposedly the largest market for the Brazilian deep offshore oil – and Mexico got close to end PEMEX monopoly over oil, facts that appealed to American companies such as Chevron and Exxon Mobile. Petrobras, whose participation in the auction was imperative due to legal requirements, lost the financial capacity to act due to the continued losses it faced by subsidising gasoline and incompetence in closing business deals (due to increasing levels of corruption). In 2015, enthusiasm for the deep offshore oil reached its lowest point and there were changes in relative prices of gasoline/ethanol. When governmental interference in domestic oil prices was reduced, also in 2015, ethanol regained competitiveness against gasoline. It is necessary to wait and see if the situation will consolidate before making any predictions regarding future increase in investments in ethanol production.

In fact, it is uncertain how ethanol production will develop in the next few years. At the same time that an important share of ethanol producers closed their business in recent years, research regarding ethanol production and improvements in technology are still very much alive and advancing. Brazilian public opinion favours the biofuel, but few groups are mobilized or understand the bigger picture – a clear example is the positive connotation federal government subsidies to gasoline had to the average Brazilian population. Other important obstacles to greater ethanol production are the influence of oil and gas groups in the government and the misunderstanding that it implies a further commoditisation of Brazilian economy. In the international market, the competition with ethanol from corn, produced by the US, and a possible ethanol from cellulose, developed by the US and Europe, undermines Brazilian ethanol's possibilities to be vastly exported. If there is no change in the next few years to create strategic reserves to avoid price volatility and reliability of supply or to invest in the production of pure ethanol vehicles, it is likely that ethanol will remain a fuel of second importance in Brazil. Another possibility is a major progress of second-generation ethanol from energy cane (a hybrid variety derived from sugar cane) produced for some cutting-edge corporations like Greenbio.

Despite its obvious contribution to reducing energy use, therefore GHG emissions, energy efficiency is not a political priority in Brazil. Yet, except during the 2001 electricity supply crisis, when Brazilian society was capable of rapid and efficient response to electricity scarcity, energy efficiency is not a priority in the country: by comparing 2003 and 2014 total energy supply and total energy use by end-use sectors, it can be seen that energy losses increased 19.40 per cent in the period (Table 9.7).

The topic of energy efficiency has been on the agenda since the 1980s. Brazil has developed a regulatory framework and mandatory programmes of energy efficiency: the national electrical energy conservation programme (in Portuguese, PROCEL) aims at reducing the consumption of electricity; the national programme for the rational use of oil and natural gas by-products (in Portuguese, CONPET) targets use of oil and derivatives; the Brazilian labelling programme (in Portuguese, PBE) classifies house appliances, devices and light utility vehicles according to their energy use. Law nr. 9991/2000 mandates electricity companies to minimum investment in energy efficiency Research & Development. Yet, public debate about energy efficiency and response to the initiatives were minimal.

The 2001 electricity supply crisis was an exception. Electricity was rationed, and surprisingly, Brazilian society was capable of rapid and efficient response to electricity scarcity. Following that, National Policy for Conservation and Rational Use of Energy (Law nr. 10295/2001) was enacted, and created a cross-Ministries committee to manage and enhance initiatives related to energy efficiency. Labelling became much more stringent; industries were obliged to increase energy efficiency so as not to face economic losses. After the crisis, though, energy waste resumed its high levels.

Electricity transmission also performs very poorly. Brazilian electricity is mostly transmitted through the SIN. An integrated transmission system has

Table 9.7 Brazilian energy efficiency, 2003-2014 (10^3 tep (toe))

TPES 2003	End use 2003	Losses	TPES 2014	End use 2014	Losses	δ losses 2003–2014
200,875	169,073	31,802	287,841	249,869	37,972	19.40%

Source: Own elaboration, using data from: *Balanço Energético Nacional Séries Completas 1970-2013*, available at ,<https://ben.epe.gov.br/BENSeriesCompletas.aspx> e *Matriz energética nacional 2014*, available at <https://ben.epe.gov.br/BENRelatorioSintese.aspx?anoColeta=2015&anoFimColeta=2014>, both accessed 07 July 2015.

supply security advantages, as long as the transmission lines are quality built and well maintained; unfortunately, the criteria in Brazil is short-term costs, not reliability and efficiency: lines have inferior quality and little maintenance – even cheap electronic leak detectors are not yet in place in the country. In addition, expansion of electricity supply and grid improvements are not coupled: for instance, UHE Belo Monte connection to the SIN will be concluded two years after it starts producing electricity; several wind farms are disconnected as well. The lack of a smart grid prevents consumers becoming energy 'prosumers' – by injecting electricity produced by solar panels and wind farms into the grid – and the intelligent use of renewable energy potential according to the characteristics of the Brazilian regions, hindering the joint exploitation of complementary renewable energy sources (solar and wind with hydropower).

The Brazilian Association for Energy Efficiency (in Portuguese, ABEE) and other NGOs are in favour of energy efficiency and smart grid in Brazil, but they are not strong enough to change the picture. Government propaganda is always based on quantity, not quality; recent policies – reducing prices for electricity, based on a distorted calculation that benefited companies that do not invest in productivity and energy efficiency – were clearly energy populist measures targeting the 2014 elections. Energy efficiency requires massive investments in long-term quality and reliability, which will never be achieved through increasing economic interventionism and reduction of macroeconomic predictability.

Inefficiency is also a concern in the automotive industry. Since the 1970s, the programme to control air pollution from automobiles (in Portuguese, PROCONVE) has been in force, and since 2012 the programme has encouraged technological innovation and densification of the productive chain of automobiles (in Portuguese, INOVAR-AUTO). Yet, in general the automotive industry accepts only vague energy efficiency labelling: Brazilian branches of European and US companies lobby against strict energy efficiency labelling and sell in Brazil out-dated and inefficient models no longer commercialized in their home countries.[8] Most consumers still take into account only short-term costs when purchasing a car. Even the flex-fuel technology was not developed due to pure environmental concerns, but as a means to employ a boosted ethanol production.

Agriculture and cattle grazing: an emerging low carbon agribusiness?

Agriculture and cattle grazing are in the third place, after LULUCF and energy, among the Brazilian GHG emissions drivers. The sector answered for 36 per cent of total net Brazilian GHG emissions in 2013 (Observatorio do Clima 2015b, 09), and its emissions increased 45.67 per cent between 1990 and 2013 (Table 9.4). Brazilian agribusiness emissions are directly related to demand for agricultural commodities in the international market, where Brazil is a major exporter. The global demand for agriculture commodities, especially soy, corn and beef. These items have been rising in recent decades, driven by China, and growth is expected to remain high in the following decades, due to increasing human population.

Eighty-four per cent of total GHG emissions of the Brazilian agribusiness sector are due to beef production (Observatorio do Clima 2015d). In the process of enteric fermentation, ruminants expel an important amount of GHG in the atmosphere, especially CH_4, methane, a gas of extremely high capacity to influence climate change. The digestive process of animals is not the only source of emissions from livestock farming: the decomposition of animal manure – and in this case, swine and chicken production are also causes of concern, as Brazil is a great exporter of both – and their use as fertilizers are also relevant. To date, there is not much to do about the methane expelled by ruminants in their digestive process, even if some research is being done regarding how changes in their diet could reduce emissions (e.g., FAO 2013) – this is the reason why reducing the consumption of beef is considered climate friendly, although it is not the object of this chapter to enter this discussion. The manure, however, can be treated in composting units and anaerobic digesters to reduce emissions: the treatment in digesters, e.g., avoid the spontaneous emission of gases that would result from the action of aerobic bacteria and also produce biogas, that later can be used as a source of electricity (Observatorio do Clima 2015d).

Emissions from crops can be reduced by employing best practices regarding soil preparation, fertilization and planting. Brazil is one of the top world consumers of fertilizers, and their use is directly related to GHG emissions. Their consumption could be reduced through more efficient use – it is proved that at least half of all fertilizer is lost in transport, before application in the field (Muller et al. 2014). Other best practices are genetic improvement of crops, better soil preparation, rotation of crops, integrated management of pests and diseases; the biological fixation of nitrogen and the practice of direct planting (without adding fertilizers at first hand) can also contribute (Observatorio do Clima 2015d).

Following the National Climate Change Programme, the Brazilian federal government has enacted the Low Carbon Agriculture Plan (in Portuguese, Plano ABC). The Plan defines climate change mitigation and adaptation targets in agriculture and livestock farming, and indicates several initiatives to be conducted, detailed in subprogrammes: restoring degraded grazing areas; integrating crops and forest; the practice of direct planting; biological fixation of nitrogen; afforestation and treating animal manure. Although comprehensive and an important indication that changes in the sector are necessary to implement a Brazilian low carbon economy, the Plan is not being implemented as fast as needed. Among the implementation problems, there is (i) lack of knowledge about specific credit lines to low carbon agriculture and associated interest rates (not much different from those in other lines); (ii) not enough monitoring of low carbon practices; and (iii) lack of communication about the programme and excessive bureaucracy (Observatorio do Clima 2015d). The economic crisis has hit the Plan hard: the amount of resources destined to the programme in the 2015/2016 period has been reduced by 33 per cent compared to the amount available in the 2014/2015 period; interest rates increased from 4.5–5 per cent to 7.6–8 per cent (Observatorio do Clima 2015d, 39). In 2015 the amount of public credit for conventional agriculture (50 billon dollars) was 80 times the equivalent for low carbon agriculture (600 million dollars).

Conclusion: what can be expected from Brazil?

This chapter has argued that in order to understand Brazilian wavering positions in the climate regime it is necessary to analyse the trajectory of Brazilian domestic policies and politics in the GHG emitting sectors, especially LULUCF, energy and agriculture, given their weight in total Brazilian GHG emissions.

From 1992 to 2005 Brazil was a conservative player in the climate regime, employing the radical interpretation of the 'common but differentiated responsibilities' (CBDR) principle, where the idea of historical responsibilities of countries is a foundation. This position was deeply connected with the idea that deforestation, which answered for the largest share of Brazilian GHG emissions, was a problem beyond control. Between 2006 and 2010, a period in which deforestation started to be successfully tackled, the Brazilian position moved towards moderate conservative: in 2009, the country abandoned the discourse of historical responsibilities and pledged a 36.1 to 38.9 per cent reduction in the projected emissions by 2020, a weak target, but better than the recalcitrant standings defended until then. Successfully tackling deforestation, however, meant that emissions from other sectors, mainly energy, agriculture and cattle grazing, became more relevant in total Brazilian emissions. Thus, climate change mitigation started to depend on a broader set of policies, such as support for low carbon energy sources, energy efficiency and low carbon agriculture, which are in their infancy in Brazil. Adding to this picture (i) the international boom for agriculture commodities, propelling agriculture exports and, consequently, their emissions, (ii) the domestic control of oil prices in Brazil, undermining the ethanol industry, and (iii) the use of thermal power plants as back up systems of new hydropower plants and as base systems of new electricity production in times of water scarcity, it was no surprise that, between 2011 and 2015, the Brazilian position in the climate regime retrogressed to conservative.

Brazilian INDC marked new progress towards moderate conservative: the country pledged to cut 37 per cent of its GHG emissions by 2025, and 43 per cent by 2030, compared to 2005 levels, the first absolute GHG reduction target compared to a base year ever presented by Brazil. The pledge has several inconsistencies that are deeply connected with the current contentious situation of climate change mitigation in Brazil and that disallow the INDC to be considered a step towards reformism. In fact, it is by looking at the pledge for specific sectors, the outlook of domestic policies and politics in those sectors and the political and economic landscape that future Brazilian standings can be envisaged.

Targets for LULUCF are an accurate representation of the difficulties in further tackling deforestation in Brazil. Further reducing deforestation, from an average 5,000 km² yearly to zero, requires stricter law enforcement and better monitoring, but also facing structural issues that privilege short-term private gains over long-term common benefits: land speculation; deficient registers of land ownership; deficient tax collection; lack of economic incentives to forest conservation and to invest in sustainable use of forest resources. The 2030 deadline to zero illegal deforestation in the Amazon indicates a paralysis in deforestation policies – particularly when considering the strong opposition developed in public opinion during 2015 against corruption and illegal activities in general, and the lack of planning to tackle structural issues raises concern over deforestation in Brazil.

In the energy sector, targets are not ambitious. As seen above, the targets for biofuels in the transport sector and renewables in the energy mix and in electricity production will be met easily and earlier than 2030 in a business as usual scenario. Having a more low carbon energy matrix than world average is no justification for lack of ambition in a sector that is key for reducing world GHG emissions and for Brazilian development. Enhancing targets would require dealing with structural issues in energy politics, especially the links between energy corporations, large-scale contractors and the federal government, in which Brazilian crony capitalism is based.

Targets for agriculture and cattle grazing are positive, but they represent only the start of change in the sector. The greatest issue is that measures to allow their implementation are insufficient. Brazil is a great agriculture exporter, and its production is based in large-scale monoculture fields that use great amounts of fertilizers and agrochemicals to sustain high outputs. A big push is needed to adapt the system to low carbon agriculture, and, to date, both the institutional framework and financial support for it are lacking.

Brazilian standing at COP 21 reflected the ambivalences of its INDC: during most of the time, Brazil sided as member of BASIC and G77 and two days before the end of the Conference decided to become member of the High Ambition Coalition lead by the European Union, United States and the Small Island Alliance, contributing somehow to a more progressive outcome of the Conference. The following months will show whether this was just a temporary and opportunistic move or if it implies Brazil shifting away from BASIC and G77.

The greatest sign of concern, however, comes from the current dynamics in Brazilian politics. Facing and tackling the structural issues that are deeply rooted in Brazil require leadership and balancing short- and long-term goals in planning and policy-making. The new government coalition formed after impeachment in May 2016 – lead by the weak and unpopular former vice-president Michel Temer – is very heterogeneous. The economic team lead by Ministry of Treasury Henrique Meirelles has recognized competence in mainstream economic, but is insensitive to low carbon development.

The Minister of Foreign Affairs Jose Serra has a more internationally cooperative approach to climate change than all the previous from the era Lula/Rousseff and the Minister of Environment Sarney Filho is more committed to climate policy than the previous one. However, urgent problems of fiscal deficit, growing public debt, resistant inflation and major reforms will take practically all the energy of government marginalizing climate policy. In order to re-build the economy that has contracted 7% in 2015–16 it would be needed some major reforms: fiscal, pensions, labor, and political parties. Most members of Congress are pursuing short-term particular interest or just trying to avoid prison because of corruption, none having the long-term collective interest at heart.

The excellent performance of the Public Prosecutor, the Federal Police and sectors of the Judiciary in the last two years has dramatically changed the previous situation of impunity – political class, big business, and political appointees in civil service – in relation to corruption. Still more advance is needed in order to consolidate a modern market democracy. The consolidation of the Brazilian position as moderate conservative moving towards reformism in the climate regime has a key pre-condition: the capacity of Temer government to survive and go ahead with major reforms. An upgrade in Brazil climate policy will not happen at least until a new elected President/Congress start to govern in 2019.

Notes

1   The definition covers manufacturing inputs such as silicon and specialized machinery, intermediate products such as solar cells, and final products such as wind turbines, heat pumps and biofuels.
2   Due to fears that the international financial resources that would be attracted to the Amazon region by CDM projects would undermine the national sovereignty of the region.
3   Different activities generate GHG emissions; they are usually divided into five categories: energy, industrial processes, waste, agriculture and cattle grazing and LULUCF. The latter refers to land use changes, especially deforestation, and is very relevant in countries with large forest cover, such as Brazil and Indonesia.
4   Own calculation based on data from Table 9.1.
5   Own calculation based on data from Table 9.2.
6   The full title of REDD is UN Collaborative Program on Reducing Emissions from Deforestation and Forest Degradation in Developing Countries.
7   The first auction including solar energy took place on 31 October 2014; solar sold 1,048 MW (installed capacity), and is expected to produce around 889.7 MW starting in 2017; in 28 August 2015, 1,043 MW (installed capacity) were sold, expected to be producing 833.8 MW by August 2017.
8   Japanese and Korean companies have been exceptions to this rule.

References

Arima, E., Walker, R. T., Perz, S. and Souza, Jr. C. (2015) 'Explaining the fragmentation in the Brazilian Amazonian forest', *Journal of Land Use Science*, DOI: 10.1080/1747423X.2015.1027797.

Brazil (2015) *Intended Nationally Determined Contribution*. www4.unfccc.int/submissions/indc/Submission per cent20Pages/submissions.aspx. Accessed 30 September 2015.

CAIT Climate Data Explorer (2015) Washington, DC: World Resources Institute. Available at: http://cait.wri.org. Accessed 17 December 2013.

Carvalho, A. C. (2012) Expansão da fronteira agropecuária e a dinâmica do desmatamento florestal na Amazônia Paraense, PhD thesis Universidade Estadual de Campinas, UNICAMP. www.bibliotecadigital.unicamp.br/document/?code= 000862229 Accessed 17 December 2013.

Eletrobras (2012) *Potencial hidrelétrico brasileiro por bacia – Dezembro 2012* (2012). www.eletrobras.com/elb/data/Pages/LUMIS21D128D3PTBRIE.htm   Accessed 17 January 2014.

EPE (Empresa de Pesquisa Energetica) (2015a) *Resenha Energética Brasileira*, exercício de 2014   www.mme.gov.br/documents/1138787/1732840/Resenha+Energética+-+Brasil+2015.pdf/4e6b9a34-6b2e-48fa-9ef8-dc7008470bf2   Accessed   15   July 2015.

EPE (Empresa de Pesquisa Energetica). (2015b) *Balanço Energético Nacional 2015* https://ben.epe.gov.br Accessed 15 September 2015.

FAO (Food and Agriculture Organization) (2013) *Tackling climate change through livestock: a global assessment of emissions and mitigation opportunities* www.fao.org/docrep/018/i3437e/i3437e.pdf Accessed 15 September 2015.

Imazon (2015) *Deforestation report for the Brazilian Amazon*. Available at http://imazon.org.br/publicacoes/deforestation-report-for-the-brazilian-amazon-july-2015-sad/?lang=en Accessed 15 September 2015.

INPE. (2015) *PRODES Project* (database on deforestation rates). Available at www.obt.inpe.br/prodes/prodes_1988_2015.htm Accessed 11 December 2015.

IEA (International Energy Agency). (2005) *Key World Energy Statistics 2005* www.iea.org/publications/freepublications/publication/KeyWorld2014.pdf Accessed 1 May 2015.

IEA (International Energy Agency). (2014) *Key World Energy Statistics 2014* www.iea.org/publications/freepublications/publication/KeyWorld2014.pdf.   Accessed   1 May 2015.

Martins, H., Fonseca, A., Souza, J., Sales, M. and Verissimo, A. (2013) *Forest Transparency, Brazilian Amazon* (Bulletin of July 2013), http://imazon.org.br/publications/?lang=en Accessed 15 September 2015.

MME – Ministerio de Minas e Energia (2009a) *Boletim mensal de monitoramento do sistema elétrico brasileiro*, Jan 2009, www.mme.gov.br/web/guest/publicacoes-e-indicadores/boletim-de-monitoramento-do-sistema-eletrico Accessed 29 August 2015.

MME – Ministerio de Minas e Energia (2009b) *Boletim mensal de monitoramento do sistema elétrico brasileiro*, Jul 2009, www.mme.gov.br/web/guest/publicacoes-e-indicadores/boletim-de-monitoramento-do-sistema-eletrico Accessed 29 August 2015.

MME – Ministerio de Minas e Energia (2015) *Boletim mensal de monitoramento do sistema elétrico brasileiro*, Jul 2015 www.mme.gov.br/web/guest/publicacoes-e-indicadores/boletim-de-monitoramento-do-sistema-eletrico Accessed 29 August 2015.

Muller, N. D., West, P. C., Gerber, J. S., MacDonald, G. K., Polasky, S. and Foley, J. A. (2014) 'A tradeoff frontier for global nitrogen use and cereal production', *Environmental Research Letters 09* 02–08.

Nogueira, L. A. H. and Costa, J. C. (2012) 'Opções tecnológicas em energia: uma visão brasileira', *Fundação Brasileira para o Desenvolvimento Sustentável* http://fbds.org.br/fbds/IMG/pdf/doc-531.pdf Accessed 17 December 2014.

Nye, J. (2011) *The Future of Power*, New York Public Affairs.

Observatorio do Clima (2015a) *Proposta para INDC brasileira* www.observatoriodoclima.eco.br/wp-content/uploads/2015/06/proposta-indc-oc.pdf Accessed 15 September 2015.

Observatorio do Clima (2015b) *Análise das emissões de GEE no Brasil (1970–2013) e suas implicações para políticas públicas, documento síntese* www.gvces.com.br/analise-das-emissoes-de-gee-no-brasil-1970-2013-e-suas-implicacoes-para-politicas-publicas?locale=pt-br Accessed 15 September 2015.

Observatorio do Clima (2015c) *Evolução das emissões de efeito estufa no Brasil (1990–2013), setor de mudança de uso da terra*, http://seeg.eco.br/biblioteca/ Accessed 15 September 2015.

Observatorio do Clima (2015d) *Evolução das emissões de efeito estufa no Brasil (1990–2013), setor de agropecuária*, http://seeg.eco.br/biblioteca/ Accessed 15 September 2015.

ONS (Operador Nacional do Sistema) (2015) *Histórico de operação, geração de energia* (www.ons.org.br/historico/geracao_energia.aspx) Accessed 15 September 2015.

The Pew Charitable Trusts (2012) *Who is winning the energy race*, www.pewenvironment.org/news-room/reports/whos-winning-the-clean-energy-race-2012-edition-85899468949 Accessed 12 July 2013.

Victor, A., Souza, Jr. C. and Verissimo, A. (2014) *Forest Transparency, Brazilian Amazon* (Bulletin of July 2014) http://imazon.org.br/publications/?lang=en Accessed 15 September 2015.

Viola, E. (2002) 'O regime internacional de mudança climática e o Brasil', *Revista Brasileira de Ciências Sociais 17* (50), 25–46.

Viola, E., Franchini, M. and Ribeiro, T. L. (2012) 'Climate governance in an international system under conservative hegemony: the role of major powers', *Revista Brasileira de Política Internacional 55*, 09–29.

Viola, E. and Franchini, M. (2013) 'Brasil na governança global do clima, 2005-2012: a luta entre conservadores e reformistas', *Contexto Internacional 35* (1), 43–76.

Viola, E. and Basso, L. (2015) 'Brazilian Energy-Climate Policy and Politics towards Low Carbon Development', *Global Society 29* (3) 427–446.

World Bank 2014 Statistics (database) <http://data.worldbank.org/indicator> Accessed 25 January 2014.

WWF; Roland Berger (2012) *Clean Economy, Living Planet: The Race to the Top of Global Clean Energy Technology Manufacturing*, www.rolandberger.com/media/publications/2012-06-06-rbsc-pub-Clean_Economy_Living_Planet.html Accessed 10 July 2013.

10 From environmental information to precaution in the face of environmental risks

An analysis of Brazil's National Policy on Climate Change and rulings by higher courts

Carlos José Saldanha Machado and
Rodrigo Machado Vilani

Brazil's environmental and climate policies have been characterized by alternating strides forward and setbacks (Machado 2014; Machado and Vilani 2015). In the phase since the Conference of the Parties (COP-21) held in December 2015 in Paris, what real contribution is Brazil making to reduce greenhouse gas emissions and strengthen actions that lead to a global climate agreement? For COP-21 held in September 2015, Brazil defined its *Intended Nationally Determined Contribution (INDC)*, or environmental goals. They can be considered ambitious (more than those defined by the United States) and unachievable, taking into account on one hand the illegal deforestation scenario in Brazil (Rajão and Soares-Filho 2015) and on the other hand, the climate contradictions discussed below in this chapter. As its main contribution, the Federal government set a 37 percent reduction of greenhouse gas emissions by 2025, plus 43 percent as an indicative reference value by 2030 – based on 2005 emission levels (Brazil 2015a). On December 12, 2015, at the COP-21, the new global climate agreement was signed, following two weeks of negotiations and discussions on the new standards to guide efforts towards a possible future. The main uncertainty on global climate governance lies precisely in knowing what efforts will actually be made by the signatory countries, which will require a permanent revision of the goals. In the coming years, due to the intense negative effects on human societies and ecosystems, will we see adequate treatment of climate risks? The answers to these questions belong to the future, but in the Brazilian case we can identify the main challenges at present for anticipating abstract risks like those related to the climate. Modern risks and abstract risks are treated here as synonymous, meaning "new risks [with] a global dimension (both in spatial and in temporal or multidimensional terms), [...] invisible (that can only be accessed by expert scientists), irreversible [...], and irrevocable [...]" (Brunet, Delvenne and Joris 2011, 181).

In relation to the INDC proposed by the Brazilian government:

> Recent studies indicate that, in the absence of mitigation efforts, the current Brazilian energy mix will continue on a trend of increasing carbon intensity, with natural gas and coal gaining importance in the power sector, and the sugar-alcohol sector undergoing a severe crisis that has caused the closure of several ethanol distilleries.
>
> (Spencer and Pierfederici 2015, 35)

The uncertainty thus lies in the materialization of proposed measures, undersigned repeatedly in documents and commitments. The time will come when consequences of the uncertainties will be definitive for human existence on the planet (Nalini 2010); the technical political, legislative, legal, institutional, social, cultural, and economic discussion should fight this uncertainty.

We thus prioritize enforcement of the precautionary principle rather than that of prevention, since these two fundamental institutes for life in society differ in terms of scientific certainty as an element in the causal nexus. While prevention applies to certain risk, precaution is directed to abstract risk (Carvalho 2008; Varella and Platiau 2004). Precaution is "prevention based on probabilities or contingencies" (Kiss and Shelton 2007, 95).

Meanwhile, our biocentric reading of society and Brazil's legal and institutional framework, based on the 1988 Constitution (Article 225), favors extending the time horizon by government policies and actions (Machado and Vilani 2015; Vilani and Machado 2010): a struggle against the short-term approach to decision-making in the country's public policies in defence of ecological equilibrium for present and future generations.

The precautionary principle was adopted as the structuring element for the analysis proposed here. The principle is essential for orienting the debate on climate risks, characterized by uncertainty, because "precaution means preparing for potential, uncertain, or even hypothetical threats" (Kiss and Shelton 2007, 95). Rulings by Brazil's highest courts (Federal Supreme Court – STF – and especially the High Court of Justice – STJ) have increasingly cited precaution in their rulings on environmental matters. Anticipation of future harm has thus gained space in court rulings. Finally, the Legislative Branch also emphasized precaution when it established the National Policy on Climate Change, under Law 12.187 of December 29, 2009, explicitly requiring enforcement of the precautionary principle, among others.

For analysis and interpretation of the National Policy on Climate Change (Law 12.187/2009) and rulings by the higher courts, we have chosen a multidisciplinary approach, focused on verification of the adoption (or lack thereof), by various government bodies, of the technical, scientific, and legal instruments and knowledge to mitigate or avert environmental risks to Brazilian society.

This chapter is organized into three main sections. The first identifies and analyzes three central contradictions in the energy sector's political and institutional framework in relation to Brazil's INDC. As we will demonstrate, the contradictions lies in the way the production of environmental information has been treated in the national legislation, the Federal government's option to expand oil and natural gas exploration and production, and stimulus for the use of individual automobiles. We will see that the origin of these contradictions lies in non-enforcement of the precautionary principle, particularly by the Brazilian Executive. Next, the above-mentioned contradictions, inherent to Brazil's environmental policy, will be analyzed on the basis of the precautionary principle, emphasizing the gap between the three contradictions and the scientific evidence on climate change. Finally, we offer remarks on the short-term, pork-barrel bias of the country's decision-making process versus the potential democratic strengthening of institutions and social actors, linked in cooperation, participation, and solidarity.

The three institutional contradictions related to measures by the energy sector in the INDC

The first contradiction in Brazil's environmental policy lies in the production of environmental information. Scientific evidence has triggered a global demand for the reduction of anthropogenic actions responsible for climate change. The human role in global warming has become uncontested, particularly since the fourth report of the Intergovernmental Panel on Climate Change (IPCC), in which researchers from the entire planet published the scientific evidence on the climate that has oriented the principal multilateral debates. For the IPCC, the greenhouse effect, intensified by CO_2 emissions from burning fossil fuels (petroleum, natural gas, and coal), that is, the result of human activity, is responsible for the average global temperature increase in the last 50 years (Machado and Vilani 2015). Notwithstanding the importance of environmental information, how has it been treated in Brazil's legal framework?

The political approach to the issue has several historical characteristics. According to Alvarenga, Castro and Magalhães Júnior (2005), in parallel with the historical lack of environmental information, some political players have seized such information for their own benefit during the decision-making process on the country's public policies. The resumption of large hydroelectric projects illustrates this issue: despite the scientific evidence on social and environmental impacts of hydroelectric dams like Belo Monte (Cunha and Ferreira 2012), this energy policy has been maintained, to the benefit of mining groups and construction companies responsible for building and operating the projects, often facilitated by public funds from the National Economic and Social Development Bank (BNDES) and with no participation by the affected communities, notably indigenous peoples (Fleury and Almeida 2013; Franco and Feitosa 2013; Pinto 2012). Milaré (2001) relates efficient action by society to the right to information, which helps curb decision-making abuses and privileges. Alvarenga, Castro and Magalhães Júnior (2005, 158) criticize the reproduction of this "*archaic structure*, marked by the prevalence of hegemonic sector interests based on political and economic influence." This context, exemplified by the Belo Monte hydroelectric dam, can be related to two symbols of Brazil's delegative democracy (Frey 2000): limited institutional density, marked by elitist influence in the decision-making process (Frey 2000), and state omission in the face of the need to collect or requisition scientific data for the public administration (Machado 2006). Along this same line, Sarlet and Fensterseifer (2014, 134) tie access to environmental information to "the full exercise of *participatory ecological democracy*" (authors' emphasis), which they thus affirm as a fundamental right.

Despite the democratic, social, and scientific relevance of environmental information, Law 10.650/2003 rules on access to environmental data and information, but limits it to mere repositories of such information without requiring its production by the environmental agencies of the SISNAMA (see article 1 of the Law concerning action by agencies of the National Environmental System, SISNAMA, established under the National Policy for the Environment, Law 6.938, of August 31, 1981). This created a mismatch in the legal system, since article 9, XI of the National Policy for the Environment explicitly requires that the government produce environmental information when it is lacking. The National Policy on Water Resources, enacted by Law 9.433/97, also explicitly provides for the production of data and information as one of the principles of its information system. According to the laws' wording, nonexistent information is not limited to information that is unavailable in the environmental agency (Machado 2006). Thus, the obligation to produce such information assumes a duty for government-backed scientific research to fill the gap of scientific knowledge needed for decision-making. Importantly, environmental information and its public availability should not be interpreted in a narrow sense, limited to environmental licensing procedures, as a hasty interpretation of article 225, paragraph 1, IV, of the 1988 Federal Constitution might suggest.

Machado (2006) emphasizes the contribution by environmental information in an analysis of the Aarhus Convention on Access to Information, Public Participation in Decision-making, and Access to Justice in Environmental Matters, of June 1998. The contribution of the Convention lies in the structuring of (and access to) environmental databases, overcoming the "policy of secrecy and obscure decisions that still exist in numerous Administrations" (Machado 2006, 163). For environmental law in particular, clarity of rulings is a primary objective. The precautionary principle, one of the principal pillars for protection of the environment and safeguards for the rights of future generations, is present in the legal frameworks of various countries and in the main international debates, such as the Convention on Biological Diversity (CBD) and the Rio Declaration on Environment and Development, of 1992 (Machado 2013; Sarlet and Fensterseifer 2014). The central thrust of the precautionary principle is decision-making based on scientific evidence concerning the environmental risks of a given anthropogenic intervention (Milaré 2001; Machado 2013; Sarlet and Fensterseifer 2014). When dealing with climate risks, anticipation of caution becomes increasingly relevant, since the severity of climate events in the future could make human existence on the planet extreme or even impossible.

Nobre (2014), discussing the future climate of the Amazon based on existing scientific evidence, concludes that we are at a decisive moment for reversing a potential scenario of even more serious natural disasters in Brazil. The author highlights that "the urgent decision [to act], already late" (Nobre 2014, 29) is based on the available scientific knowledge and refers to "a 'war' on ignorance" (Nobre 2014, 31) to curb further postponement of the necessary measures to protect the Amazon and thus to avoid the disappearance of water throughout the country.

In the fight against climate ignorance, the Brazilian state needs to overcome the first contradiction and launch a process of production of scientific evidence, linked to the country's research institutions, encouraging social and institutional maturity on the theme. We favor the wide dissemination of research results in Portuguese and reports with appropriate structure and language for different social actors.

The second contradiction refers to the Federal government's decision to expand exploration and production activities in oil and gas beginning in 2008. We will analyze several of this contradiction's multiple facets. First, the veto by the President of Brazil, backed by the Ministries of Finance and Planning, Budget, and Management and the Federal Attorney-General (AGU), on article 4, III of the National Policy on Climate Change (PNMC). The article vetoed by the President had proposed "incentive for the development and use of clean technologies and the gradual abandonment of energy sources that use fossil fuels." The article was vetoed because it purportedly ran counter to the public interest, and on grounds of unconstitutionality. Machado and Vilani (2015) challenge the definition of "public interest" claimed by the President as grounds for the veto. The authors claim that the veto failed to consider the public interest written into the Federal Constitution, articles 3 and 225, which determine that such interest primarily involves healthy quality of life for the population in an ecologically balanced environment. Therefore, the public interest claimed as grounds for the Presidential veto clashes with these vital and ecological premises for full human dignity, the scope of which demands the reduction of air pollution to ensure a healthy environment for present and future generations (Machado and Vilani 2015).

The claims for the veto are that "the country's current energy policy has already prioritized renewable energy sources in its matrix and achieved widely acknowledged strides in the use of clean technologies. One of this policy's pillars is the rational use of the various available energy resources, making inadequate a provision focused on abandoning the use of fossil fuels" (Brazil 2009). The Federal Executive Branch concludes by stating that the strategy "for the sector should meet the principles and objectives set out by Law 9.478 of August 6, 1997, which combines protection of the environment with other relevant values for energy policy and security" (Brazil 2009). Law 9.478/97, known as the "Petroleum Law," obviously makes no mention of reducing fossil fuels and thus does not make such reduction mandatory. From this analysis emerges the second aspect of this contradiction. Petroleum has been Brazil's greatest public rallying cry in recent decades, and oddly enough, it has gained force during the current political and financial crisis in Petrobras, the result of the embezzlement of money from the state-owned company to fund politicians and political parties in the National Congress. According to the official propaganda, "Petrobras is ours!" and "Its energy comes from rising to the challenge!" The problem is that this fossil energy entails human, environmental, and climatic consequences that are harmful, insurmountable, permanent, and irreversible.

The institutionalization of the developmentalist model in Brazil occurred together with transformations in legal structures and in the state's own political organization, which has emphasized and increasingly prioritized the use of fossil energy. Such institutional shortsightedness appears clearly in the Ten-Year Plan for Energy Expansion by 2023 (PDE) of the Ministry of Mines and Energy – MME (MME 2014). The document makes the following forecasts for investments in the energy sector from 2014 to 2023, totaling 1,263 billion *reais* (MME 2014, 409): petroleum and natural gas – BRL 879 billion (69.6 percent); supply of liquid biofuels – BRL 82 billion (6.5 percent); the remainder, BRL 301 billion, referring to electric power supply.

Machado and Vilani (2015) call attention to the paradox of Brazil's climate governance contained in the Ten-Year Plan for Energy Expansion: only 6.5 percent of the predicted investments in the sector are focused on decarbonization of the energy matrix. Therefore, the investment pattern clashes in practice with the pretension of the country's INDC, specifically the goals of "expanding the use of renewable sources, in addition to hydroelectric power, in the total energy matrix, from 28 percent to 33 percent by 2030" and "expanding the use of domestic use of non-fossil energy sources, increasing the share of renewable energies (in addition to hydroelectric power) in the supply of electric power" (Brazil 2015a, 3).

As a logical consequence, in 2023 transportation will be the main sector responsible for emissions in Brazil, with 306 million tons of carbon equivalent ($MtCO_2eq$), or approximately 46 percent of total emissions! This legacy for future generations began in the 1950s, when highway transportation achieved priority status on the government agenda, and to this day it contributes to air pollution by burning fossil fuels from petroleum, besides increasing the cost of harvests from the Central-West Region due to the high transportation costs of farm production (Reis, Fadigas and Carvalho 2005).

This shortsighted rationale has oriented government actions and given rise to the *third contradiction*: stimulus for individual automobile use. In transportation, Brazil has never approached the principal global climate change debates with effective action. The country has confronted the official discourse on government efforts to reduce greenhouse gas emissions (Viola and Franchini 2013; Obermaier and Rosa 2013). To keep the transportation sector afloat, for example, Brazil still mainly uses fossil fuels (80.96 percent according to Ambrizzi and Araújo 2014). The Brazilian INDC showcases ethanol, a typically Brazilian product. In the document, expansion of the supply of biofuels appears as one of the additional measures for the 2 degrees Celsuis average temperature goal (Brazil 2015a). However, "ethanol diplomacy [was] central to the international strategy of President Lula in 2006 and 2007 but [was] nearly abandoned after announcement of the Pre-Salt discovery" (Viola and Franchini 2013, 71). Machado (2014, 243) describes this political characteristic as "doublethink and double-act by the Brazilian Executive" on environmental issues in the last 50 years.

Politically speaking, Brazil is clearly confronting its choice for unsustainability, given the finite nature of petroleum and natural gas. The certainty of this finitude allows defining "sustainable" development based on the use of renewable resources. Inversely, "unsustainable" is the model that depends on non-renewable natural resources, degrading its own material base for survival (Gilpin 1996).

The logic against the reduction of dependence on fossil fuels is oriented (among other issues) by the prevalence of individual transportation as official government policy. This same logic, perpetuated and sponsored by the Brazilian government, entails the third climate contradiction. The National Association of Automobile Manufacturers (ANFAVEA), founded in 1956 during expansion of the country's automotive industry, has lobbied heavily since 2002, exerting strong influence on the government's economic policy measures, meeting with "Cabinet members and heavyweight politicians" according to Lazzarini (2011, 92).

In terms of economic policy, this proximity resulted in tax cuts on industrial products (IPI) for automotive vehicles, inducing the demand for individual transportation. According to the National Transit Department, in September 2015 there were approximately 90 million vehicles on Brazil's roads, of which 50 million were automobiles, or double the 25 million circulating in 2001 (Fenabrave 2015). Notably, the national campaign for purchasing new vehicles fails to include any measure for retrofitting the country's fleet, as adopted by other countries, for example Germany (Guimaraes and Lee 2010). Given such measures, Viola and Franchini (2013, 56) issued a warning about a "standstill in the enforcement of the climate law." The authors cite two reasons: i) industry stimulus "via tax cuts on industrial products (IPI) – introduced in 2012 without any climate or environmental conditions"; and ii) eliminate the "Contribution on Economic Activities (CIDE) on importation and production of petroleum and petroleum products to avoid a new fuel price increase, and reduction of electricity tariffs by late 2012" (Viola and Franchini 2013, 57–58).

This reveals a clear mismatch with one of the purported measures in the INDC submitted to the COP-21 by the Brazilian government, namely: "in the transportation sector, to promote efficiency measures and improvements in transportation infrastructure and public transportation in urban areas" (Brazil 2015a, 4). The overloaded roads in medium-sized and large Brazilian cities have aggravated the problems with traffic jams.

To overcome the third contradiction does not require new proposals. Minimizing the use of non-renewables in favor of the transition to renewable resources, as anticipated by Meadows et al. (1972) and reinforced more recently by Meadows, Randers and Meadows (2004) and Daly (1996), is a single action, but it requires a clear political stance, as opposed to repeated vacillation by the Federal Executive, marked by a policy centered on minerals and fossil fuels for development.

The three contradictions analyzed above – the way the production of environmental information has been treated in Brazil's legal system; the Federal government's decision to expand exploration and production activities in petroleum and natural gas since 2008; and stimulus for individual use of the automobile – represent obstacles to sustainable development. This observation results from the comparison between these and three of the recently announced Sustainable Development Goals (Brazil 2015b). Among the goals, we highlight the following: Goal 11. Make cities and human settlements inclusive, safe, resilient, and sustainable; Goal 12. Ensure sustainable production and consumption patterns; and Goal 13. Take urgent action to combat climate change and its impacts.

The incentive for automobiles runs contrary to sustainable cities and sustainable production and consumption patterns. This government initiative contradicts the climate agenda for the reduction of greenhouse gases. Stimulating individual automobile use exclusively feeds a model for consumer society based on reproducing dependence on fossil fuels.

The precautionary principle gains relevance from the perspective of the Sustainable Development Goals. "The Precautionary Principle is often seen as an integral principle of sustainable development, that is, development that meets the needs of the present without compromising the abilities of future generations to meet their needs" (UNESCO 2005, 8). The next section will thus provide details on the Brazilian case concerning the principle's application by government in relation to climate risks.

The precautionary principle in the three branches of government in Brazil

There is no doubt about the relevance to climate risks of Brazil's enforcement of the precautionary principle. Its relationship to scientific certainty has already been highlighted in 1992 during the United Nations Conference on the Environment and Development, in Rio de Janeiro. The lack of scientific certainty, as observed in abstract risks, cannot be an obstacle to government action. Thus, "climate change is a global threat that satisfies many of the conditions for which a precautionary approach is indicated" (Iverson and Perrings 2011, 12).

The lack of a scientific basis and citizens' participation in the decision-making process on national public policies (as in the energy sector, see for example the Belo Monte Hydroelectric Dam – Section 2) clashes with the constitutional and general principles of environmental law. Throughout Brazil, rulings issued in disagreement with the established legal order in the past and present have caused drought, deforestation, pollution, (water, air, and soil), silting, landslides, and other social and environmental disasters. Nature in Brazil, as elsewhere on the planet, still resists, moved by the force of resilience – or the capacity of systems to respond to and recover from disturbances, reaching multiple equilibriums (McGreavy 2015). The same resilience operates as energy inducing emerging spontaneous social changes in solidarity, increasingly biocentric in the present and for the future.

Different areas of knowledge are asked to contribute to the construction of a global pact for the climate. These areas of knowledge need to extend beyond the Cartesian disciplinary perspective to a multidimensional vision that proposes possible alternatives for perennial life in the future. In this sense, "a precautionary approach is needed and this requires a number of changes in scientific culture and in the way risk assessment is performed" (Van Der Sluijs and Turkenburg 2006, 17).

Scientific evidence allows establishing policies that incorporate abstract risks. The anticipation of climate risks will be one of the precautionary principle's main contributions to future generations. With broad consensus in the legal literature, the use of this principle has become the prerogative of a new sought-after model for the decision-making process in life in society, conducted according to clear, sufficient, and uncontroversial scientific evidence.

Nalini (2010, 143) emphasizes that "expert scientific certification is not necessary to realize that the Earth is increasingly surprising its unwitting occupants." At any rate, depending on the point of view, there are persistent scientific certainties that corroborate the causal links between our actions (harmful events) and climate changes.

In this sense, the concept of precaution is not intrinsically incoherent, as claimed by Giddens (2014). Contrary to the author's analysis, the principle provides the legal order with an expanded definition of causal link, moving beyond present material damage (subject to measurement and compensation) to the danger of occurrence of harm. The precautionary principle takes a step ahead in time to prevent methods, practices, substances, or processes from annihilating the conditions for future life on the planet (Machado and Vilani 2015). Giddens (2014, p. 152) is not wrong to state that inaction carries risk, but inaction is not the purpose of precaution. We can illustrate this with a pioneering approach to precaution on environmental matters, in Rachel Carson's *Silent Spring*. In the early 1960s, Carson contended "that we have allowed these chemicals to be used with little or no advance investigation of their effect on soil, water, wildlife, and man himself" (Carson 2002 [1962], 13).

Thus, precaution does not rule out any measure definitively, but only suspends decisions that expose human and/or environmental health to risks in a scenario of scientific uncertainties until these are superseded by advances in technical and scientific knowledge. In fact, by defending risk assessment as a substitute for precaution, Giddens (2014) is not creating an alternative. In our view, risk assessment, as a technical and scientific tool, is the materialization of the precautionary principle. Corroborating this reasoning, Nobre (2014) concludes that "despite a mountain of scientific evidence, we have still been incapable of acting, and if we are too slow, we will likely have to deal with incomprehensible damage, having always received free shade and cool spring water from the great forest" (p. 36). Speed in acting prudently, regardless of the existence of a clear and incontestable causal link, is thus the objective of precaution, not the opposite. In the final analysis, precaution represents the understanding of a new formulation of the laws of nature, seen by Prigogine (1996, 153) no longer as "certitudes, but rather as possibilities." Precaution affirms that the future is not prescribed, imposed, insoluble, but can have its direction corrected by anticipation of the decision concerning the abstract risks that we should not, need not, or desire not to take on as a society – the paradoxical, unequal, and complex Brazilian society of the Anthropocene.

The state's centrality in this dynamic of solidarity between the present and future is in step with overcoming the limits of government action. This process "demands the patience of historical time" (Machado 2014, 18) to move beyond the dichotomies of the Brazilian state (social and neoliberal, elitist and inclusive, multipartisan and fragmented). Despite these contradictions and paradoxes, the Brazilian state has made social, institutional, and legal progress. Acknowledgment of the environmental variable in the broad enforcement of the precautionary principle by the Legislative and Judiciary is often found in the Brazilian legal order. The Federal Supreme Court (STF) defines the principle of sustainable development as a basic element for interpretation of the Constitution and (as a logical spinoff) for analyzing the "clash between relevant Constitutional provisions" when, rather than economic issues, it is necessary to prioritize "the right to preservation of the environment, translated as the public good, to be enjoyed by everyone, to be defended in favor of the present and future generations" (STF 2005). The wide receptiveness to the precautionary principle in Brazil's courts includes the position by the High Court of Justice (STJ), that the principle's enforcement aimed "to avoid harm to the public order generated by uncertainty concerning environmental risks" (STJ 2013).

The position by the High Court is emblematic, having ruled repeatedly in favor of environmental quality. In a ruling in August 2009, even prior to the National Policy on Climate Change, the STJ upheld the existence of a causal link between anthropogenic actions and climate change, specifically concerning post-harvest burning of sugarcane fields. The court ruled that "burning [sugarcane fields] is incompatible with the objectives of environmental protection established by the Federal Constitution and other environmental legislation. Especially in a time of climate change, any exception to this general prohibition must only be allowed by federal law, but *should be interpreted restrictively by administrators and judges*" (our emphasis) (STJ 2009a). The STJ followed the same line of enforcement of precaution by adopting a scientific basis for ruling against burning sugarcane fields. The court explicitly adopted the fight against outdated industrial and agricultural practices when it ruled that "any activity should be conducted with modern industrial instruments and technology to reduce environmental impact" (STJ 2009b).

The STJ not only called on the Judiciary to prevent and punish activities that contribute to climate change, but also required clear attention to the issue by the Federal Executive. The Brazilian legal order thus consolidates what has been called "climate change law" (Ayala 2010). Enforcement of the precautionary principle by the Brazilian Judiciary has thus reinforced the importance of scientific evidence to support court rulings and has become an important step forward in Brazil's environmental law (Machado and Vilani 2015). The Brazilian and international legislation is vast, but "the problem we are faced with is not, in fact, philosophical but legal, and in a wider sense, political" (Bobbio 1996, 12). The assertion by Bobbio (1996, 12) is entirely pertinent, since in relation to climate changes, we need to discuss and decide on "the surest method for guaranteeing rights, and preventing their continuing violation in spite of all the solemn declarations."

The increasing risks have required a broader approach to the law by the Judiciary, overcoming dogmas of traditional jurisprudence, bound to the present and limited to the reparation of definite harm. In these times of climate change, the lack of clarity on the occurrence of harm requires that the law be open to scientific knowledge, and that the jurist engage in "more in-depth study of sciences, based on solid technical reports" (Wedy 2009, 100). Only with the aid of "technical and scientific observation" will the law reach the complexity of causes and effects operating in the environment and produce the correct "translation of expert evidence" (Carvalho 2008, 106). The precautionary principle, as an important clause in the pact for climate governance, has become an important basis for the fight against public inaction in the face of scientific uncertainties and certainties that jeopardize (or may come to jeopardize) life in all its forms. The flipside of this is to act in defense of a pact for the planet's time and our existence and that of all life forms, to change our view, keep nature in mind, influence men and women one-by-one in their daily activities or their more lasting research, to complete still-unfinished clauses in a socio-environmental pact.

The work by the Office of the Public Prosecutor (MP) in this socio-environmental pact, complementing that of the Judiciary, is an important example of institutional maturity in Brazil and in the enforcement of the precautionary principle. Under the organization of the Federal Republic of Brazil, the Office of the Public Prosecutor is an essential institution for the state's jurisdictional function. According to its Constitutional attributions (Article 129, III), it functions include processing public inquiries and class action suits. Class action suits are the main legal instrument in Brazil for the repression and reparation of environmental harm, as regulated by Law 7.347 of July 24, 1985. The Public Prosecutor is literally "society's environmental advocate and negotiator" (McAllister 2004, 230). By way of example, the fourth Chamber for the Coordination and Review of the Environment and Cultural Heritage, of the Office of the Federal Public Prosecutor (MPF), publishes yearly activity reports, highlighting the number of class action suits in the country: 564 in 2014 and 623 in 2013 (MPF 2015). This emphasizes the strengthening of the Office in achieving its institutional mission, defined constitutionally as the defense of the legal order, democratic rule of law, and inalienable collective and individual interests (1988 Federal Constitution, Article 127/heading).

In the legislative sphere, implementation of the National Policy for the Environment (Law 6.938/81) witnessed the awakening of a Democratic Rule of Environmental Law, or simply Environmental State, as defined by Canotilho (1999), gaining clear shape with the enactment of the country's new Constitution in 1988. This Environmental State, with a legal basis and an open-ended timeframe, commands the government to elaborate public policies based on the delicate link between ecological equilibrium and human life in present and future generations and in cooperation according to the principle of environmental democracy (Leite and Ayala 2004; Canotilho 1999). To illustrate how the Brazilian legal order allows precaution, take Law 9.605/98, with the penalization of mere conduct (that is, the existence of abstract danger is sufficient to punish the perpetrator). Crimes of pollution are a clear example of criminal conduct not requiring proof of harm. There was also an attempt to enforce the precautionary principle by the National Policy on Climate Change, in the provision discussed in section 2. Machado and Vilani (2015) highlight the "recognition of the precautionary principle" by various pieces of Brazilian legislation. These feature: Law 11.105/2005, on genetically modified organisms; Decree 2.519/1998, ratifying the Convention on Biological Diversity – CDB; Law 9.605/1998, better known as the Law on Environmental Crimes with Administrative Sanctions; and Resolution 303/2002 of the National Council on the Environment, which provides parameters, definitions, and limits for Permanent Preservation Areas (Table 10.1).

The use of scientific and legal measures to avoid risk has thus oriented action by government in Brazil. Precaution places a limit on state action and aims to protect citizens against the abuse of power. In the Environmental State, public policies should be sufficient, reasonable, necessary, and proportional, while prohibiting excessive and arbitrary omission by government acts (Canotilho 1999; Wedy 2009). Although the long journey towards the creation of an Environmental State is still in its initial steps (since Brazil's National Constitution is still recent), we can safely state that the "environmentalization" of the law (Rocha and Carvalho 2006) at the national level continues on a steady and promising course (Machado 2014).

By pointing to climate-related contradictions, we merely intended to show the preponderant role of the Executive Branch in creating a blatant paradox between the official discourse for COP-21 and the reality of recently practiced government measures in Brazil.

There is an obvious mismatch and lack of linkage between the Executive and Legislative Branches. This obviously hinders agreement on (and optimization of) efforts to meet the climate goals in particular and the sustainable development goals in general. It has been up to the Judiciary, through

Table 10.1 Precautionary principle in prevailing Brazilian federal laws and rulings

Law	Purpose	Recognition of the precautionary principle
Law 11.105/2005	Regulates the use of genetically modified organisms	Article 1. This law establishes safety standards and inspection mechanisms for the construction, cultivation, production, handling, transportation, transfer, importation, exportation, storage, study, marketing, consumption, release into the environment, and disposal of genetically modified organisms – GMOs and their byproducts, having as guidelines stimulus for scientific advancement in the area of biosafety and biotechnology, the protection of life and human, animal, and plant health, and observance of the precautionary principle for the protection of the environment.
Decree 2.519/1998	Ratifies the Convention on Biological Diversity	Preamble: Aware of the general lack of information and knowledge regarding biological diversity and of the urgent need to develop scientific, technical, and institutional capacities to provide the basic understanding upon which to plan and implement appropriate measures,
Law 9.605/1998	Rules on penal and administrative sanctions resulting from conducts and activities harmful to the environment	Article 42. Make, sell, transport, or release balloons which *may cause* fires in forests and other forms of vegetation, in urban areas or any type of human settlement [...]. Art. 54. To cause pollution of any nature at levels that result or *may result* in harm to human health, or that cause the death of animals or significant destruction to the flora: [...] § 3° The same penalties of the previous paragraph apply to anyone who fails to adopt, when so required by the competent authority, *precautionary measures in case of risk* of severe or irreversible environmental harm. Art. 61. Disseminate a disease, pest, or species that may cause harm to agriculture, fishing, fauna, flora, or ecosystems [...] (our emphasis)
Resolution 303/2002	Provides parameters, definitions, and limits for Permanent Preservation Areas.	In view of the socio-environmental function of property provided in Articles 5, paragraphs XXIII/170, paragraph VI, 182/2, 186/II, and 225 of the Constitution and the principles of prevention, precaution, and polluter-pays...

enforcement of the precautionary principle, to fill this gap. Still, due to the checks and balances in Brazil's Republican system, such action by the Judiciary has its limits, and its rulings alone cannot provide sufficient basis for building a national development model that ensures ecological equilibrium, human dignity, and the reduction of Brazil's social and regional inequalities.

The country's political and institutional complexity cannot be overcome by a legal principle alone. The openness of the three branches of government to the precautionary principle, even with steps forward and setbacks, has provided an ample opportunity for increasing the democratic maturity of institutions and society. The Judiciary appears to have taken the first step in this process. The Legislative and Executive still follow the short-term vision of economic results, but they occasionally absorb the abstract risks. Important strides are made on such occasions, like the production of essential legal provisions for the protection of the rights of future generations.

The major challenge of the Brazilian state in the Anthropocene is to overcome the isolated nature of these advances and adopt a long-term perspective as the guiding thread for the decision-making process in the country's cross-cutting public policies, ensuring broad grassroots participation and an in-depth scientific basis for the debate on public policies and actions.

Thoughts for a possible future

By the end of this chapter, we can state, based on the analysis of government actions, that the goals announced in the INDC will not be met. That has been the story for the last 45 years! Although acknowledging legal and institutional strides in the enforcement of the precautionary principle, especially by the Judiciary, there is still an evident gap in the country's political and institutional framework. The predominance of lobbies for economic interests in the National Congress (bicameral, consisting of the Senate with 81 members and the Chamber of Deputies with 513 members) is the principal obstacle to the democratization of Brazil's national political structure. For example, the current Congress (2015-2018 Legislature) has 119 National Deputies with direct links to agribusiness, or 23 percent of the total of 513 in the Chamber of Deputies. Lobbyists for the banking, meatpacking, and mining industries can count on 197 (38 percent), 162 (31 percent), and 85 (16 percent) of the Chamber of Deputies, respectively. In the final months of 2015, the Office of the Federal Public Prosecutor denounced a racket for the sale of so-called "provisional measures" (Executive orders with the force of law for a temporary period, and that require approval by the National Congress to become fully-fledged laws), allegedly paid for by the automobile industry to favor its interests in the form of tax cuts. The gap between the interests of small groups in the present and the rights of future generations is the biggest challenge faced by the Brazilian state.

The varying and unequal degree of appropriation of scientific evidence on climate change by different government stakeholders points to the urgent need for maturity in the production, management, and dissemination of environmental information. A broad, participant, and transparent debate should already have started in Brazil on the incorporation of climate risks in the national decision-making process.

Precaution has been far removed from the short-term logic based on the Federal Executive's recurrent argument about overcoming the economic crisis by increasing automobile sales, and by unbridled reliance on oil as the instrument for economic development. The contradictions of Brazil's climate policy in the Anthropocene are many and complex, extending far beyond the apparent simplicity presented in this chapter. Despite commitments assumed during the COP-21, the country's political decision-makers have failed to attack deforestation and have avoided investments in alternative energies, to the point of vetoing wind and solar energy in the Federal Four-Year Plan (2016–2019) published in January 2016. The progress achieved in 2014, when Brazil ranked fourth in the world in wind power, appears to have been swept away by the economic stagnation aggravated in 2015.

Brazil thus faces numerous challenges in the Anthropocene: from corruption in the form of political parties in organized crime, invasion of the Amazon by the oil industry and agribusiness, to the construction of an urban society oriented by unsustainable consumption of natural resources, encroachment on indigenous lands, beef cattle raising and soybean plantations, and hydroelectric dams.

If the numerous political and institutional challenges head us toward worse social segregation and environmental degradation, the examples of advances and maturity discussed in this chapter show the potential for building a just and ecologically balanced society. It is up to the Brazilian state to play an innovative and challenging role in organizing human coexistence and overcoming its own political and institutional dichotomies that aggravate the complex scenario of risks in the Anthropocene.

Thus, we men and women of science cannot simply accept supporting roles on the public stage due to the disillusionment dominating the contemporary landscape. We must play leading roles in founding a world for all that bestows rights and duties on everyone. When human intervention is needed to change the course of events and create the new, the world is threatened by its own destructive action.

List of abbreviations

MP (*Ministério Público*): Office of the Public Prosecutor
MPF (*Ministério Público Federal*): Office of the Federal Public Prosecutor
STF (*Superior Tribunal Federal*): Federal Supreme Court
STJ (*Superior Tribunal de Justiça*): High Court of Justice

References

Alvarenga, L. J., Castro, F. V. F. and Magalhães Júnior, A. P. (2005) "Participação cidadã e informação na gestão de recursos hídricos," *Revista de Direito Ambiental 10*, 148–162.

Ambrizzi, T. and Araújo, M. (2014) eds. *Base científica das mudanças climáticas*, Universidade Federal do Rio de Janeiro, Rio de Janeiro.

Ayala, P. A. (2010) "O direito ambiental das mudanças climáticas: mínimo existencial ecológico, e proibição de retrocesso na ordem constitucional Brazileira," in Benjamin, A. H., Irigaray, C. T., Lecey, E. and Cappeli, S. eds, *Florestas, mudanças climáticas e serviços ecológicos*. Imprensa Oficial do Estado de São Paulo, São Paulo, 261–294.

Bobbio, N. (1996) *The Age of Rights: Human Rights Now and in the Future*. Polity Press, London.

Brazil (2009) Mensagem n. 1.123, de 29 de dezembro de 2009 (http://bit.ly/1zelQV5) Accessed 24 October 2015.

Brazil (2015a) Pretendida Contribuição Nacionalmente Determinada para consecução do objetivo da Convenção-Quadro das Nações Unidas sobre Mudança do Clima (http://bit.ly/1Ru0Jm3). Accessed 24 October 2015.

Brazil (2015b) Objetivos do Desenvolvimento Sustentável (http://bit.ly/1MLlm9H) Accessed 24 October 2015.

Brunet, S., Delvenne, P. and Joris, G. (2011) "O princípio da precaução como uma ferramenta estratégica para redesenhar a (sub)política. Compreensão e perspectivas da ciência política de língua francesa," *Sociologias*, *13*, 176–200.

Canotilho, J. J. G. (1999) *Estado de Direito*. Gradiva, Lisboa.

Carson, R. (2002 [1962]) *Silent Spring*, 40th Anniversary Edition. Mariner Books, New York.

Carvalho, D. W. (2008) *Dano ambiental futuro: a responsabilização civil pelo risco ambiental*, Forense Universitária, Rio de Janeiro.

Cunha, D. A. and Ferreira, L. V. (2012) "Impacts of the Belo Monte hydroelectric dam construction on pioneer vegetation formations along the Xingu River, Pará State, Brazil," *Brazilian Journal of Botany*, *35*, 159–167.

Daly, H. E. (1996) *Beyond Growth*. Beacon, Boston.

Departamento Nacional De Trânsito – Denatran (2015) Frota de veículos (http://bit. ly/1OYP8Mw). Accessed 29 April 2015.

Fenabrave (2016) Anuário 2015. O desempenho da distribuição automotiva no Brasil. Fenabrave, São Paulo.

Fleury, L. C. and Almeida, J. (2013) "A construção da Usina Hidrelétrica de Belo Monte: conflito ambiental e o dilema do desenvolvimento," *Ambiente e Sociedade, 16* 141–156.

Franco, F. C. O. and Feitosa, M. L. P. (2013) "Desenvolvimento e direitos humanos: marcas de inconstitucionalidade no processo Belo Monte," *Revista Direito GV, 9* 93–114.

Frey, K. (2000) "Políticas públicas: um debate conceitual e reflexões referentes à prática da análise de políticas públicas no Brazil," *Planejamento e Políticas Públicas*, 21, 211–259.

Giddens, A. (2014) *Turbulent and Mighty Continent: What Future for Europe?* Polity Press, Cambridge.

Gilpin, A. ed. (1996) *Dictionary of environment and sustainable development.* John Wiley & Sons, Chichester.

Guimaraes, L. E. and Lee, F. (2010) "Levantamento do perfil e avaliação da frota de veiculos de passeio Brazileira visando racionalizar as emissões de dióxido de carbono," *Sociedade e Natureza, 22*, 577–592.

Iverson, T. and Perrings, C. T. (2011) *Precautionary Principle and Global Environmental Change.* Nairobi, The United Nations Environment Programme.

Kiss, A. and Shelton, D. (2007) *Guide to International Environmental Law.* Martinus Nijhoff, Leiden.

Lazzarini, S. G. (2011) *Capitalismo de laços: os donos do Brazil e suas conexões.* Elsevier, Rio de Janeiro.

Leite, J. R. M. and Ayala, P. A. (2004) *Direito ambiental na sociedade de risco.* Forense Universitária, Rio de Janeiro.

Machado C. J. S. (2013) *Animais na sociedade brasileira: práticas, relações e interdependências.* E-papers, Rio de Janeiro.

Machado, C. J. S. (2014) *Desenvolvimento Sustentável para o Antropoceno.* E-papers, Rio de Janeiro.

Machado, C. J. S. and Vilani, R. M. (2015) *Governança climática no Antropoceno*: da rudeza dos fatos à esperança no Brasil. E-papers, Rio de Janeiro.

Machado, P. A. L. (2006) *Direito à informação e meio ambiente.* Malheiros, São Paulo.

McAllister, L. K. (2004) Environmental Enforcement and the Rule of Law in Brazil PhD Thesis. University of California.

McGreavy, B. (2015) "Resilience as discourse," *Environmental Communication, 10* 104–121.

Meadows, D. H., Meadows, D. L., Randers, J. and Behrens, W. (1972) *The limits to growth: a report for the Club of Rome's project on the predicament of mankind.* New York: Potomac Associates.

Meadows, D. H., Randers, J. and Meadows, D. (2004) *Limits to growth: the 30-year update.* Chelsea Green, White River Junction.

Milaré, É. (2001) *Direito do ambiente.* RT, São Paulo.

Ministério de Minas e Energia – MME (2014) *Plano Decenal de Expansão de Energia 2023.* MME/EPE, Brasília.

Ministério Público Federal – MPF (2015) *Relatório de Atividades* (http://bit.ly/1JZB03v) Accessed 29 May 2015.

Nalini, J. R. (2010) "As mudanças climáticas perante o direito" in Benjamin, A. H., Irigaray, C. T., Lecey, E. and Cappeli, S. eds, *Florestas, mudanças climáticas e serviços ecológicos.* Imprensa Oficial do Estado de São Paulo, São Paulo, 143–160.

Nobre, A. D. (2014) *O Futuro Climático da Amazônia.* ARA, São José dos Campos.

Obermaier, M. and Rosa, L. P. (2013) "Mudança climática e adaptação no Brazil: uma análise crítica" *Estudos Avançados, 27*, 155-176.

Pinto, L. F. (2012) "De Tucuruí a Belo Monte: a história avança mesmo?" *Boletim do Museu Paraense Emílio Goeldi, 7*, 777–782.

Prigogine, I. (1996) *The End of Certainty: Chaos, Time, and the New Laws of Nature.* The Free Press, New York and London.

Rajão, R. and Soares-Filho, B. (2015) "A encruzilhada das emissões do desmatamento" Rio de Janeiro, *Observatório do Clima* (http://bit.ly/29IMKta). Accessed 2 March 2016.

Reis, L. B., Fadigas, E. A. A. and Carvalho, C. E. (2005) *Energia, recursos naturais e a prática do desenvolvimento sustentável.* Manole, Barueri.

Rocha, L. S. and Carvalho, D. W. (2006) "Policontexturalidade e direito ambiental reflexivo," *Revista Seqüência*, *53*, 9–28.

Sarlet, I. W. and Fensterseifer, T. (2014) *Princípios do direito ambiental*. Saraiva, São Paulo.

Spencer, T. and Pierfederici, R. eds. (2015) *Beyond the numbers: understanding the transformation induced by INDCs*. Paris, IDDRI – MILES Project Consortium.

Superior Tribunal de Justiça – STJ (2013) AgRg na SLS 1419/DF (http://bit.ly/1IlrE2I). Accessed 2 March 2015.

Superior Tribunal de Justiça – STJ (2009a) REsp 1000731/RO (http://bit.ly/1Lv4RR5). Accessed 15 April 2015.

Superior Tribunal de Justiça – STJ (2009b) AgRg nos EDcl no REsp 1094873/SP (http://bit.ly/1EVBXJL) Accessed 2 March 2015.

Supremo Tribunal Federal – STF (2005) ADI 3540 MC (http://bit.ly/1JEvbWy). Accessed 12 March 2015.

UNESCO (2005) *The Precautionary Principle*. UNESCO, Paris.

Van Der Sluijs, J. P. and Turkenburg, W. (2006) Climate Change and the Precautionary Principle in Fisher, E., Jones, J. and Von Schomberg, R. eds, *Implementing the Precautionary Principle: Perspectives and Prospects*. Edward Elgar Publishing, Oxford 245–269.

Varella, M. D. and Platiau, A. F. B. eds. (2004) *Princípio da precaução*. Del Rey, Belo Horizonte.

Vilani, R. M. and Machado, C. J. S. (2010) "A competência da União para elaboração de plano nacional das atividades de exploração de petróleo e gás natural no Brazil," *Ambiente & Sociedade*, *13*, 187–206.

Viola, E. and Franchini, M. (2013) "Brazil na Governança Global do Clima, 2005–2012: A Luta entre Conservadores e Reformistas", *Contexto Internacional*, *35*, 43–76.

Wedy, G. (2009) *O princípio constitucional da precaução*. Fórum, Belo Horizonte.

11 Shaping up Brazil's long-term development considering climate change impacts

Sérgio Margulis and Natalie Unterstell

Tackling climate change risks is one of the greatest development challenges nations face today. Climate change is already under way and it will increasingly affect all elements of life – access to water, food production, health, and the environment. 'If left unchecked, climate change could cause significant economic and ecological dislocations' (IPCC 2007). Even if the global economy is fully decarbonized, some degree of global warming from past emissions is already locked in, posing a serious challenge to social and economic development in all countries. For example, climate change is expected to bring greater water stress and scarcity in dry regions, while more intense floods and inundations are to threaten urban areas already vulnerable to these events. Therefore, it is imperative that we, societies, adapt to the already changing climate.

Still, the issue of climate change can seem remote compared with immediate problems such as poverty, disease and economic stagnation. A typical public investment dilemma arises from these competing priorities: should a country wait for the climate impacts to become stronger? Or should it address them now, since we already know the potential impacts? Since development risks being seriously undermined by climate change, forward planning is critical to avoid creating expensive future risks.

Growth policies should begin to assess climate risks as a matter of course. Efforts to increase agricultural productivity or develop coastal zones, for example, should not come at the expense of higher susceptibility to climate shocks. The design of new infrastructure, crucial for growth and development, will also have to be amended to make these structures fit for climate change. Climate change risks will need to be considered systematically in development planning at all levels in order to build in adaptation measures. Particular attention should also be paid to policies and projects with long-term consequences. These include, in particular, large-scale infrastructure projects, transport networks, land use planning, urban development master plans and others, which play a key role in underpinning economic development and poverty reduction. Although a range of activities contribute to reducing vulnerability to many climate change impacts, in some cases, development may increase vulnerability to climate change. For example, coastal zone development plans that fail to take into account sea level rise will put people, industries, private property and basic infrastructure at risk and prove unsustainable in the long term. In addition, climate change considerations may raise the importance of supporting such sectors as agriculture, rural development, water resource management and urban infrastructure.

Growth does not automatically reduce vulnerability, only the right kind of growth does. For example, investment in skills and access to finance, indeed reduce vulnerability to climate change. However, investment in infrastructure and efforts to stimulate entrepreneurship and competitive markets must take more of a risk management perspective and recognize climate risks (Bowen et al. 2011). Adaptation measures play a role in adjusting the 'right kind of growth,', if it serves to develop the 'right kind of infrastructure' in the 'right places.'

Coping with changing climate conditions

The mean global temperature increased by almost 0.74 degrees Celsius over the course of the twentieth century, following a sharp increase in GHG concentrations since the 1950s. Most of the warming took place in the last few decades. For Brazil, the observed increase has been greatest in the South and Southeast regions, as suggested in the IPCC's Fifth Assessment Report (AR5).

The IPCC estimates, on the basis of scenarios of future GHG emissions and projections from computer models of the climate, that the Earth's average surface temperature will increase by between 1.1 degrees Celsius and 6.4 degrees Celsius (relative to 1990) by the end of this century. During the same period, global mean sea levels are projected, according to the IPCC, to rise by at least 18 cm and perhaps by 59 cm.[1]

The rise in global temperature also affects other climate variables, especially precipitation. Such increase will not be spread evenly, neither across regions nor within countries. In 2010, the Economics of Climate Change study in Brazil (ECCB) (Margulis and Dubeaux 2010) utilized the Hadley Center ES2 Global Circulation Model to evaluate potential impacts of climate change in Brazil. Based on the assessment of impacts on average values, ECCB pointed to a potentially significant increase in the variability of flows in wet and dry periods in the North region, precisely in places where mega hydropower-plants are being built, and to a reduction in the average flow particularly in the Northeast of Brazil, a region already affected by water shortage.

While changes in average climate conditions are important, societies are especially vulnerable to extreme conditions, such as floods, droughts, heat waves and cyclones. Climate change is expected to amplify extremes, which can have much more significant consequences on society than average increases in temperature or sea level. The IPCC's AR5 indicated that although the average value of events such as rainfall and streamflow has not changed significantly over time, they have become more frequent.

To cope with historical climate and climate variability, agricultural risk information systems and emergency response systems have been put in place, together with the construction of heavy infrastructure such as ditches and reservoirs. These and other mechanisms along with social capital and cultural norms all contribute to and can either increase or diminish a society's 'coping capacity' (e.g., Adger et al. 2007).

However, a number of the coping strategies historically employed are coming under increasing pressure from multiple non-climatic stresses that may make them less effective over time. For example, fire fighting in the Northern and Center-Western regions – already limited in their capacity to protect vegetation and properties against extreme dry weather now – will become even less effective against future fires when subject to higher temperatures and less frequent rainfall. This means that certain regions, sectors, and populations are becoming more vulnerable to climate variability and change, even at present.

Managing the risks from climate change is not something that individual landowners can do on their own. While mitigation of greenhouse gases emissions may require changes in individual attitudes and behaviour, adaptation will have to involve collective action. Infrastructure experts are used to dealing with risks and there is ample scope for maintaining a dialogue in terms of climate risks.

The missing link: future climate scenarios and current investments

The need for infrastructure investment over the coming decades is enormous in Brazil. Inadequate infrastructure, and the corresponding lack of investment, has long weighed on the productivity of the largest Latin American economy (LatinFinance 2015). Brazil scores low on a large variety of qualitative indicators of infrastructure adequacy (Frischtak and Moreira 2014). Based on overall infrastructure quality, Brazil ranked 120 out of 144 countries surveyed by the World Economic Forum in 2014, with particularly poor results for roads and ports. Moreover, the recent drought episode has underscored vulnerabilities from the high dependence on hydropower for electricity generation (Garcia-Escribano et al. 2015).

Climate change does not alter this need but may increase its costs. Making infrastructure resilient to climate change is an important and early adaptation challenge. Infrastructure assets are long-lived and have the potential to lock-in development patterns for a long time (Margulis and Narain 2010). A rapid assessment of policy documents of the Federal Government of Brazil shows that no investment nor budget plans in the 2011–2015 policy cycle have taken any sort of climate change scenario into consideration.

Climate-proofing is not cheap: infrastructure adaptation tends to dominate adaptation cost estimates (Margulis and Narain 2010). The need for adaptation in Brazil's energy sector by 2050, coupled with the need for expansion of the energy system, was estimated to exceed US\$ 10 billion, and another almost US\$ 100 billion of investment (Margulis and Dubeaux 2010).

A higher probability of extreme events may make fiscal sustainability both more important and more difficult to achieve (Lis and Nickel 2009; Williges et al. 2015). Government budgets may come under pressure if more funding is required for emergency services, reconstruction and climate proofing existing infrastructure. There may also be setbacks in terms of social returns on policy, like in the extraordinary decrease of inequality achieved by Brazil in the last 30 years. Fiscal pressure may be compounded by a temporary fall in revenues in the aftermath of a disaster and by the risk of moral hazard if private actors rely on public emergency coverage (Heipertz and Nickel 2008).

Climate change may also affect where infrastructure is built and how it is designed, as the current geography of agricultural production may be completely transformed, creating new transportation needs and perhaps making current infrastructure idle. The same possibility exists in the energy sector, where the development of the hydropower potential may be challenged by changing precipitation patterns and fragile technologies. Furthermore, the energy sector planning assume that there will be no significant changes in rivers runoff volumes in the coming decades, leaving a significant optimistic bias on prospects for the production of energy from hydropower plants (80 per cent of power in Brazil comes from hydropower).

There may be a need for additional infrastructure, dedicated to climate protection, such as sea defences and flood protection. In order to identify and quantify these demands, the generation of climate-related information is necessary. It also requires significant mainstreaming of climate issues in current planning in both energy and water management planning, given how little future climate effects have been considered in current policy and investments planning in Brazil. To this end, starting this process is important.

Climate data and projections are crucial to this end. The public sector also needs to make sure that end users understand and use this information properly. Sectors are happy to stay within their comfort zones and conduct risk assessments, but not to move beyond these diagnostics unless more reliable and consistent information is available. There is a need to educate planners and investors regarding the use and interpretation of climate data, and the scientific community can be of great help, if also invited to the process.

Thus there is an opportunity – and urgency – to advance forward-looking infrastructure planning and development strategies, in order to build the 'right type of infrastructure' in 'the right places'. What would they be? Infrastructure that maintains resilience to future expected climate effects, avoiding concentration in areas that are highly exposed to climate change or else, incorporate in the design phases new resilience parameters. This means that infrastructure regulation must ensure that service providers incorporate resilience to climate effects, avoiding future climate risks.

Starting up the future adaptation of the Brazilian economy to climate change

The Brazilian Government set up an inter-ministerial group consisting of the ministries directly interested or related to climate change issues in order to prepare a national adaptation plan. The Secretariat of Strategic Affairs of the Presidency (SAE-PR)[2] dedicated itself, from 2013 to 2015, to prepare an encompassing and ambitious technical-economic study that aimed at advancing the adaptation undertakings in the national long-term development planning. The study aimed to serve as a main technical support document to the preparation of the national adaptation plan.

The study was known as the 'Brazil 2040' – due to its planning horizon – and considered seven critical sectors of the Brazilian economy: water, agriculture, energy, urban infrastructure, coastal infrastructure, health and transportation. A steering committee, consisting of 15 decision-making institutions of the federal government plus three civil society representatives, was formed. Eight of the most respected research institutions in Brazil, focusing on climate-related issues, were hired to develop expert assessments on each of the above areas. In total, 25 partners took part in the preparation of climate scenarios, sector analysis and adaptation plans, including the following: Ministry of Science, Technology and Innovation, Special Secretariat for Ports, Ministry of Cities, Ministry of Transport, Energy Planning Company, Ministry of Integration, Ministry of Planning, Navy, National Water Agency, Ministry of Mines and Energy, Cabinet of Institutional Security of the Presidency, Ministry of Agriculture and Livestock, Logistics Planning Enterprise, Ministry of Environment and others.

The study had a relatively simple and intuitive logic behind it, consisting of four steps. The first was to have a set of different climate scenarios for Brazil, with a minimum and adequate level of downscaling. Based on these scenarios, the second step consisted of the identification and assessment of the likely impacts of each climate scenario on water resources. This is crucially necessary because of the high significance of the climate-induced impacts of water resources on people, ecosystems and economic sectors – such as droughts, floods, sea-level rise, irrigation and water supply and sanitation. For Brazil, in particular, the impacts of water runoff on hydropower generation are extremely significant.

Based on the projected variations on climate and on water resources, the third step consisted of analyzing in quantified terms the impacts on people, ecosystems and economic sectors. This step involved mainly relating the effects of changes in climate and water variables – such as temperature, precipitation and water runoff – on the productivity of economic sectors, existing infrastructure, people's health, etc.

Lastly, based on the identification of the expected impacts, the fourth step would consist of identifying appropriate adaptation measures. Such measures could involve expensive infrastructure (such as dams to store water, or construction of dykes in coastal areas), but also simple and much less expensive measures such as risk alert systems, changes in agricultural practices, informing and strengthening social networks, etc.

Assumptions and technical considerations for the Brazil 2040 study

The scenarios used in the Brazil 2040 study were based on the IPCC's AR5, which adopted Representative Concentration Pathways (RCPs). RCPs refer to different possible trajectories of the concentration of greenhouse gases (GHG) in the atmosphere, resulting from different scenarios of emissions of such GHGs. Higher concentrations of GHGs correspond to more radiation being trapped in the atmosphere and heating up the Earth.

In the RCP 8.5, GHG emissions rise continuously throughout the twenty-first century, resulting in three times the current CO_2 concentrations in the atmosphere in 2100. The projected average temperature increase is 3.7 degrees Celsius and the average sea level rise about 0.63 meters.

In the RCP 4.5, GHG emissions reach a peak in 2040, and then they start to decrease due to the mitigation of emissions. The atmospheric concentration of GHGs is stabilized around 2060 and reaches 550 parts per million by the end of the century (representing an increase of 50 per cent over the concentration of the pre-industrial era). The increase of average temperature in this case would be 1.8 degrees Celsius and average rise in sea level of 0.47 meters in 2100.

For the Brazil 2040 study, SAE decided to use two global circulation models (GCM) downscaled to the Brazilian conditions at a sufficient scale to project the most relevant impacts at a national level. Previous studies, such as The Economics of Climate Change in Brazil (ECCB) and other assessment reports (such as those from the Brazilian Panel on Climate Change) were solely based on one global model, and downscaled to a resolution of only 40 sq. km x 40 sq. km. INPE (National Institute for Space Research), who was responsible for modelling the climate scenarios, identified two GCMs as suitable for the study, one capturing wet conditions and the other dry conditions for the Brazilian Amazon and the Northeast regions.

The two models were the (i) Model for Interdisciplinary Research on Climate (MIROC) from Japan and the (ii) Hadley Centre Global Environment Model (HADGEM) of the British Met Office. The downscaling of INPE resulted in four scenarios with higher resolution and monthly values for temperature and precipitation until 2100. The use of an odd number of scenarios was avoided, as a way to prevent that a moderate scenario would dominate the analysis. So an even number of four scenarios of global warming was adopted.

These four scenarios were then used by the FUNCEME (Ceará Foundation of Meteorology and Water Resources) as inputs to their physical hydrological models, which were calibrated to the spatial scale of the hydrological stations of the hydropower plants that currently are the basis for the operations of the

ONS (National Electric System Operator). FUNCEME's models adjusted equations governing phenomena such as rainfall, evaporation, infiltration and storage in aquifers so that, from historical rainfall and temperature data, the model could approximate the historical inflow to dams. The methodology involved calibrating the model according to a different set of rainfall and temperatures, simulating and verifying that the resulting flows fitted the observed (past) data. The hydrological modelling was applied on a monthly basis to the 195 hydropower plants, as per the ONS configuration, from 2010 to 2100.

For the sectorial analyses, each of the five sector teams adjusted parameters to evaluate impacts under the conditions of each of the two climate models and the two RCP scenarios. In the case of the energy sector, the climate-water models were used as inputs to two energy sector models that identified the new mix of energy sources in Brazil capable of supplying fixed levels of demand at minimum costs. The analyses of coastal infrastructure depended fundamentally on scenarios of sea level rise, and not on the climate scenarios themselves, and for this reason they were given exogenously to the climate models.

Climate change scenarios: Temperature, rainfall and water flow projections for Brazil in 2040, 2071 and 2100

Brazil is set to become warmer and drier in all scenarios generated by INPE for the Brazil 2040 study. The values presented in the climate maps below (Figure 11.1) correspond to the minimum and maximum differences in temperature between the measured period and the pattern observed from 1961 to 1990.

The figure indicates that warming is maximum in the Center-West Region, in every month of the year. Average temperatures in the hottest months of the year could rise by 3 degrees Celsius over current averages in that region. Between 2071 and 2100, the maximum warming extends to the Northeast, North and South regions. At the end of the century, an increase of eight degrees Celsius is projected for the Northern region, where the Amazon rainforest is located, under the RCP 8.5 scenarios.

With regard to changes in precipitation, this is where the projections between the two climate models differ most. Reduction in rainfall in the rainy season (summer) is projected in most of the country, with a maximum reduction in the Midwest Region and Southeast areas. The reduction appears more intense in the first 30 years, then less intense from 2040 to 2070, and returns to be quite intense in the last 30 years of the century.

These projections reveal positive rainfall anomalies in the extreme South of Brazil, whereas other regions would present negative anomalies. Simulations of the HadGEM2-ES model, which is a 'more intense model,' projects a reduction of precipitation throughout the century, especially in the coastal areas. Comparing the two scenarios, RCP8.5 appears drier in most parts of Brazil, especially on the coast of the Northeast and Southeast regions of the country. The Southern region tends to become rainier, while the Southeast, the Midwest

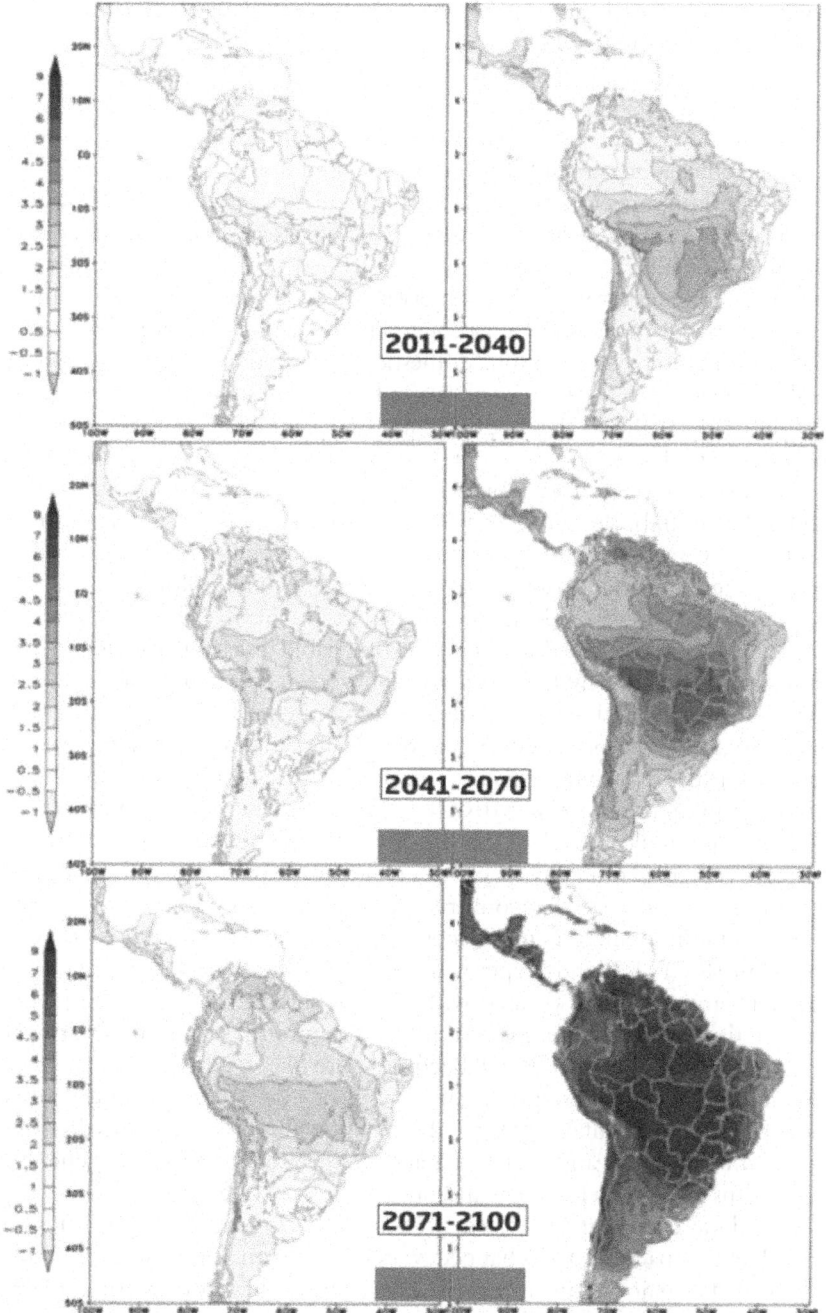

Figure 11.1 Minimum (left) and maximum (right) temperature change, 2010–2040, using two models (MIROC-ETA and HadGenES2-ETA) and two RCPs (4.5 and 8.5)

and parts of the North and Northeast have projected reductions in rainfall, especially in the summer months.

Impacts on the power grid

The projected changes in rainfall patterns presented in the previous section are a cause of significant concern in Brazil. The immediate consequence of changes in precipitation is changes in the amounts of runoff water in the major river basins in Brazil. In general, the projections agree on a downward trend in the average water flows compared to historical observations – between 38 per cent and 57 per cent relative to the historical average. The impacts on hydropower, on the availability of water for irrigation, domestic consumption, industries, watering animals, as well as, no less important, water for ecosystems – are all concerning.

FUNCEME – the Federal University of Ceará State – carried out the projections of changes in water runoff as a consequence of changes in precipitation. The spatial distribution of the projected water runoff for the period 2011–40 in both RCP scenarios is shown in Figure 11.2. The simulated scenarios have in common a positive trend in water flows in the Southern region and reductions in most of the Midwest, North and Northeast regions.

Measurements were made in the reservoirs of the major power plants in Brazil because they hold the longest and most reliable historical data. Reductions in virtually all Brazilian hydrological stations were accounted for in projections derived from the HadGenES model, while the MIROC5 model showed increase of flows in some stations in the Southeast and South regions and reductions in the Northern region. The models results diverge about the rest of the Southern region: the MIROC5 projected increase in flow rates of various basins, while the HadGEM2-ES model shows flow rates below the historical average as much as 40 per cent in some basins in the three periods. The results suggest that the country's four largest power stations – Itaipu, Furnas, Sobradinho and Tucuruí – could experience declining flows between 38 per cent (RCP 4.5) and 57 per cent (RCP 8.5).

Higher temperatures associated with climate change will place considerable strain on the power sector as currently configured. Across the North and Northeast, drought conditions will become more likely, whether due to greater evaporation as a result of higher temperatures or – in some areas – less rainfall or more sporadic rainfall. Extended droughts clearly reduce hydropower output, while droughts and heat waves add stress to transmission and generation systems, thus reducing efficiency and raising the cost of electricity.

The Federal University of Rio de Janeiro (UFRJ) carried out the study that analyzed the impacts of the changed water runoff on the power generation capacity in the major basins in Brazil. The team evaluated how the operation and the expansion of the National Interconnected System (SIN) could be affected by climate change scenarios. The exercise breaks new ground assessing climate impacts on the Brazilian power system precisely because, surprisingly, they are currently not taken into account in the conventional energy planning.

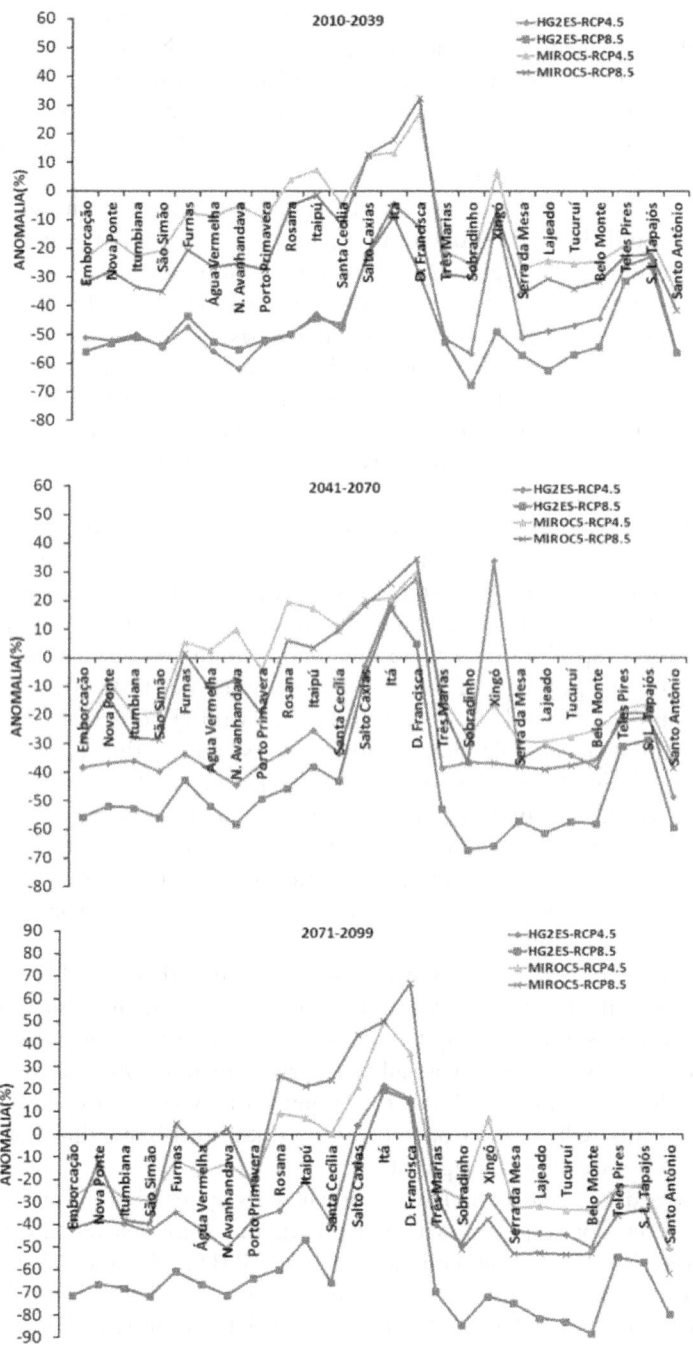

Figure 11.2 Average projected change in water runoff in selected hydropower reservoirs, 2011–2039 (FUNCEME 2015)

In all climate-hydrology scenarios that have been simulated, the losses in terms of power generation and the operation of the National Interconnected System (SIN) are immense. The estimated costs range from US$ 3 to 280 billion. In the worst scenarios (RCP 8.5), the hydropower potential can get 8 to 20 per cent lower by 2040. In the best case, the capacity still decreases between 4 and 15 per cent.

Most of the hydropower plants in Brazil were built without regulating reservoirs, and that has reduced the capacity of the national power system to sustain prolonged droughts (by releasing accumulated water in reservoirs). This is of course more critical in the dry seasons and in drier climate conditions – which are in fact projected for so many basins. As for new hydropower investments, since the runoff projections are the same as for the existing plants, it is crucial that early decisions are made regarding the need and appropriateness to build more regulation reservoirs. This may in fact dictate the feasibility of these investments and also the level of reliability of the hydropower system in Brazil.

The projections for the new strategic areas are also of reduction, or significant reduction, of water runoff, such as São Luiz do Tapajós in the Northern State of Pará, which will see a flow reduction of at least 20 per cent. Belo Monte, an icon in investments already being made in the Amazon, also shows a decrease in generation capacity under various climatic and hydrological scenarios.

In terms of operation of the national system, the study suggests that the climate impacts would be very significant, leading to a disruption of the system and unacceptable deficit probabilities, above the 5 per cent level currently admitted. In the scenario of milder impacts (RCP 4.5), the risk of a deficit would, on average, be above 10 per cent, reaching 17 per cent in some years. In the extreme impact scenario (RCP 8.5), the deficit would go above 90 per cent. This leads to dramatic increases in costs associated with such deficits as well as very high operating costs (ranging from 3.5 to 16.7 times relative to baseline).

The explanation for increased operational costs is that the SIN depends on both hydro and thermal power generation. The thermal plants tend to be cheaper in terms of cost of capital but present higher operating costs and lower efficiency, so they operate basically in periods of unfavourable hydrology to offset hydropower plants. Under the projected new conditions, however, the thermal plants will have to operate on a continuous basis, thus charging the system with greater operating costs.

The speed that Brazil will have to enhance its hydro, thermal and alternative power generation capacity will depend on the scenarios of climate change evolution. The worst-case scenarios, such as the RCP 4.5 and 8.5, will call for more accelerated expansion of capacity because the SIN will not be able to cope with such high levels of risk. The impacts of climate change in Brazil coupled with the country's extreme dependence on hydropower will force the country to increasingly rely on thermal power, thus significantly increasing its GHG emissions. This appears a little odd, but emissions from energy use in

Brazil have always been significantly low precisely because of hydropower. In the absence of more significant movements towards decarbonization, climate change will force Brazil to increase its GHG emissions from energy use. In contrast, in a world where carbon costs are internalized, the increase in emissions due to less water availability could be significantly lower.

The trajectory of climate change (i.e., the RCP scenario that will really occur), will indicate the need for, and the speed of adaptation measures in the energy sector. In the case of adaptation via system expansion, the cost difference between the RCP 8.5 and RCP 4.5 is US$ 76 and 122 billion, considering the MIROC or the HadGEM model, respectively. That is, if an international cooperative effort to restrict emissions is as good as to reach an RCP 4.5, the avoided adaptation costs for the Brazilian power grid would be of this order of magnitude.

This reinforces the need to use these results as an 'urgent alert' to the authorities who must respond both to increased investment in research in this area (even more if one considers the ongoing high investment in hydropower plants in Amazon region), as well as the establishment of a set of preventive measures that should be quickly incorporated into energy planning by the government.

A last point to note is that the adaptation measures proposed by the energy team were the least costly ones. They even involve the use of coal and other (cheap) thermal sources that are extremely polluting and sources of GHGs. This strategy clearly confronts Brazil's pursuit of a low-carbon economy. The numbers thus provided are lower bound estimates of the real costs of adaptation in the energy sector. The same applies to other forms of adaptation, that may either involve higher emissions of GHGs – which is clearly a non-starter – or that may collide with one another. For example, more intense use of irrigation in agriculture, to counter higher temperatures, may clash with increased demand in urban, industrial and other water uses. This calls for integrated approaches – both between mitigation and adaptation measures as well as between economic sectors.

Impacts on agriculture

The extent and the magnitude of how agriculture can be affected by climate change was evaluated by a joint team formed by EMBRAPA (agronomic analyses) and Agroicone (economic analyses), which looked into some of the most important crops in Brazil – soybeans, corn (summer and winter crops), beans (summer and winter crops), rice, cotton, wheat and sugarcane. They also considered cattle ranching and a few industrial products from agriculture (oil and soybeans meal, sugar and ethanol, beef, pork and poultry and milk).

They assessed climate impacts over those individual crops, evaluated change in production possibilities and estimated changes in land prices. Impacts are related to change in water availability in the different regions. Since the water supply in Brazil for agricultural crops, pastures, forests and orchards comes almost exclusively from the rain, the most important climate risk factor derives from the water conditions during cultivation.

The study indicated that agriculture can be particularly hard hit by a warming climate in Brazil, as the agricultural risks are increased for almost all crops in all the scenarios, in comparison with today's level of agricultural risk. An increase in the number of municipalities classified as high risk in all agro-climatic simulated scenarios is expected, negatively affecting the current areas of production.

Some of the country's main crops could suffer a serious decline in the areas already under cultivation. The most severely affected is precisely the country's main export crop – soybeans – that could face losses of up to 67 per cent of the area planted in the Southern region by 2040. This could translate into significant financial losses, since soybeans currently bring in US $20 billion in export earnings every year. Area losses projected for other crops include corn (28 per cent), beans (26 per cent) and rice (24 per cent).

Rice is a culture that has negative impacts in all scenarios and variables for the family and not family farms. For soybeans and corn, there is significant vulnerability. Investing in alternatives of production, through improved varieties, alternative technologies of production and developing mechanisms for risk management in these cultures is suggested as a priority by the researchers.

The Mid-West and Northern regions could offset some of the above-mentioned production losses through relocation of production. Pasturelands also lose with a warming climate. Not necessarily because the new climate conditions will become detrimental to the pasture, but because both degraded areas as well as other areas appropriate for agriculture will likely be occupied by relocated crops. The livestock area is projected to fall by 6.5 per cent. Yet, the study expects there to be an intensification of production, so that the beef production could avoid significant impact, falling by 'only' 2.6 per cent in 2040 compared to the baseline scenario.

Sugarcane appears to be a production alternative, as it experience gains of productivity and risk profile in scenarios of higher temperature.

The impacts at a national level obscure local impacts that can be very relevant, especially for the Southern region, which can lose most in terms of production value, employment and income. Some municipalities are projected to fail to produce soybeans in the future, as they become classified as located in high climatic risk.

Regions such as the Semi Arid, North, Midwest and Southeast can become riskier in terms of production of current suitable crops and may suffer from higher depreciation of land. In the Northeast, there might be greater pressure on smallholders dependent on government transfers, pressing public policy to make a direct link between climate variability and social policies. Smallholders and non-commercial farmers deserve special attention because they have less structure for relocation of cultures.

The impacts suggest problems of social nature and leads to special attention to the rural middle class in vulnerable regions, especially family farming enterprises. Maize and beans are pointed out by the EMBRAPA and Agroicone researchers as important alternative products for this type of producer.

In general, the states of the South region tend to experience a net gain of alternatives for production. In the Northeast, there is a possibility of gains in regions hit by rain, near the coast. This situation is quite uncertain, as there is uncertainty on the impacts of climate change on rainfall distribution. Much of the semi-arid regions North, Northeast and Midwest have losses in terms of increased climate risk for the agricultural production. This result, compared to changes in land values, shows trends similar to the impacts.

The value of the land reflects the worsening expectations for production. Expected reduction in land prices is due to the expectation of lower returns of production and the elevation of the climatic risks.

Impacts on coastal zones: cities and ports

The assessment of coastal risks conducted by ITA (Institute of Aviation Technology) was rather limited by a lack of data. Among the vulnerable metropolitan areas that are economically very important are Rio de Janeiro and Santos, considering sea level rise, storm surge risk, landslides and floods. The two cities had a minimum level of data allowing for a preliminary analysis of vulnerability risks for the four climate change scenarios. The cities were categorized into areas of very low to very high vulnerability. Vulnerability was assessed based on an overlapping of existing infrastructure (roads, sewerage treatment plants, hospitals, metrolines) with existing risk maps of climate events.

In Rio de Janeiro, the researchers crossed the vulnerability map with real estate valuation. In the mildest scenario, RCP 4.5, an area worth R$ 109 billion was deemed threatened in the next 25 years. In the worst scenario, the figure has risen to R$ 124 billion. In Santos, around 60 per cent of the city fell into the highly vulnerable category, including critical infrastructure like hospitals, which have to maintain reliability of their services in disasters and extreme weather events.

With regard to ports, most of them already suffer from the rising sea level through reduction of freeboard. The Port of Santos, for example, which is designed to have a freeboard of 1.18 m, gets just 0.95 m at low tide times. It may fall to 0.72 m by 2040 in the worst climate change scenario.

Freeboard areas are already below the recommended levels for the ports of Belem, Recife, Maceió, Niteroi, Rio de Janeiro, São Sebastião, Santos and Paranaguá. Projections of tide elevation, for the years 2030 and 2050 imply a further reduction of free board of the docks (for example, the port of Recife, which features a freeboard area of 0.97 m in 2015, would have this value reduced to 0.87 m in 2030 and 0.70 m in 2050). In addition, the average trend by 2050 for all ports is: (a) a 33 per cent average increase in sedimentation of the external access channels; and (b) the need to increase the width of the external access channels, due to the increasing size of vessels and the increase in the height of the waves.

The main impacts of sea level rise and its immediate consequences on the waves systems and storm surges on ports are: the partial drowning of mangroves and marshes; the reduction of the freeboard area of the wharfs; flood of port and road systems; the deterioration of the defence of coastal and estuarine ports such as jetties, breakwaters and jetties' guides-currents; the massive overtopping of defence works from the coastal ports; and silting of the bar channels or external channels.

In terms of adaptive measures, the team suggested both green and grey infrastructure measures: strengthening of piers with massive armour and breakwaters; artificial blocks (Fortaleza, Recife and Maceió); heightening of the massive piers and breakwaters (Fortaleza, Recife, Maceió, Tubarão, Imbituba, Laguna and Rio Grande); and annual increases of dredging the bars (all ports except Santos) are of immediate concern. Further, recognizing the many benefits that natural infrastructure provides, the coastal and ports expert team recommended immediate management of mangroves (Recife, Maceió, Niterói, Rio de Janeiro and Santos).

Measures to be completed by 2030 should include heightening of vestment and quay micro drainage (Belem, Recife, Maceió, Niterói, Rio de Janeiro, São Sebastião, Santos, Paranaguá and Antonina); macro drainage without storm surge barriers (Belém, Maceió, Niterói, Rio de Janeiro, São Sebastião, Paranaguá and Antonina); and reinforcing piers of massive armor and break-waters, artificial blocks (Pinto, Barra do Riacho, Tubarão and Praia Mole).

For 2050, the team recommended macro drainage without storm surge barriers (Praia Mole); macro drainage with storm surge barriers (Recife, Tubarão and Santos); reinforcing piers of massive armor and breakwaters, artificial blocks (Imbituba, Laguna and Rio Grande); and heightening of the massive piers and breakwaters (Barra do Riacho and Praia Mole).

Compared to the RS$ 2.7 billion of investments in ports included in the PAC (the national infrastructure program called the Growth Acceleration Programme) for the period 2007–2010, estimates of costs of the suggested adaptation measures are (in 2007 values): (a) immediate action: R$ 13 million; (b) up to 2030: R$ 1.665 billion; and (c) between 2030 and 2050: R$ 7.255 billion (including three storm surge barriers). These investments are perfectly compatible and feasible in the very long horizon of 35 years, considering the strategic importance of the port system for the country's development as well as the contribution in synergy that these works can offer to the metropolitan areas around it.

Impacts on road transportation

The analysis of the impacts on road infrastructure was deemed relevant based on Margulis et al. (2009), who suggested that the costs of adaptation in this sector are among the highest of all infrastructure sectors. The study was carried out by a joint team of the Federal University of Rio de Janeiro (UFRJ) and the Military Institute of Engineering (IME). They constructed a Vulnerability

Index of Road Infrastructure – IViR – to evaluate hotspots in terms of temperature and precipitation. The criterion used to define hotspots was to consider surface temperature climate on determining the upper and lower limit of pavement resistance of highways (only paved roads were considered because of a lack of data and maps of unpaved roads in Brazil, which constitute a vast majority of roads in the country).

The results showed that much of the road network is located in the hotspot temperature, triggering the need for engineering interventions on existing roads and the need to adapt the designs for the planned roads. Yet, the expected impacts in terms of changes in precipitation are more significant. The hotspots were defined based on the overlap of the future climate precipitation maps and the annual maximum rainfall intensity recorded in one day, considering the time of recurrence five years, the minimum scale used for drainage structures.

The results indicate that the areas likely to face climate variations effects have reduced magnitude. Segments with high vulnerability in the present were located in the states of Acre, Alagoas, Paraíba, Piauí, Rio Grande do Norte, Rondônia and Tocantins, while 14 other states and the Federal District showed potential future vulnerability. In fact, one can infer that the main problem is an already existing problem of adaptation deficit in the poorer areas in Brazil. The roads with highest vulnerability to climate events coincide with the poorer regions in Brazil.

Adaptation measures in the road transportation sector include improvements in the road infrastructure per se – and these vary from more simple operations and maintenance intensification to more expensive protection measures such as pavement rutting, new standards, etc., and up to total replacement – as well as mobility behaviour measures, such as encouraging freight trucks to travel at night.

Impacts on urban infrastructure

Changes in storm intensity associated with climate change have the potential to overburden urban drainage systems across much of the Brazilian cities, particularly in areas where storm intensity is expected to increase significantly. Cities with older infrastructure that have not made sufficient or appropriate maintenance and expansion investments, may already experience floods at times of more intense rainfall. Significant investment in urban drainage infrastructure may be necessary to prevent the exceeding of system capacity.

One significant problem in evaluating such potential problems is the lack of data about the existing network of drainage systems. This is a widespread problem throughout Brazil. We do not know of any studies carried out for any cities in Brazil that have used storm-water modelling to project likely impacts of increased rainfall on their drainage systems. Because of the great heterogeneity among cities with respect to their current urban drainage systems, the team in charge of these analyses (Instituto Tecnológico de Aeronáutica – ITA) employed a generic impact assessment approach intended to be applicable to big cities in Brazil.

The analyses were carried out for the two largest cities in Brazil – Rio de Janeiro and São Paulo. Neither municipality currently considers climate scenarios in their planning scope. Micro drainage systems were not analyzed, but the macro drainage systems, specifically the channels excluding detention basins were analyzed. To project the runoff, the curve number method was applied to the selected basins, which are part of the National Plan for Risk Management and Response to Natural Disasters 2012–2014. The macro drainage plans generally assume stationary climate series and reproduce their history into the future, focusing on investments in the extreme stochastic recurrence times and the risks assumed from these projections.

The modelling exercise conducted for São Paulo pointed to a significant reduction in the amount of runoff, as a result of the conditions of use and occupation imposed by the simulated scenario, providing smaller surplus rainfall, that is, higher infiltration rates. This means that the current infrastructure would be sufficient to cope with the changing climate, at least in terms of average changes. However, extreme weather conditions remain to be evaluated. If the pattern of expansion of the São Paulo Metropolitan Region is maintained as per historical trends, in 2030 the urban area will be approximately 38 per cent bigger than the current one, increasing the risks of natural disasters such as floods and massive landslides on slopes, reaching the population as a whole and especially the most vulnerable (Nobre 2011).

For the Rio de Janeiro analysis, the team adopted the Mangue Canal basin as the study area. Regarding the scenario HadGEM2 RCP 8.5, there was a 15 per cent decrease in peak flow compared to the current situation. In the MIROC5 RCP 8.5, this peak was increased by 16 per cent. The big difference in behaviour between the models makes it difficult to infer a future pattern of climate change. It was observed that the conditions of use and occupation of land, imposed by the simulated scenario, provide greater infiltration rates and the consequent reduction in the amount of runoff generated in the basin, both for the current situation, as for the simulations the climate models used.

The projections of the climate models pointed to a reduction in the average annual rainfall, but possibly expanding the range of extremes. If on one hand this reduction involves less pressure on the drainage networks, the variation of the extremes – and possibly increased frequency of these events – point to the need for more resilience of the infrastructure. Also, it was found that the current model of use and occupation of these cities could compromise the ability to quickly drain the rainwater causing floods. Depending on the patterns of future land use and occupation, the results could be more drastic in terms of future needs of expanding the drainage infrastructure. The master plan of the cities must include zoning of permeable areas, preferably on each lot/household. Lowland areas and larger riverbeds need to be saved from intense occupation through the implementation of parks and/or recreational areas.

Conclusions

In the context of planned adaptation to climate change, decision-makers in the private and public sectors first need to become aware of the potential impacts and risks, and how these risks may affect them or their specific business and management responsibilities. This awareness needs to be coupled with a fuller understanding and capacity to analyze such information in order to develop policy initiatives, strategies and plans. Echoing Moser and Luers, this ability and resulting understanding can, but may not suffice to, provide the necessary motivation and willingness to act (as per Rayner et al. 2005; UKCIP 2003). Rapid assessments done by the authors revealed that in Brazil, in a majority of cases, decision makers currently are neither aware of the potential risks nor are prepared and equipped to address them.

The overarching message emerging from the Brazil 2040 study is that the Brazilian economy, its population, the built assets, and its ecosystems are rather vulnerable to a range of impacts projected by different climate scenarios: the situation is very concerning for some specific sectors and regions. The government in general must first realize and understand this threat. This means that the model results still need to be translated into information that is understandable and salient to decision-makers. For example, while sea-level rise projections are valuable as a general indicator to raise awareness of future coastal risks in a general sense, permitting officers who determine setback distances to site new buildings need to know how these projections translate – together with possible changes in storm activity – into future coastal erosion rates. The expected result is the mainstreaming of climate risk into public policy in the country. Brazil is still rather distant from this point, and discussions have only recently begun at the highest levels in the Brazilian federal government. This suggests that it will take even longer for the existing knowledge to affect actions on the ground. In order to move forward with this agenda, the following minimum set of questions must be addressed:

- What level of climate change (or risk of change) is society willing to accept?
- What goals should adaptation achieve, e.g., preserving the status quo, actively managing change toward new conditions, promoting deeper societal changes required for sustainability?
- What is an acceptable level of individual vs. public risk and how should the responsibility be shared?
- How to initiate (and provide adequate funding and staff to arrange) public forums to discuss climate change risks and response options; forums could be agency-specific or location-specific, for the private sector, public officials, or the general public?
- How to promote integrated resource and hazard management plans that promote or require incorporating climate risks among other multiple and interacting stressors, like deforestation?

Multi-year investments in adaptation planning are very complex because they require a great deal of sophisticated information that needs to be translated into accessible and useful information to a large range of stakeholders. Given the complexity, more than an exchange of fixed political views or technical data is required: commitment to collaboration and joint problem-solving are essential to address the climate change threats. Adaptation to climate change is indeed a core development issue, and choices to act on it are as complex as the development challenges themselves. And the challenge is made more difficult because of the complexities and uncertainties of the climate change phenomena.

While the Brazil 2040 study involved a good degree of individual and independent sector analysis, it is clear that its most interesting aspect would be to bring all pieces together in an integrated and consistent manner. The various hypotheses and assumptions made by each sector team were in principle consistent with one another. This includes not only the climate projections, but also projections about population and economic growth, changes in levels of demand, etc. Just like any plan of any other sector or theme, a climate adaptation plan requires all proposed policies and actions to be comparable to one another and thus to be as quantified as possible. While full benefit–cost analyses of all such policies and actions would never be entirely possible, valuation provides an impartial view of trade-offs involved in choices and decision-making. Different national and international agencies have started undertaking initiatives on this aspect, but very few concrete measures have been implemented in Brazil so far.

The Brazil 2040 study identified some specific entry points of climate change adaptation needs. However, a larger effort aiming at removing barriers, informing people and decision-makers, creating incentives and modernizing policies in order to encourage investments, practices and partnerships that facilitate resilience to climate impacts remains to be undertaken.

Notes

1   Values of sea-level rise significantly larger cannot be excluded as certain processes leading to sea-level rise are still not well understood (Oppenheimer et al. 2007). For example, in 2014, Quaternary Science Reviews published an extensive expert assessment held with 90 experts from 18 countries, who were among the most active scientific publishers on the sea level topic in recent years, projecting the rising sea level to be probably about 0.44 to 0.6 meters by 2100 and 0.6–1.0 meters by 2300. The most extreme scenarios were 0.7–1.2 meters by 2100 and 2.0–3.0 meters up to 2300.
2   Extinct by the Ministerial Reform undertaken in October 2015. Its functions were transmitted to the Ministry of Planning and Budget.

References

Adger, W.N., Agrawala, S., Mirza, M.M.Q., Conde, C., O'Brien, K., Pulhin, J., Pulwarty, R., Smit, B. and Takahashi, K. (2007) 'Assessment of adaptation practices, options, constraints and capacity', in M.L. Parry, O.F. Canziani, J.P. Palutikof, P.J. van der Linden and C.E. Hanson, (eds), *Climate Change: Impacts, Adaptation and Vulnerability* – Contribution of Working Group II to the Fourth Assessment Report of the Intergovernmental Panel on Climate Change Cambridge University Press, Cambridge, UK, 717–743.

Bowen, A., Cochrane, S. and Fankhauser, S. (2011) *Climate change, adaptation and economic growth*. Springer.

Frischtak, C. and Moreira, M.M. (2014) *Where Is Brazil Going? Taking Stock of Recent Trends in Industrial and Trade Policies and the Challenges Ahead* (http://ssrn.com/abstract=2479536 or http://dx.doi.org/10.2139/ssrn.2479536) Accessed 1 November 2015.

FUNCEME (2015) 'Brasil 3 Tempos' Relatório IV PNUD (http://www.sae.gov.br/imprensa/noticia/brasil-2040-cenarios-e-alternativas-de-adaptacao-a-mudanca-do-clima/) Accessed 1 March 2016.

Fundação Elizeu Alves (2014) 'Brasil 3 Tempos' *Relatório II PNUD* (www.sae.gov.br/imprensa/noticia/brasil-2040-cenarios-e-alternativas-de-adaptacao-a-mudanca-do-clima/) Accessed 1 November 2015.

Fundação Elizeu Alves (2014) 'Brasil 3 Tempos' *Relatório III PNUD* (www.sae.gov.br/imprensa/noticia/brasil-2040-cenarios-e-alternativas-de-adaptacao-a-mudanca-do-clima/) Accessed 1 November 2015.

Fundação Elizeu Alves (2014) 'Brasil 3 Tempos' *Relatório IV PNUD* (www.sae.gov.br/imprensa/noticia/brasil-2040-cenarios-e-alternativas-de-adaptacao-a-mudanca-do-clima/) Accessed 1 November 2015.

Fundação Elizeu Alves (2015) 'Brasil 3 Tempos' *Relatório IV PNUD* (www.sae.gov.br/imprensa/noticia/brasil-2040-cenarios-e-alternativas-de-adaptacao-a-mudanca-do-clima/) Accessed 1 November 2015.

Fundação Elizeu Alves (2015) 'Brasil 3 Tempos' *Relatório V PNUD* (www.sae.gov.br/imprensa/noticia/brasil-2040-cenarios-e-alternativas-de-adaptacao-a-mudanca-do-clima/) Accessed 1 November 2015.

Garcia-Escribano, M., Goes, C. and Karpowicz, I. (2015) 'Filling the Gap: Infrastructure Investment in Brazil', *IMF Working Papers*, July.

Heipertz, M. and Nickel, C. (2008) 'Climate change brings stormy days: case studies on the impact of extreme weather events on public finances' *European Central Bank Working Paper*.

Horton, B.P., Rahmstorf, S., Engelhart, S.E. and Kemp, A.C. (2014) 'Expert assessment of sea-level rise by AD 2100 and AD 2300'. *Quaternary Science Reviews*. Volume 84, Elsevier 15 January.

IPCC (2007) 'Climate Change 2007: Synthesis Report. Contribution of Working Groups I, II and III to the Fourth Assessment Report of the Intergovernmental Panel on Climate Change' Core Writing Team, Pachauri, R.K and Reisinger, A. (eds.). *IPCC*, Geneva, Switzerland, 104 pp.

LatinFinance (2015) BRAZIL INFRASTRUCTURE: Bridging the gap.

Lis, E.M. and Nickel, C. (2009) 'The impact of extreme weather events on budget balances and implications for fiscal policy', *Working Paper 1055*, Frankfurt: European Central Bank.

Margulis, S. and Narain, U. (2010) 'The costs to developing countries of adapting to climate change: new methods and estimates'. *The global report of the economics of adaptation to climate change study*. Washington, DC: World Bank (http://documents. worldbank.org/curated/en/2010/01/12563514/costs-developing-countries-adapting-climate-change-new-methods-estimates-global-report-economics-adaptation-climate-change-study) Accessed 1 November 2015.

Margulis, S. and Dubeaux, C. (eds). (2010) *Economics of Climate Change in Brazil: Costs and opportunities*. São Paulo SBD/FEA/USP.

Margulis, S., Narain, U., Pandey, K., Cretegny, L., Bucher, A., Schneider, R., Hughes, G. and Essam, T. (2009) Economics of Adaptation to Climate Change. World Bank. Washington, DC.

Moser, S.C. and Luers, A.L. (2008) 'Managing climate risks in California: the need to engage resource managers for successful adaptation to change', *Springer Science + Business Media* B.V. January.

Narain, U., Margulis, S. and Essam, T. (2011) 'Estimating Costs of Adaptation to Climate Change'. *Climate Policy*. Vol. 11 (3), pp. 1001–1019, Routledge.

Neumann, J.E. and Price, J.C. (2009) 'Adapting To Climate Change: The Public Policy Response: Public Infrastructure.' *Final report prepared for Resources for the Future*, June (www.rff.org/News/ClimateAdaptation/Pages/domestic_publications. aspx) Accessed 1 November 2015.

Nobre, C. (2011) (ed.) *Vulnerabilidades das megacidades brasileiras às mudanças climáticas: região metropolitana de São Paulo: relatório final / –* São José dos Campos, SP: INPE, 192 p.

Oppenheimer, M., O'Neill, B.C., Webster, M. and Agrawala, S. (2007) 'The Limits of Consensus'. *Science* 14, September 2007: Vol. 317, no. 5844: pp. 1505–1506. (www. sciencemag.org/content/317/5844/1505.summary) Accessed 1 November 2015.

Rahmstorf, S. (2006) 'A Semi-Empirical Approach to Projecting Future Sea-Level Rise', *Science* 19: Vol. 315 no. 5810, pp. 368–370.

Ranger, N., Millner, A., Dietz, S., Fankhauser, S., Lopez, A. and Ruta, G. (2010) *Adaptation in the UK: A Decision Making Process*, London: Grantham Research Institute on Climate Change and the Environment, London School of Economics and Political Science.

Rayner, S., Lach, D. and Ingram, H. (2005) 'Weather forecasts are for wimps: why water resource managers do not use climate forecasts', *Climate Change* 69: 197–227

Schwing, R.C. and Albers, W.A. (eds.) (1980) *Societal Risk Assessment:* How Safe is Safe Enough? Springer Science + Business Media LLC.

Secretaria de Assuntos Estratégicos da Presidência da República do Brasil (2013) Brasil 2040: documento de trabalho (http://issuu.com/sae.pr/docs/brasil2040). Accessed 1 November 2015.

Smith, M.S., Horrocks, L., Harvey, A. and Hamilton, C. (2010) 'Rethinking adaptation for a 4°C world', *Philosophical Transactions of the Royal Society* A. 369, 196–216. (http:// rsta.royalsocietypublishing.org/content/369/1934/196) Accessed 1 November 2015.

Susskind, L. (2010) 'Policy & Practice: Responding to the risks posed by climate change: Cities have no choice but to adapt', *Town Planning Review*, Vol. 81, Issue 3, pp. 217-235 (E- Journal).

Susskind, L., Rumore, D., Hulet, C. and Field, P. (2015) *Managing Climate Risks in Coastal Communities: Strategies for Engagement, Readiness and Adaptation*, Anthem Press.

United Kingdom Climate Impacts Group – (UKCIP) (2003) *Climate Change and Local Communities: How Prepared Are You?* UKCIP, Oxford. (www.ukcip.org.uk/resources/publications/documents/Local_authority.pdf) Accessed 3 March 2016.

Unterstell, N. (2015) In Paris, adaptation should go global (www.nivela.org) Accessed 1 November 2015.

Williges, K., Hochrainer-Stigler, S., Mochizuki, J. and Mechier, R. (2015) 'Modelling the indirect and fiscal risks from natural disasters for informing options for enhancing resilience and building back better', Background Paper prepared for the *2015 Global Assessment Report on Disaster Risk Reduction* UNISDR, Geneva, Switzerland.

12 Pathways to a low carbon economy in Brazil[1]

Emilio Lèbre La Rovere, Claudio Gesteira,
Carolina Grottera and William Wills

Introduction

Brazil occupies a unique position among the major greenhouse gas (GHG) emitting countries due to its low per capita energy-related GHG emissions (2.4 tons CO_2 in 2014), attributable to Brazil's abundant clean energy sources. The sources of major emissions have historically been concentrated in agriculture, forestry, and other land use (AFOLU), and are related mostly to deforestation, crop growing and livestock. Recently, deforestation in Brazil has slowed considerably, to the point where forestry has ceased to be the major source of emissions. Thanks to reduced deforestation, Brazil has reduced its overall GHG emissions by 41 percent from 2005 to 2012, and its total GHG emissions per capita decreased from a high in 2004 of 14.4 tCO_2e to an estimated 6.5 tCO_2e in 2012.

Brazil achieved this reduction in emissions per capita through recent governmental policies combining command-and-control tools (enforcing laws and regulations, such as the Forest Code, through inspecting rural properties and roads spotted by satellite imagery) and economic instruments (requiring agricultural and cattle-raising projects to demonstrate compliance with environmental regulations to be eligible for public bank soft loans that supply most of the credit to this sector).

Before this recent decline, earlier agriculture and livestock emissions growth was driven by the expansion of the agricultural frontier that pushed crop and cattle-raising activities into the *cerrado* (savannah) and Amazon biomes. Brazil is one of the world's most important suppliers of commodities such as soybeans and meat, with 210 million heads of cattle in 2010. Deforestation in the Amazon peaked at 2.8 Mha (million hectares) in 2004. Governmental efforts have succeeded in bringing Amazon deforestation down to 0.7 Mha in 2010 and 0.5 Mha in 2014. On the other hand, emissions related to fossil fuel combustion for energy production and consumption have continued to increase significantly, in parallel with the growth of the Brazilian economy. Fossil fuel combustion for energy production and consumption have reached nearly the same level of those from agriculture plus cattle breeding, and due to this fast growth rate are expected to become the dominant source of GHG emissions over the next decade (La Rovere et al. 2013).

Brazil faces the challenge of building upon its historically low energy-related GHG emission levels through new decarbonization strategies, while pursuing higher living standards for its population. Average annual income per capita in 2005 was only $4,767. Inequality, as evidenced by Brazil's uneven income distribution, is a major problem. In 2005, the poorest 16 percent of the population had an average income per capita of $481 per year, or less than two times the national annual minimum wage. Meanwhile 60 percent of the population had an average income per capita of $1,819 per year, the equivalent of 2 to 10 times the national annual minimum wage. The richest 24 percent of the population had average annual income per capita of $10,848, or more than 10 times the annual minimum wage (IES-Brasil 2015). Brazil has made some progress in reducing income inequality in the last decade, thanks to the government consistently increasing the minimum wage faster than the inflation rate and social transfer programs (e.g. *Bolsa Família*). They decreased the Gini coefficient from 0.57 in 2005 to 0.53 in 2013. But inequalities are still a leading concern: in 2013, 15.5 million people in Brazil were living below the poverty line, of whom 6.2 million were living in extreme poverty (Brasil 2015). Inequality between regions is also a problem; reducing these is the object of some regional incentive programs.

Electricity access is very high and increasing (99 percent of urban households and 90 percent of rural households were electrified in 2010). Access to a clean water supply is also high (93 percent of the population in the largest 100 municipalities, in 2013). But important challenges remain in providing citizens access to basic services, most notably, Brazil has a housing deficit. Brazil has, in total, 65 million households, but 5.4 million houses are "missing." Some 2.7 million households lived in a residence with more than one family. In addition, nearly half the population had to build its own house. Water is expensive and subject to shortages, as was the case during the recent drought. The coverage of sanitation services is still poor: only 48 percent of the population benefits from sewage collection networks and only 39 percent of the population had its sewage properly treated in 2012. Building a low-carbon infrastructure to meet all these demands remains a huge challenge.

At the time of the 1973 oil shock, Brazil was strongly dependent on oil imports. It relied on imported oil for 83 percent of domestic needs, mostly for the industrial and transportation sectors (oil products are not used significantly in electricity generation or the residential sector; ambient heating is needed only sparingly in the south of Brazil). Oil imports have been important, particularly to fuel the on-road transportation modes that dominate urban and long-distance travel – freight as well as passenger. In 1980, after the second oil shock, more than half of Brazilian hard-currency earnings from exports were used to pay its oil import bill. Brazil has found large off-shore oil reserves during the last four decades, allowing for a substantial increase in oil production, and sharply reducing the country's dependence on oil imports (as of 2014, imported oil accounted for 6 percent of domestic oil consumption; while 44 percent of natural gas domestic consumption was imported, 75 percent of coal's and 5 percent of electricity's, with 87 percent of overall energy use secured by domestic production).

More recently, the discovery of large offshore oil reserves in the pre-salt layer has created expectations that Brazil will become a major oil exporter, since the size of the reserves exceed the country's own consumption needs. Current government plans envision doubling domestic production by 2030 (from the 2014 level of 2.25 billion barrels oil/day), and using half of future production for export. Congress has approved a law to use 75 percent of the oil rent to fund education and 25 percent for health. This assumes that the pace of growth of domestic energy consumption will be kept moderate, through continuing to produce renewable energy. Already, renewables accounted for 39.5 percent of energy consumption in 2014. Renewables include hydropower, sugarcane products (ethanol used as liquid biofuel in transport and bagasse for cogeneration of heat and power), and more recently, the fast growth of wind energy.

Brazil is not endowed with large coal reserves. Its small reserves are of a low-grade variety, with its demand limited to the few industries that use it for specific processes (e.g. coke for steel mills, ceramics and cement) and some complementary electricity generation. The volume of natural gas produced in the country is equivalent in 2014 to 24 Mtoe (net of losses and reinjection). It has not traced the rapid growth in demand, 9.5 percent from 2013 to 2014, mainly for power generation and industrial use, creating a need to import gas through the pipeline from Bolivia or as liquefied natural gas (LNG) from other countries. Natural gas imports represent 41 percent of domestic supply in 2014, but it is expected to sharply decline, or to be eliminated entirely in the future as recent discoveries are fully exploited. If the country embarks on a low-carbon path, it will allow the country's natural gas to be diverted from power generation towards its highest use, as an industrial feedstock.

Brazil is endowed with a huge renewable energy potential that makes a growth trajectory with low energy emissions appear entirely technically feasible, with a wide spectrum of options. In 2014, hydropower provided 65 percent of the country's electricity needs and hydropower's full potential is still untapped, although not all of it will be used due to concerns over local environmental impacts in the Amazon region. Brazil also has an abundance of land that can be sustainably used to produce biofuel feedstocks, especially sugar cane for ethanol. Since the launch of the Brazilian Ethanol Program in 1975, all the gasoline used in the country is blended with 22–25 percent ethanol. Domestically produced ethanol is also used for the pure ethanol and flex fuel engines of light-duty vehicles. A learning curve over the last 40 years has allowed Brazilian producers to increase yields from 4,000 to 7,000 liters/ha.year of ethanol from sugarcane, while production grew from 0.7 billion liters in 1975 to 27 billion liters in 2010. The first new plants producing second-generation ethanol from cellulosic materials (e.g. sugarcane bagasse) have already reached 25,000 liters/ha.year, illustrating the huge potential for increasing ethanol production. Biodiesel production reached 3 billion liters/ year in 2010, mainly obtained as a byproduct of soybean oil production. Overall, diesel oil used in the country includes a 7 percent blend of biodiesel. Brazil also has important wind energy potential. Initial estimates of its potential for installing 150 GW have yet to be updated; this figure will grow due to technical progress. In 2014, wind power installed reached 4.9 GW, with power generation increasing from 1.2 TWh/year in 2008 to 12.2 TWh/year in 2014 (EPE 2015). Solar energy is also widely available at high levels throughout the country. Therefore, keeping a low energy-emissions growth trajectory appears technically feasible, with a wide range of options of renewable energy sources.

In the period 2004–2012, Brazil's GDP increased by 32 percent and more than 23 million people were lifted out of poverty, while emissions dropped 52 percent, delinking economic growth from emission increase over the period (Brazil 2015). However, this was only possible thanks to a dramatic cut in Amazon deforestation, as energy-related GHG emissions have increased in the same period. The challenge now is to decouple economic growth and social gains from the use of fossil fuels, ensuring a sufficient supply of renewable energy to fuel economic growth and increase the living standards of all the population (La Rovere et al. 2013). The good news is that the long-term deep decarbonization of the Brazilian economy will receive a boost from a relatively advanced starting point, an already comparatively low-carbon energy system, and the country's huge potential to further expand production of renewable energy.

GHG emissions: current levels, drivers and past trends

Brazilian GHG emissions increased from 1.4 billion metric tons CO_2 equivalent ($GtCO_2e$) in 1990 to 2.5 $GtCO_2e$ in 2004, followed by a substantial reduction (by half) to 1.25 $GtCO_2e$ in 2010, thanks to the sharp fall of deforestation (see Figure 12.1).

As a consequence of the lower rate of deforestation, the share of CO_2 in the GHG emissions mix has declined sharply, from 73 percent to 57 percent between 2005 and 2010. The recent upturn in GHG emissions has been driven, notably, by methane emissions from the enteric fermentation of Brazil's large cattle herd (numbering 213 million heads in 2012). Also, the share of fossil fuel combustion in total GHG emissions has been steadily increasing in recent years, from 16 percent in 2005 to 32 percent in 2010. Fossil fuel combustion ranked second, after agriculture and livestock, in 2010 (see Figure 12.1). Among fossil fuels, oil is by far the dominant source of CO_2 emissions, followed by natural gas, and coal (in 2010, 220, 62 and 44 $MtCO_2$, respectively).

The Brazilian population increased from 145 million to 191 million people from 1990–2010. Population growth rates have declined, to a rate 0.9 percent per year today (from 1.6 percent per year in 2000). Economic growth has been an important driver of increased energy-related CO_2 emissions, as GDP nearly

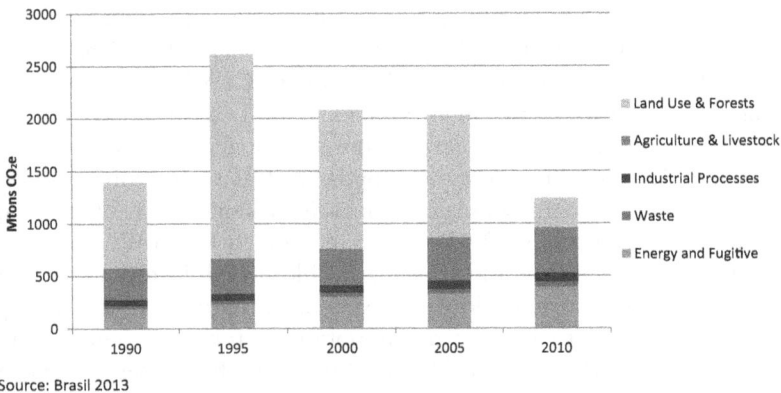

Source: Brasil 2013

Figure 12.1 GHG emissions by source: 1990–2010

Source: Brasil 2013.

doubled from 1990 to 2010 (an 89 percent increase, in real terms). The carbon content of the energy supply has followed the ups and downs of the development of renewables: the share of renewables in total energy mix fell from 49 percent in 1990 to 41 percent in 2000 due to a slowdown in the deployment of hydropower and ethanol from sugarcane. Then it rose, after 2000, to reach 45 percent of total energy supply in 2010. In 2010, transportation was the largest energy-related CO_2 emissions source (162 $MtCO_2$), followed by industry (112 $MtCO_2$), electricity generation (31 $MtCO_2$) and buildings (21 $MtCO_2$). Roads play an overwhelming role in Brazil, both in intercity freight transport and intercity passenger traffic, and the growth of air travel and car ownership are also important drivers of energy-related GHG emissions. Energy-intensive industries (iron and steel, pulp and paper, cement, petrochemicals) are important drivers of GHG emissions, but their share of GDP in the Brazilian economy is declining now.

Brazil's future development pathway: an overview

The general methodological approach in designing the Deep Decarbonization Pathway Project (DDPP) scenario for Brazil was to highlight the implications of deep decarbonization strategies – embedded in a pathway of rapid economic and social development.

Demographic assumptions up to 2050 are based on IBGE ("Instituto Brasileiro de Geografia e Estatística," a federal body) projections of Brazilian population changes. From 2015–2050, Brazil will experience a major shift in its demographic profile, mainly because of a decrease in fertility. From 200 million people in 2015, the population is expected to grow to a peak of around 225 million between 2030 and 2040, before slowly falling to about 221 million in 2050. This shift will bring about its own challenges, including the projected rise in the already significant deficit of the public retirement pension system.

Development assumptions are based on government plans, particularly the Energy Long-Term National Plan, PNE 2050 (EPE 2014). Economic growth is assumed to be very strong through 2050, with a quadrupling of average GDP per capita to reach about $19,000 (2005 US dollars) by 2050.

Our macroeconomic reference scenario follows the government plans (the governmental plan scenario, or GPS). We build on it, with some assumptions to complement and extend the GPS to 2050.

A Deep Decarbonization Pathway then, is designed to include mitigation policy actions and measures targeted to bring Brazilian GHG emissions per capita to 1.7 tCO_{2e} per year, of which 1.2 tCO_{2e} comes from energy-related emissions, consistent with a world average required to limit global warming to 2 degrees Celsius.

Methods – modeling methodology and economic consistency

The construction of the Deep Decarbonization Pathway scenario uses a framework that aligns national development goals with the 2 degrees Celsius global climate target. The 2050 GHG emission target for Brazil is set at 367 $MtCO_2e$, a 70 percent reduction compared to 2010 (1,214 $GtCO_2e$). To reach this goal, the designed pathway includes a number of mitigation actions in various sectors.

Initially, the Deep Decarbonization Pathway (DDP) follows the most ambitious scenario up to 2030 designed in the IES-Brasil project (Economic and Social Implications of GHG Mitigation Scenarios in Brazil up to 2030),[2] outlined by a Scenario-Building Team (SBT) made up of experts from the government, academia, private sector and civil society. This IES–Brazil scenario considers additional mitigation measures that go beyond an extension of current government plans. Hence the DDP from 2010 to 2030 considers assumptions virtually identical to the IES–Brasil project, reaching a 1,009 $MtCO_2e$ emissions level in 2030 (483 $MtCO_2$ from energy-related emissions). After 2030, a number of additional mitigation actions, which are foreseen to be economically feasible by that date, are introduced in the energy and transport sectors, to achieve a 367 $MtCO_2e$ emissions level in 2050.

In the energy sector, these actions include a strong expansion of solar power, its share of total energy generation growing from close to zero to 11.3 percent in the period, and a doubling of biomass-based thermopower, to producing almost 20 percent of all electricity in 2050. The growth of solar and biomass power sources results in a lower reliance on hydropower, down to a share of little more than 50 percent, although it continues to expand in absolute terms. Until 2050, the complete replacement of natural gas by biofuels, the last fossil fuel still in use for power generation by 2030, is completed. This allows for fully emissions-free electricity generation by the end of the period.

Rail and water represent about 60 percent of total freight transport in ton-kilometers, in 2030. After 2030, reliance on rail and water increases to reach more than 70 percent of total ton-kilometers in 2050. In addition, better geographical distribution of production, consumption and import/export hubs create logistical efficiency gains. Those improvements permit a decoupling of freight transport needs from production (14 percent decrease of ton-kilometers from 2030–2050, in parallel with 80 percent increase of GDP). The modal shift, combined with reduction of transport activities, translates directly into a sizable drop in road transportation activities, reducing overall energy needs. Freight transport emissions decrease further through the increased use of biodiesel. Together, these measures permit Brazil to cut sectoral emissions by almost 50 percent from 2030–2050.

Passenger transportation, by contrast, sees a growth in activity levels over the period, by about 30 percent, given the continued trend of the expansion of Brazil's urban centers. Urban growth triggers a 15 percent rise in energy needs from 2031–2050, essentially because of increased public transportation needs. The DDP assumes the electrification of passenger transport, through a shift from the use of passenger cars to rail, and from fossil-fuel run vehicles to electric cars. The result is that, although passenger transportation activity increases over the period by about 30 percent, emissions fall, more than offsetting the activity growth and decreasing emissions by about 30 percent from 2030–2050. A comprehensive list of all mitigation actions per sector can be found in Table 12.2 on page 262.

The investment required for these transformations is assessed through a one-way soft-link in sectoral modules for which a series of mitigation actions are associated. Each mitigation action presents, for a given level of specification, a cost and an energy-use profile. For example, energy-efficiency actions show a reduction in energy in their energy-use profile, whereas mitigation actions related to biofuels consist of switching from fossil to renewable energy. Mitigation actions that are not related to energy demand or supply in the AFOLU and waste sectors are assessed directly through their associated emissions.

These sectoral results provide inputs to the CGE model IMACLIM-BR. In this model, the technical coefficients of the DDP are calibrated according to the percentage variation of energy use compared to the reference scenario, the governmental planning scenario (GPS). Monetary values are the total investment requirements for all mitigation actions considered, per sector.

The IMACLIM-BR model is also used to simulate the introduction of a carbon tax on burning fossil fuels. The tax level increases linearly, from $0/tCO$_2$e in 2010 to $112/tCO2e in 2030 and then to $168/tCO$_2$e in 2050.[3] The tax revenues are fully recycled through lower social security taxes on labor, so that fiscal neutrality is ensured and the overall tax burden is kept at the same level as before the carbon tax.

The model ensures macroeconomic consistency between the sectoral modules and the IMACLIM-BR framework through aligning of some key variables, such as population, GDP, GDP structure and final energy consumption. Most of the mitigation actions considered are cost-efficient and their effects on GDP and total investment rate over GDP are minor (especially if compared to the substantial long-run uncertainties) so that we adopt, as an approximation, neglecting the effect on energy demand of economic feedback from IMACLIM-BR.

Decarbonization strategy

Given the huge potential of natural resources in Brazil, there is a wide range of possible decarbonization strategies that may be proposed. The analysis conducted in this report starts from the Deep Decarbonization Pathway presented in the DDPP 2014 report (La Rovere and Gesteira 2014) and further explores the possibility of an earlier and more pronounced peak of GHG emissions, in order to make it more consistent with a 2 degrees Celsius-compatible global emissions trajectory. Through 2030, this Brazilian Deep Decarbonization Pathway assumes that a majority of the economy-wide emission reductions will be realized through actions outside of the energy sector. However, actions will need to be taken in the near-term to set in motion the major infrastructure changes that would allow energy-related emissions to fall significantly after 2030, thanks to major investments in renewables, energy efficiency and low carbon transportation. Thus, Brazil's energy-related emissions are expected to grow in the immediate future, peak around 2030, and then decline through 2050. This report outlines a Deep Decarbonization Pathway of the energy system that would be achieved through efficiency gains and fuel switching, as well as new technologies such as electric vehicles and energy storage for intermittent sources. Clean power generation would be provided by hydropower, complemented by bioelectricity (to ensure reliability) along with emerging onshore and offshore wind, as well as solar photovoltaic energy. In the productive sector, increased use of green electricity and biomass coupled with an interim substitution of natural gas for coal and petroleum products would be required.

Since Brazil has sizable biological CO_2 sinks, which are expected to increase until 2050 through substantial reforestation and afforestation efforts, the decarbonization strategy will be strongly complemented by initiatives promoting CO_2 sinks to compensate for energy-related GHG emissions.

The following sections describe the three main pillars of this Deep Decarbonization Pathway strategy: Agriculture and Livestock, Forestry and Land Use (AFOLU); biofuels and hydropower. Table 12.2 includes a list of all sectoral mitigation actions up to 2050 in the Deep Decarbonization Pathway (DDP).

Agriculture and livestock, forestry and land use

According to Strassburg et al. (2014), "Brazil's existing agricultural lands are enough to sustain production at levels expected to meet future demand (including both internal consumption and exports) for meat, crops, wood and biofuels until 2040 without further conversion of natural habitats." The cattle breeding subsector has the greatest mitigation potential, because its emissions account for approximately half of all Brazilian GHG emissions (Bustamante et al. 2012). The total area reserved for pasture lands comprises 170 million hectares, versus 60 million for crops. However, current productivity (94 million animal units) is 32–34 percent of the estimated carrying capacity, which accounts for 274 to 293 million animal units. It is envisageable to increase pasture productivity to 49–52 percent of the carrying capacity, while maintaining the present geographical patterns of production, allowing to produce more food from the same area with lower environmental impact.

Insofar as agriculture (including livestock) is currently Brazil's most important source of GHG emissions, the DDP assumes the extension of the policies and measures of the Plan for Consolidation of a Low Carbon Emission Economy in Agriculture (Brasil 2012), launched to meet the voluntary goals set by the Brazilian government for 2020. It thus assumes mitigation actions, such as the recovery of degraded pasture land. Moreover, both the Plan above and the DDP assume there will be an increase in land used by agroforestry and intensive cattle-raising (integrated agriculture/husbandry/forestry activities), while the planted area under low tillage techniques would also be expanded. In addition, areas cultivated with biologic nitrogen fixation techniques will be increased, replacing the use of nitrogenous fertilizers, and there would be greater use of technologies for proper treatment of animal wastes.

In forestry and land use, the DDP assumes the extension of the policies and measures of the Action Plan for Prevention and Control of Deforestation in the Amazon (Brasil 2004) and of the Action Plan for Prevention and Control of Deforestation and Fires in the Savannahs (Brasil 2009) launched to meet the voluntary goals set for 2020. These action plans include a number of the initiatives, combining economic and command-and-control policy tools that have succeeded in bringing down the rate of deforestation in recent years (see Figure 12.1).

Moreover, the proposed decarbonization pathway assumes the successful implementation of afforestation and reforestation activities, which would lead to a dramatic increase of forest plantations using eucalyptus and pine trees, not only for the pulp and paper industry but also for timber as well as the charcoal used in the production of pig iron and steel. In fact, huge areas of degraded land are available in the country where these afforestation programs would be developed, achieving both environmental and economic benefits. Given the likelihood that such initiatives will continue and expand in the coming decades, it is expected that as early as the mid-2020s, land-use change and forestry will become substantial net carbon sinks, and will, by 2050, be capable of offsetting a substantial share of the emissions from the energy sector.

The waste management system will require large investments in sewage pipelines, waste disposal facilities and industrial effluents treatment units, with methane capture and burning facilities that may curtail emissions. The capture of methane creates a renewable fuel source, and biogas would be used to replace some fossil natural gas.

The Transition Strategy

Up to 2030, the DDP suggests that a number of available technologies can be deployed at a larger scale than proposed in governmental plans, bringing about a further lowering of carbon emissions. From 2005 to 2030, the Deep Decarbonization Pathway is based upon the same assumptions as the most ambitious mitigation scenario of the IES–Brasil project, regarding additional mitigation actions and their investment requirements. Thus, the DDP includes all mitigation actions approved or already under implementation by the government (per the governmental plan scenario or GPS), extended up to 2030 at higher penetration rates. It includes, as well, a set of extra mitigation actions, leading to further decarbonization.

From 2030 to 2050, new mitigation technologies will become available, allowing for the deeper decarbonization of the Brazilian economy. Given the uncertainties about winners and losers in the technological race by 2050, the deployment of mitigation actions included in this period in the DDP must be seen as an illustration of what a deeply decarbonized domestic energy system could look like at that time horizon. The sectoral mitigation measures up to 2050 are presented in Table 12.2.

Results and discussion

The quantitative results from the storyline of the Deep Decarbonization Pathway (DDP) described in the previous section are presented and discussed below.

Emissions pathways

In most countries, the key challenge to a deep decarbonization pathway is the combustion in fossil fuels. In Brazil, this is also true (see La Rovere et al. 2013) but the country is right now in a transition from deforestation to energy as the main source of GHG emissions, as explained in the Introduction. Moreover, carbon sequestration from reforestation and afforestation schemes has the potential to play an important role in the long-term decarbonization of the Brazilian economy. Therefore, in this section we first present the GHG emissions from AFOLU and its share of overall GHG emissions, before detailing the results for energy-related emissions in different sectors of the economy.

Non-energy related GHG emissions

The DDP anticipates a deepening of already successful strategies in AFOLU (as previously described), ensuring the continuous decrease of non-energy related GHG emissions until AFOLU becomes a net sink before 2050. Those strategies – exerting control over deforestation, coupled with an increase in forest restoration and afforestation – compensate for the growth of energy-related GHG emissions until 2030. Thanks to the policies and measures under

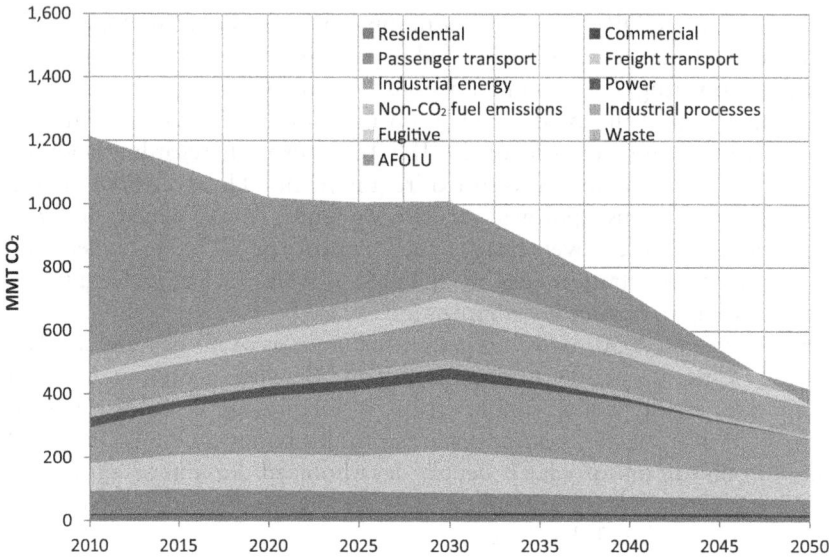

Figure 12.2 DDP: GHG emissions per sector 2010–2050

Source: La Rovere et al. 2015

implementation to meet the voluntary targets announced in Copenhagen, overall GHG emissions decline until 2020, stabilize through 2030 and resume a downward path from 2030 to 2050, as shown in Figure 12.2.

Energy-related GHG emissions

By 2050, under the DDP scenario, total final energy needs increase considerably, due primarily to economic development and secondarily, population growth. Renewables and biomass become the dominant source of primary energy and are used to meet the majority of these energy needs, notably through the direct use of biomass and zero-carbon electricity generation. Brazil has a strong potential for energy efficiency; the government recently introduced several energy-saving initiatives that will be extended across the board, such as the replacement of incandescent light bulbs, energy efficiency labeling of home appliances and incentives for establishing less energy-intensive urban mobility infrastructure.

Under the DDP scenario, energy-related CO_2 emissions peak by 2030 and decline thereafter, to reach in 2050 262 $MtCO_2$ a lower level than in 2010. Opposite trends drive these evolutions. On the one hand, the main driver of emission increase is the strong growth of GDP per capita and, to a lesser extent, population growth – which stops when the population stabilizes by 2040. On the other hand, a downward push on emissions is exerted by a substantial shift towards a renewable energy supply (especially through the increased use of hydropower, wind, solar energy and biomass to produce electricity and industrial heat, as well as the use of biofuels and electricity for transportation) and a decrease in final energy intensity per unit of GDP, induced by structural shifts in the transportation and industrial sectors. The transportation and industrial sectors will be responsible for the bulk of emissions. Transportation emissions will dominate in the near future, and industry will become the dominant source in the mid-term; both sectors reach 120 $MtCO_2$ in 2050. Emissions from buildings (22 $MtCO_2$ in 2050) and power generation remain very low throughout the 2010–2050 period.

Final energy and energy demand

Industry has historically been the most important source of energy demand, corresponding in 2010 to almost half of final energy consumed, followed by transportation (about one-third) and buildings (less than one-fifth). Energy demand from buildings is supposed to increase slowly but steadily up to 2050, due to improved overall living standards, which more than offset efficiency gains. Energy demand from the transportation sector is expected to peak in 2030, and the industrial sector in 2040. The reversal of the energy-consumption trends happens later in industry because of a greater rigidity in some of its processes, which do not allow for fuel replacement and structural changes as easily or thoroughly as the transportation sector does. By 2050, buildings would be consuming more energy than transportation, while the largest demand will continue to come from industry.

Final energy, currently consumed mostly in the form of liquid fuels (almost 50 percent), will undergo a shift towards electricity and biomass, both with a more than threefold increase up to 2050, with electricity becoming the prevailing energy source. Natural gas consumption will also increase considerably, albeit less than electricity and biomass, while coal consumption will decrease.

Primary energy increases about proportionately to final energy, pushed by two opposite drivers. On the one hand, losses in electricity distribution drop as a result of technological improvements and the better spatial distribution of generation (notably, lower dependence on hydropower produced far from consumption centers). Both of these improvements contribute to proportionally reducing primary energy needs. On the other hand, a greater reliance in biomass, with less than 100 percent transformation efficiency (as opposed to hydro), will increase the amount of primary energy required to meet the base load.

The emissions impact of primary energy will be curtailed strongly by the considerably higher share of renewables and biomass.

Transportation

The DDP projection sees demand for transportation services increase considerably over the period 2010–2050. Passenger transportation increases by a factor of 2.6 due to increased urbanization, which not only results in more people relying on urban transportation but also (given the geographical expansion of metropolitan areas) longer commuting distances. Freight transportation is also expected to increase considerably.

In the transportation sector, the reliance on renewables, especially ethanol, will increase. Regular gasoline sold in the country will continue to have the current 27 percent mandatory ethanol content (anhydrous ethanol), and most new cars manufactured for the domestic market will continue to be flex fuel, capable of running solely on ethanol (hydrated ethanol with up to 5 percent water content). An ambitious biofuel program will increase the production of ethanol from sugarcane and biodiesel and biokerosene from a combination of sugarcane and palm oil. This would allow renewable ethanol to substitute for a significant amount of gasoline as it begins to fuel most of the light-duty vehicle fleet (along with some natural gas, used mostly by taxicabs in major cities). The amount of biodiesel blend in diesel, used by trucks and buses, would be further increased to 25 percent (the government has just regulated an increase from 5 percent to 7 percent). Through these combined measures, more than half the total energy used in transportation would be renewable.

In addition, electric vehicles will be an alternative/complementary option that grows in importance over time, notably given these vehicles' benefits aside from lower GHG emissions – such as producing less urban air pollution and noise. By 2050, almost half of the light-vehicle fleet is expected to be electric-powered and 30 percent of the bus fleet is expected to be electrified. Altogether in 2050, the carbon intensity of fuels used in transportation per unit of energy would be reduced by almost half.

Higher national standards for energy efficiency standards would be used to increase the fuel economy of all vehicles (cars, buses and trucks) and current tax incentives, for cars with smaller motors and lower fuel consumption, will be strengthened and expanded, further increasing the already large current share of one-liter motors. At the same time, a shift towards railways and waterways would be promoted (wherever possible) for a deep decarbonization of the transportation sector. For deep decarbonization of freight transport, the current share of transport by trains, ships and barges could be intensified, so it rises from a level now below 50 percent to 70 percent in 2050. The DDP scenario also includes the significant extension of urban mass transportation infrastructure (subways and trains, bus rapid transport systems, etc.), which should correspond to close to 50 percent of all the passenger transportation needs in 2050, a significant increase from the current level of below 10 percent.

Buildings

Demand for energy in buildings rises strongly in the DDP scenario, reflecting 16 percent population growth, 267 percent economic growth, and the drive for social inclusiveness. While the analysis envisions Brazil pursuing energy efficiency, it has only moderate impacts on energy demand (compared to countries with colder climates), given Brazil's almost non-existent heating needs. Fuel shifts in household energy consumption focus on increasing solar thermal for hot water, with some replacement of LPG by natural gas, ethanol and grid electricity. The adoption of solar photovoltaic panels in residences would be stimulated by a proper regulatory framework and smart grid infrastructure, allowing for an increasing share of photovoltaic power.

The share of lighting in household electricity consumption will be reduced from the current level of about 20 percent to about 5 percent to 10 percent by 2050, thanks to more intensive adoption of compact fluorescent light bulbs (CFLs) and light-emitting diodes (LEDs). At the same time, consumption from electronic equipment and electric appliances would increase by a factor of more than five. More specifically, the share of residential electricity consumed by refrigerators and freezers, already used in more than 90 percent of residences, will fall from its current level of 33 percent because of efficiency improvements. By contrast, air conditioners, used now by only about 15 percent of households, will be more widely adopted (by up to 75 percent of houses), and despite the addition of more efficient technologies (such as split units, central air conditioning and heat pumps), their total consumption of electricity will increase roughly four-fold.

Electric showers, now present in about 72 percent of residences and accounting for 27 percent of residential energy consumption on average, will be replaced mostly by solar heaters, so their share falls to no more than 10 percent of household consumption by 2050. For cooking purposes, households would transition from LPG to natural gas in urban settings, and from firewood to LPG in rural areas. Although this last transition goes against decarbonization, it is necessary because of the social benefits of replacing firewood as a domestic cooking fuel.

However, by far the highest growth in energy consumption would come from household uses not specified above, such as entertainment and telecommunications equipment (TVs, computers, internet links, cellphones) and other electric appliances (hairdryers, microwave ovens, toasters, washing machines, vacuum cleaners, water purifiers, etc.). These appliances, now found in only 44 percent of households, will be present in almost all residences by 2050, and their relative use is likely to be intensified so they become the source of more than half of residential energy demand.

In the commercial sector, decarbonization measures are similar to those in the residential sector, with more weight given to energy efficiency in air conditioning installations, which are used extensively in modern shopping centers and malls, office buildings, hospitals, universities, etc.

In both the residential and commercial sectors, the Deep Decarbonization Pathway includes increasing the energy efficiency of all LPG uses (cooking and water heating), and greater energy efficiency in all electricity uses to reduce the growth in demand. In the end, energy use in buildings is expected to triple and the bulk of this increase will be satisfied by low-carbon electricity.

Industry

Several industrial sectors have essentially domestic markets, which are expected to grow considerably in response to economic growth. Notably, an expansion in housing should be the main driver of demand for cement, and an important one for steel. All the non-agricultural sectors, which depend more on exports, are assumed to grow roughly at the same rate, slightly below GDP growth, increasing 224 percent from 2010–2050. The industrial sector's overall share of GDP would drop by 4 points (from 34 percent of GDP in 2010 to 30 percent of GDP in 2050), mostly due to a relative decrease of manufacturing and mining. This shift captures a structural change in the Brazilian economy towards the service sector, the share of agriculture and livestock in GDP remaining constant at about 5–6 percent of GDP.

In most major industrial sectors, the share of electricity in total energy demand is expected to rise considerably by 2050. For generating heat in industry, biomass would partially replace fossil fuels, wherever possible. The share of natural gas to meet industry needs is expected to increase, while fuel oil and other oil products would be phased out. In steel manufacturing, a greater share going to charcoal would imply lowering the demand for coke from coal. And in the cement industry, emission reductions would be achieved through changes in the industrial process, partially substituting clinker by blast-furnace slag and/or fly ash (both abundantly available in Brazil), which at the same time reduces process emissions and the associated energy needs.

In absolute terms, the energy demanded by industry would more than double by 2050, despite the assumption of widespread adoption of energy efficiency measures, such as more efficient furnaces. As a consequence, the share of industrial sectors as a whole in energy-related emissions would grow through 2040. Under the Deep Decarbonization Pathway, the growth in industrial energy emissions will be tempered by a reduction of both the energy intensity of industrial products and the emission factors, especially in the steel and cement industries. This will be permitted by a substantial rise in energy efficiency. In the aggregate across the industrial sector, this will result in a decrease of energy intensity per unit of value added of 21 percent in 2050, compared to 2010. In steel manufacturing, a greater share going to charcoal would imply lowering the demand for coke from coal. And in the cement industry, emission reductions would be achieved through changes in the industrial process, partially substituting clinker by blast-furnace slag and/or fly ash (both abundantly available in Brazil), which at the same time reduces process emissions and the associated energy needs.

The non-energy emissions of industrial processes will also increase, given the inflexibility of some of those processes.

A substantial effort will be required to reduce CH_4 and CO_2 fugitive emissions from the oil and gas production system (platforms and transport facilities), as the huge resources of the pre-salt layer are exploited. With the deployment of new infrastructure and some technical progress, in addition to the Petrobras program already under implementation, it is expected that the rate of natural gas venting and flaring can be reduced. Under this assumption, much higher levels of production can be obtained, with overall fugitive emissions reaching their peak of 64 $MtCO_2$ in 2030, and then falling back to around 25 $MtCO_2$ by 2050, equivalent to the 2015 level.

Energy supply

Brazil's Deep Decarbonization Pathway includes further expanding the country's hydropower, tapping the potential for more than doubling the installed capacity with environmentally acceptable projects, along with an expansion of bioelectricity, wind and solar photovoltaic generation. Nuclear energy currently provides only 2.7 percent of total electricity in Brazil, and no further increase of this output level is considered in the DDP, aside from the operation of the Angra 3 plant that is already under construction. This is because of its high operational and investment costs compared to other electricity generation options, especially hydropower, and its uncertain social acceptance.

Further utilizing Brazil's hydropower potential requires improving the design and construction of hydropower plants with reservoirs, while simultaneously meeting local environmental concerns. In recent years, hydropower plants have been constructed with minimal reservoirs (i.e. mostly run-of-the-river plants), limited energy storage capacity and without dispatchable generation. Improved designs are needed to improve the reliability of this intermittent resource. In addition, the DDPP analysis includes using the huge potential for renewable biomass, mainly from wood and the sugarcane byproducts of ethanol production (i.e. bagasse, tops and leaves and stillage).

This renewable electricity mix can be designed to match the country's variable electricity demand by exploiting the complementarity between the renewable resources. Offshore wind farms may become a relevant option, given the abundance of offshore sites, thanks to a potential synergy with the huge effort on offshore oil and gas drilling that would help reduce its costs (construction and operation of off-shore wind farms would strongly benefit the infrastructure and logistics in place for oil and gas platforms).

Advanced batteries, as they become available, together with bio-thermoelectricity, could help overcome the challenge posed by non-dispatchable, intermittent renewable power sources such as solar and wind. If the challenge is overcome, renewable sources could replace natural gas as the base load supply, thereby further reducing GHG emissions from power generation. The resulting scenario leads to a completely decarbonized power sector in 2050.

Macroeconomic implications

This section, an analysis of the macroeconomic and social implications of the Deep Decarbonization Pathway, was prepared based on IMACLIM-BR runs especially calibrated for this study. As mentioned above, IMACLIM-BR is a hybrid CGE model, developed to assess the macroeconomic and social implications of climate policies in the medium and long term.

For this DDP analysis, in order to limit GHG emissions, a carbon tax on the burning of fossil fuels was simulated, growing linearly from $0/tCO$_2$e in 2015 to $112/tCO$_2$e in 2030, and then to $168/tCO$_2$e in 2050. This carbon tax would stimulate the introduction of a number of mitigation measures, carefully chosen to compose the DDP. They represent investments adding up to almost $2.8 trillion (2010 US dollars) from 2015–2050 (see details on page 000). The IMACLIM-BR assessment of macroeconomic trajectories demonstrates that those investments will help to deeply reduce GHG emissions without harming the country's economic growth potential.

Table 12.1 presents key macroeconomic and social indicators related to the DDP.

Table 12.1 Deep Decarbonization Pathway: key macroeconomic and social indicators

	2010	DDPP-2030	DDPP-2050
Population (millions)	191	223	221
GDP (trillion 2010 US$)	2.14	4.53	8.64
GDP growth per year (2010–2030; 2030–2050) (%)		3.81	3.28
Investment rate (% of GDP)	19.50	20.84	25.16
Total investments (trillion 2010 US$)	0.14	0.94	2.17
Number of full time jobs (million)	94.1	128.0	115.9
Unemployment rate (%)	6.70	3.81	5.49
GDP per capita (thousand 2010 US$)	11.2	20.8	39.1
GINI	0.53	0.42	0.33
Income share of 16% poorest households	2.1	2.9	5.7
Income share of 60% middle class households	28.7	35.1	40.0
Income share of 24% richest households	69.2	61.9	54.3
Accumulated Price index		1.17	1.31
Trade Balance (billion 2010 US$)	20.3	35.9	44.6
Trade Balance (% GDP)	0.95	0.79	0.52
Exchange rate (BrR$/US$)	1.76	2.42	2.42
International oil price (2010 US$/barrel)		95.20	95.20
Carbon tax (2010 US$/tCO$_2$e)	0	112	168
% of Agriculture in GDP	5.3	5.6	6.1
% of Industry in GDP	28.1	26.5	24.2
% of Services in GDP	66.6	68.0	69.7

Note: for monetary figures of this table, 1 2005 US$ = 1.12 2010 US$, and 1 BrR$2005 = 1.23 BrR$ 2010; therefore, 2010 US$ 95.20/barrel = 2005 US$ 85/barrel

Source: La Rovere et al. 2015.

From 2015–2050, Brazil will experience a major shift in its demographic profile, mainly because of a decrease in the fertility rate. Growing from 200 million people in 2015, Brazil's population is expected to hit a peak between 2030 and 2040, and slowly fall to about 221 million in 2050. This will bring other kinds of challenges, such as the projected growth of an already significant deficit in the public retirement pension fund.

Brazil's GDP is expected to grow at an average rate of 3.5 percent per year from 2015–2050. GDP will grow from \$2.14 trillion in 2010 to \$4.53 trillion in 2030 to \$8.64 trillion in 2050 (all in 2010 dollars). GDP per capita will also increase significantly, starting from \$11,240 in 2010 to \$20,800 in 2030 to \$39,100 in 2050 (all in 2010 dollars).

In the Deep Decarbonization Pathway a partial but important decoupling between GDP growth and GHG emissions for Brazil takes place from 2005 to 2050, enabling the country to reach in 2050 a level of 30 percent of 2010 GHG emissions.

Carbon revenues collected by the government are used to reduce payroll taxes, to stimulate the creation of new jobs and offset the recessive effect of tax-induced price increases. The number of full-time jobs in the Brazilian economy under the DDP is expected to grow from 91 million in 2005 to 128 million in 2030, and then to experience a decrease to 116 million in 2050, notably due to continued labor productivity gains. As a result, unemployment rates decrease from 7.0 percent in 2005 to 3.8 percent in 2030, and then start to grow very slowly, to 5.5 percent in 2050, because of a combination of fewer full-time jobs and the demographic trend of a larger active population.

In terms of trade, oil exports from the pre-salt layer are expected to grow until 2030, and then slowly decrease until 2050. During this period, industry will improve its efficiency and increase competitiveness thanks to importation of capital goods, so the trade balance remains positive. However, the trade surplus experiences a relative decrease, from 0.95 percent of GDP in 2010 to 0.8 percent in 2030 and to 0.5 percent in 2050.

Prices rise in the DDP scenario, with the carbon tax and all the investment in mitigation measures, but even considering the higher price index, families experience a significant rise in real consumption. For example, Brazil's lower-income class of up to two minimum wages (24 percent of population in 2005) increases real consumption by a factor of 3.7, the mid-income class of 2–10 minimum wages (60 percent of population in 2005) has a 2.6 increase, while the highest income class above 10 minimum wages (16 percent of population in 2005) increases real consumption by a factor of 2.1, indicating a significant reduction in income distribution inequality by 2050. It is clear that the poorest households' consumption increases much faster than the richest. This happens due to an explicit government policy aimed at reducing income distribution inequality in Brazil through better public education for poor families, translating into better productivity of those workers, and thus higher salaries. Also, public transfers to poor families are expected to follow GDP growth, remaining at a level of about 0.5 percent of GDP from 2010–2050. In aggregate, this reduction of income distribution inequality, allows for a fall of the GINI index from 0.53 to 0.334.

In conclusion, it is possible to significantly reduce emissions in the coming decades without jeopardizing Brazil's strong economic growth and social development.

Economic implications and co-benefits

The DDP analysis includes assessing the investment requirements for a series of mitigation actions. Different sectors present a range of possibilities, notably the AFOLU sector, a historically high emitting sector due to the deforestation associated with cattle raising and agriculture.

When it comes to mitigation actions and the associated investment requirements to 2030, from the base year 2010, the DDPP analysis considers circumstances virtually identical to those in the most ambitious scenario posited by the IES–Brasil project. This scenario includes all mitigation actions already agreed upon and foreseen by the government, and considers higher penetration rates, as well as a set of extra mitigation actions, leading to great decarbonization efforts. Since the DDP analysis takes into account a longer timeframe (up to 2050), and a high growth of Brazilian economy throughout this period, a lower discount rate of 4 percent per year was chosen, instead of 8 percent per year used in the IES–Brasil study.

After 2030, saturation levels on existing actions are considered, when applicable. For example, the modernization of existing refineries takes place before 2030, therefore it is not necessary to continue investing in these improvements after 2030, even if refineries are still in operation. Also, there is a threshold for commercial forests areas: above approximately 2.5 million hectares, their rate of expansion is slowed.

Furthermore, some extra mitigation actions are implemented:

- Transportation sector: Light electric vehicle use grows steeply, reaching 46 percent of the total fleet in 2050, as their costs gradually come to equal those of fuel-powered cars.
- Energy sector: Lithium-ion batteries used to store energy were considered for offshore and onshore wind, as well as for solar energy. As more intermittent sources come online, energy storage becomes essential to assure the supply. The amount of energy guaranteed through batteries grows proportionally to the share of solar and wind power, reaching up to six days add up in 2050.

To increase power generation from sugarcane bagasse, storing some of the biomass is also necessary. This share will be used during the off-season, and hence must be dehydrated. The analysis considered that from 2031–2050, the share of sugarcane bagasse that requires dehydration gradually increases. Moreover, the analysis considered that an increasing part of the sugarcane straw would be used in cogeneration. This becomes possible due to the growing mechanization of sugarcane crops and efforts to abolish the practice of burning the straw in the fields.

Table 12.2 depicts all mitigation actions considered, including their year of implementation, penetration rate in 2050 and investment levels. Monetary values are in 2010 constant US dollars, with a 2015 present value at a 4 percent annual discount rate.

Assessing co-benefits from mitigation actions makes an even stronger case for climate action. Co-benefits can also justify some of the high investment levels the analysis finds will be required for a few mitigation actions. Subways are an iconic case: even though investment requirements are high, shifting from cars to urban rail improves air quality and mobility in general. Expanding bicycle lanes also generates health benefits.

Other actions bring about positive co-benefit impacts, for example, in the energy sector: Increasing the share of renewables in electricity generation guarantees energy security and may help improve the balance of payment terms as less fuel must be imported. The preserved biodiversity resulting from action in the AFOLU sector ensures the provision of environmental services.

Implementing a deep decarbonization pathway in Brazil

As the long-term starts today, it is of utmost importance to discuss the requirements for a transition from current policies to the enabling conditions of a deep decarbonization in Brazil. This section addresses the need for connecting short- and long-term concerns.

Challenges and enabling conditions

A fundamental, society-wide transformation is implied in decarbonizing the country's economy, which will certainly have its winners and its losers. Some preconditions will be necessary to obtain the political resolve necessary to muster the forces for change. The first precondition is solid public awareness of the potential dangers of climate change – and the dangers of inaction. Brazil will clearly benefit from a decarbonized world, given the abundance of non-fossil natural resources in the country.

The main risk is the temptation to channel the recently discovered huge offshore oil and gas resources to expand domestic use, through a low pricing policy aimed at helping curb inflation. So far, the announced government policy, confirmed by Congress, points in the opposite direction; the stated objective is exporting the bulk of the oil resources and channeling oil revenues to finance government investments in health and education. It is imperative, if a low-carbon future is to be feasible in Brazil, to stick to this policy, avoiding the wrong use of the newfound oil resources. Such wrong use would undermine current and future efforts to foster energy efficiency and the use of renewable energy sources.

Table 12.2 Mitigation actions, year of implementation, penetration in 2050 and investment levels

Sector	Mitigation action	Year of implementation	Penetration in 2050	Total investment level (2015–2050) – million 2010 US dollars at 4% p.y. discount rate
Services	Efficient light bulbs (services)	2015	100%	32,520
Residential	Efficient LPG stoves	2015	100%	
	Efficient light bulbs (residential)	2015	97%	10,231
	Thermosolar water heating	2015	45%	31,553
AFOLU	Planted forests	2021	average 5.9 million ha	4,000
	Biological nitrogen fixation (corn crops)	2021	average 5.3 million ha	874
	Agroforestry systems	2026	average 1.8 million ha	122
	Degraded pasture recovery	2026	average 4,782 properties	204
	Atlantic forest restoration	2015	average 16.0 million ha	9,040
	Swine waste management	2021	average 16.0 million ha	90,231
Transportation	Bicycle lanes	2015	n/a	22
	Increased consumption of ethanol	2015	n/a	673
	Traffic optimization	2015	n/a	343
	Heavy electric vehicles	2020	n/a	37,265
	Light electric vehicles	2031	n/a	661,983
	BRT systems (Bus Rapid Transit)	2015	n/a	14,261
	VLT systems (Light Urban Train)	2015	n/a	15,738
	Railways and waterways	2015	n/a	41,032
	Increased consumption of biodiesel (15% in the diesel mix)	2020	n/a	169,533
	Subways	2015	n/a	199,450
	Energy efficiency – light vehicles	2021	n/a	2,139
	Energy efficiency – heavy duty vehicles	2017	n/a	58,252

Energy	Additional hydroelectric generation expansion	2021	n/a	194,932
	Additional onshore wind generation expansion (inc. storage)	2021	n/a	239,977
	Additional offshore wind generation expansion (inc. storage)	2031	n/a	0,245
	Additional solar photovoltaic generation expansion (inc. storage)	2021	n/a	293,041
	Additional sugarcane bagasse generation expansion (inc. dehydration)	2021	n/a	297,089
	Improvements in refineries – energy integration and heat reduction	2021	n/a	56,040
Waste	Methane destruction in landfills	2015	100%	1,996
	Methane destruction in dumpsites and controlled or remediated landfills	2015	59%	18,165
Industry	Carbon intensity reduction by 2% – steel	2015	100%	1,257
	Eucalyptus incorporation for charcoal – steel	2015	100%	66,621
	Carbon intensity reduction and increased co-processing – cement	2015	100%	35,784

Source: La Rovere et al. 2015.

The main technological challenges for the country are designing and building a new generation of hydropower plants in the Amazon that would avoid disrupting ecosystems, and using dispatchable bioelectricity to replace fossil-fuel generation.

Many of the strategies will require structural changes and high upfront costs. The barriers to their implementation are related to pricing, funding and vested interests, especially in two fields: power generation and transportation (long distance transportation and urban mobility). The huge upfront costs and long construction times involved in tapping the hydropower potential and building low-carbon transportation infrastructure will require substantial financial outlays and upgraded institutional arrangements (e.g. public/private partnerships) to provide adequate funding. The financial flow will need to come largely from outside, given the low savings capacity of the Brazilian economy.

Internationally, a set of technical and policy actions with a realistic chance of delivering on the promise of a climate-stable planet, together with a convincing argument for the perils of inaction, will be required to mobilize the resources needed for crucial initiatives. These actions will include: accelerated research to develop safe, energy-dense renewable fuels; research on industrial processes and materials that will bring down the investment costs of renewable power sources; and establishing mechanisms for technology transfer. It will also be crucial for governments worldwide to adopt carbon taxation schemes and to cut fossil fuel subsidies.

Near-term priorities

There are a number of immediate policy and planning measures that can be recommended to engage Brazil in a deep decarbonization process. Reinforcing the initiatives aimed at curbing deforestation is one such measure, to ensure there will be no major deviations from a trajectory that ends in zero deforestation within a decade, at most. Another policy priority should be substantially expanding forest plantations on degraded land, with appropriate financial schemes to meet the upfront costs. Another measure must be expending effort to pass legislation so the net effect of the taxes and subsidies on energy markets favor widespread adoption of renewable energy and energy-efficiency options. To this end, in the near-term, it is essential to cut subsidies on gasoline and diesel, and to restore the financial health of the electricity generation sector.

Extending the coverage of current incentives to invest in renewable energy in order to encompass other types of equipment, such as photovoltaic and solar heaters, can produce short-term returns. Prompting electricity providers to adopt smart-grid technologies and drafting a detailed and economically meaningful plan for restructuring long-distance transport in Brazil is another. This will involve prioritizing an infrastructure that allows for the most energy and emissions-efficient modes of transportation, such as railways and waterways. This could both cut emissions and respond to the business community's concerns. A similar initiative should also be undertaken, in collaboration with local authorities, with respect to urban mobility – an aspect of Brazilian infrastructure that needs urgent improvement and is thus currently high on the political agenda.

Acknowledgements

The authors thank the financial support provided by SDSN and IDDRI to the research performed (La Rovere et al. 2014 and 2015).

Notes

1   This work is based on the second report about Brazil to the Deep Decarbonization Pathways Project, jointly coordinated by Columbia University and IDDRI (Sciences Po, Paris) in 2014-2015.
2   IES–Brasil (Implicações Econômicas e Sociais de Cenários de Mitigação no Brasil até 2030 / Economic and Social Implications of GHG Mitigation Scenarios in Brazil up to 2030) is an initiative of the Brazilian Forum on Climate Change, mandated by the Brazilian Minister of Environment, in collaboration with the MAPS Programme. More can be found in: www.mapsprogramme.org/.
3   These values correspond to 100 US\$/tCO2e and 150 US\$/tCO2e in 2005 values, respectively.

References

Brasil, Ministério da Ciência, Tecnologia e Inovação – MCTI (2013) *Estimativas Anuais de Emissões de Gases de Efeito Estufa no Brasil.*
Brasil, Ministério do Desenvolvimento Social – MDS (2015) *Data Social 2.0* http:// aplicacoes.mds.gov.br/sagidata/METRO/metro.php?p_id=4   Accessed   24 September 2015.
Brasil, Ministério do Meio Ambiente – MMA (2012) *Plano ABC – Agricultura de Baixo Carbono*   www.mma.gov.br/images/arquivo/80076/Plano_ABC_versao_final_13 jan2012.pdf Accessed 24 September 2015.
Brasil, Ministério do Meio Ambiente – MMA (2009) *Plano de Ação para Prevenção e Controle do Desmatamento e das Queimadas no Cerrado – PPCerrado* www.mma.gov. br/florestas/controle-e-prevenção-do-desmatamento/plano-de-ação-para-cerrado-ppcerrado Accessed 24 September 2015.
Brasil, Ministério do Meio Ambiente – MMA (2004) *Plano de Ação para a Prevenção e Controle do Desmatamento na Amazônia Legal – PPCDAm* www.mma.gov.br/ florestas/controle-e-prevenção-do-desmatamento/plano-de-ação-para-amazônia-ppcdam
Bustamante, M.M.C., Nobre, C.A., Smeraldi, R., Aguiar, A.P.D., Barioni, L.G., Ferreira, L.G., Longo, K., May, P., Pinto, A.S. and Ometto, J.P.H.B. (2012) "Estimating greenhouse gas emissions from cattle raising in Brazil," *Climatic Change,* 115 pp. 559–577.
EPE (2014) *Plano Nacional de Energia/Energy Long-Term National Plan* PNE 2050.
EPE (2015) *Balanço Energético Nacional* https://ben.epe.gov.br/downloads/Relatorio_ Final_BEN_2015.pdf Accessed 24 September 2015.

IES–Brasil Project Team (2015) *Implicações Econômicas e Sociais de Cenários de Mitigação no Brasil até 2030/Economic and Social Implications of GHG Mitigation Scenarios in Brazil up to 2030*, Rio de Janeiro, report to the Mitigation Action, Plans and Scenarios (MAPS) Programme.

La Rovere, E.L., Dubeux, C., Pereira Jr., A. O. and Wills, W. (2013) "Beyond 2020: From deforestation to the energy challenge in Brazil," *Climate Policy – Special issue: Low carbon drivers for a sustainable world*, v.13, n.01, 70–86.

La Rovere, E.L. and Gesteira, C.M. (2014) *Pathways to deep decarbonization in Brazil* DDPP 2014 report SDSN-IDDRI-COPPE/UFRJ.

La Rovere, E.L., Gesteira, C.M., Grottera, C. and Wills, W. (2015) *Pathways to deep decarbonization in Brazil*, DDPP 2015 report SDSN-IDDRI-COPPE/UFRJ.

Strassburg, B.B.N., Latawiec, A.E., Barioni, L.G., Nobre, C.A., Silva, V.P., Valentim, J.F., Vianna, M. and Assad, E.D. (2014) "When enough should be enough: Improving the use of current agricultural lands could meet production demands and spare natural habitats in Brazil," *Global Environmental Change*, Volume 28 September.

Part IV

Environmental governance

13 Financing sustainability

Where has all the money gone?

Ladislau Dowbor

The overall challenge we are facing is not difficult to define: on one hand, climate change, the destruction of wild life, overfishing in the oceans, the cutting down of rainforests and so many other slow motion catastrophes demand deep-pocket investments to save the planet; on the other hand, studies on inequality and social dramas are showing with very clear figures that the economic governance is out of control. In short, we are destroying the planet for the benefit of the few. This doesn't work. And it is not getting fixed: the earth does not vote, nor do the poor or the next generations that will bear the disasters we are creating. It is a lopsided democracy.

Yet we have the necessary financial resources and the knowledge to face both the environmental and the social inclusion issues. The world presently produces 73 trillion dollars of goods and services for 7.2 billion inhabitants. In rough figures, this means over 3,000 dollars per month for every four-person family. Total in the accumulated personal wealth (houses, cars, savings etc.) and social capital (roads, dams etc.), and we see every human being could live a reasonably decent life. From this point of view, Brazil is situated exactly at the world average. But of course, some people are more average than others, and this goes for countries and cities.

Money is just paper. But those who have access to it in large figures, can buy a villa in Nice, fly a private jet, invest in a productive startup, or just buy other papers, such as stocks, derivatives, foreign currency etc. Overall, the villa in Nice and the private jet won't make a big difference. Productive investment does make a difference, because invested money becomes growth-inducing capital: it creates goods and services, and generates jobs, which in turn will expand demand and markets for the created goods and services. The wheels will turn. These were the (relatively) good times. But nowadays buying other papers such as stocks, derivatives, foreign currency etc. is the real thing where big money is made. Commenting on the 2015 Addis Ababa Conference on Financing for Development, Joseph Stiglitz sets the key issue clearly: 'There is no shortage of money waiting to be put to productive use. The problem is that the world's financial markets, meant to intermediate efficiently between savings and investment opportunities, instead misallocate capital and create risk'.

As we entered the 1980s, investing in papers started to generate more profit than investing in production. Centuries after the Venice bankers, we rediscovered the golden path: *pecunia pecuniam parit,* money begets money. This is somewhat difficult for the English speaking world to grasp, since buying dollars or options in derivatives, gambling on what will go up or down, is called investment just like building a dam or creating a factory. In French the distinction is clear: *investissements* is one thing, *placements financiers* another. In Brazil, *investimento* and *aplicações financeiras. The Economist,* at a loss to properly address money clearly going into speculation, such as gambling on the variation of different financial papers without creating any new wealth, called them *speculative investments.* But the basic issue is that getting rich and investing in production are now very much divorced. This is all that the 'world's financial markets', in Stiglitz's words above, is about. As the UNEP 2015 Report has it, 'the job of ensuring that the financial system is fit for sustainable development has just begun.' Meanwhile, *pecunia pecuniam parit.*

The world is awash with information on what should be done. From *Our Common Future* to *The Future we Want,* from the *Millennium Goals* to the *Sustainable Development Goals,* and the Paris COP-21 Summit, the road we must take has never been so clear. But in the present rules of the game, when putting money in the world's financial markets earns big returns, while putting money in clean production or productive inclusion of the poor earns warm smiles, things will just not work. Not because people are evil, particularly at the top, but because an institutional investor managing a huge pension fund is bound to maximize returns or lose his job. And as it seeks maximum returns, it will not hesitate to squeeze the real-economics productive firm to the limit.

The engineers at Samarco, a Brazilian mining company which provoked the major environmental disaster in Brazil, when two mining dams collapsed in 2015, know how to build a dam and understand risk evaluation, but Samarco is controlled by Vale and Billiton, which manage huge financial systems where financial returns are way above environment or social concerns in corporate governance. This disaster is not very different, in terms of how decisions are made at the top, from the Volkswagen scandal on toxic emissions, or the already forgotten Enron scam, the irresponsible speculation frenzy at Lehman Brothers, or the Libor and other fraud involving practically every member of the top financial institutions club.

The basic challenge is how to build new rules of the game, so that the convergence between environmental, social and economic interests can be restored. In Brazil, private pension funds have roughly 200 billion dollars to work with. They seek a basic return of 5.5 per cent a year to cover future pensions payments. This money could be invested in the small and medium enterprise to stimulate production and jobs, or in new less disruptive technology in agriculture and so forth. This would stimulate economic growth and better cover our needs in the future, as well as improve sustainability, income and social stability, the so-called triple bottom line. As it is, they can simply put the money in government bonds which in 2015 pay 14.25 per cent a year, ensuring total liquidity, no risk, good earnings. The government pays this interest from the taxes it collects from the population. Thus the public pays for private pension funds with public money, and the wheel turns without any money going into what would be useful for society.

This simple example shows that the issue is not with the lack of financial resources to face population challenges, but with what we can call the governance, or the decision process, concerning their management. The financial flows have been cornered to serve financial intermediaries, instead of serving sustainable development. If we do not face this challenge, no amount of discussions will help. It is not a question of sequestering the villa in Nice, but of generating rules of the game where the staggering amount of unproductive money is put back to work for society, and for the earth, and for the future generations. In this chapter, we shall concentrate on the concrete example of how financial intermediaries in Brazil have stalled 20 years of progress, and thrown the country into recession.

The international dimension

But first we must have a look at the overall situation in this area of finance. No country is an island these days, and this is particularly true for money, which has been reduced to bits, travelling in fractions of seconds and huge amounts on the waves. Dematerialized money flows in the global space, while regulation is fragmented into roughly 195 countries with different governments and a diversity of regulations. Global finance flows in unregulated space, notwithstanding the strong declarations during successive G20 meetings. No global regulation means no government regulation, since if a central bank decides to regulate big money, it will simply flow somewhere else. The huge title on the IMF publication *Finance&Development* cover, 'Who's in charge?' says it all. Nobody is in charge. Actually, the financial corporations are in charge.

One of the few positive results of the 2008 ongoing crisis is that for the first time we have some solid numbers concerning the financial world. The Swiss Federal Institute of Technology (ETH) presents the first figures on how the global network of corporate control works. Basically, in the corporate world, 737 groups control 80 per cent of the system, while a nucleus of 147 corporations controls 40 per cent of the system. Of these, 75 per cent are financial intermediation groups. These are staggering figures, and the research is generating ample debate. With such a degree of concentration of economic power, we need no conspiracy theory. And the degree of concentration also generates systemic risk, because of the huge volatility and the amounts of resources involved. The sheer size of the groups involved in worldwide different activities makes rational management impossible, and the political power they wield makes democratic process a fiction.

The crisis has also highlighted the importance of tax havens and illegal money in general. James Henry, former McKinsey chief economist, estimates the overall value at between 21 and 32 trillion dollars, roughly between one

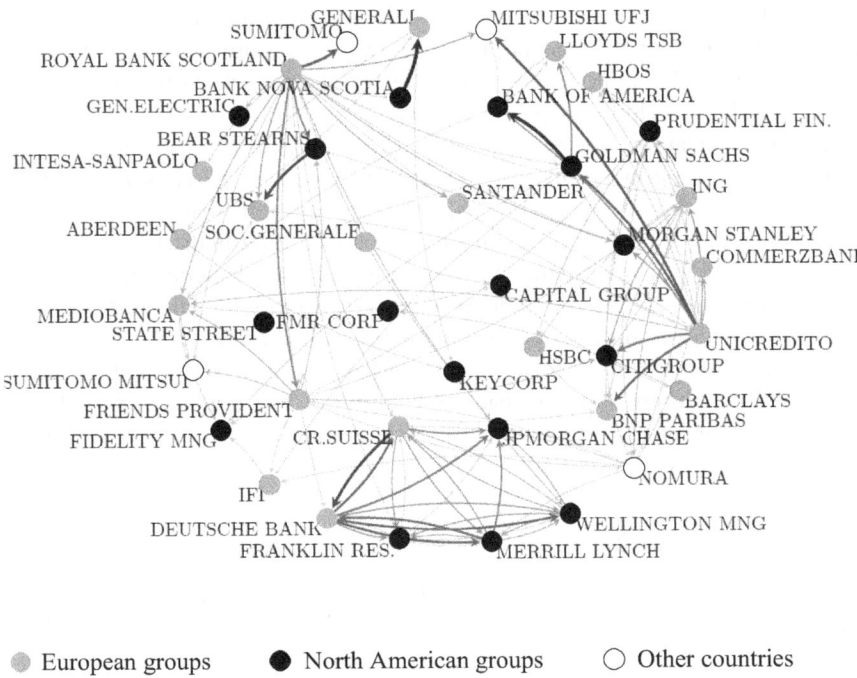

● European groups   ● North American groups   ○ Other countries

Figure 13.1 Example of a few international financial connections

Source: Vitali, Glattfelder and Battiston, (2012) http://j-node.blogspot.com/2011/10/network-of-global-corporate-control.html

third and one half of world GDP. In a special report *The Economist* settled on the probable figure of 20 trillion. Most importantly, it shows that the illegal money is not stacked in some paradisiac islands, but basically in Delaware, Miami and London, and managed by the same big banks which rig the Libor and Euribor figures, engage in money laundering and so forth. Basically the same financial groups presented in the ETH study.

The international dimension, since the 2008 crisis, if not better regulated, is now at least better documented. The crisis and the evident chaotic behavior of the financial system stimulated a collection of basic data on international finance, which interestingly, always eluded the International Financial Statistics of the IMF. In other studies we presented the detail of each of the new research carried out, with only the main results summarized here:

Research by the ICIJ (International Consortium of Investigative Journalists) has identified many names of companies and owners of fortunes, with details of instructions and transactions progressively released as they work with the immense files received. In November 2014, they published the massive tax evasion scheme of multinationals, using the tax haven that

Luxembourg has become. The amounts of evasion by Itaú and Bradesco, two big Brazilians banks, are presented in detail.

(ICIJ 2014 and Rodrigues 2014)

The study by Joshua Schneyer, systematizing Reuters data, shows that 16 international business groups control the bulk of intermediation of planetary commodities (grain, energy, minerals), mostly based in tax havens (Geneva in particular), creating the current framework for financial-commercial speculation on products that make up the blood of the world economy. Derivatives of this speculative economy (outstanding derivatives) surpass 600 trillion dollars, for a world GDP of 73 trillion dollars (BIS 2013 and Schneyer 2013).

The Credit Suisse discloses the analysis of the large world fortunes showing the concentration of ownership of 223 trillion dollars accrued (accumulated assets, not the annual income), that is to say, basically 1 per cent of the wealthiest own about 50 per cent of the wealth accumulated on the planet. The most commented figure in the 2015 Davos economic forum was that 80 persons have more wealth than the bottom half of the world population (Oxfam 2015).

The drain on productive activities, on the side of consumption as well as investment, is planetary. It is part of an international machine that since the liberalization of financial regulation, by the Reagan and Thatcher governments in the early 1980s, until the settlement of the main regulatory system, the Glass–Steagall Act, by Clinton in 1999, generated an international free-for-all.

Thus, we have a deformed planetary system, and Brazil is just playing its part in the global process of capital concentration through financial and commercial intermediaries. Information is far from complete in this shadowy area. However, two studies give us some orders of magnitude.

A study of the Tax Justice Network gives us some basic figures of capital amounts in tax havens by regions. In Brazil, the order of magnitude is of 519.5 billion dollars, which represents about a third of Brazilian GDP. (First line, sixth column of figures on Table 13.1). This is naturally the stock of Brazilian money in tax havens, not the yearly flow, but it represents a huge amount, and the government is fighting for its repatriation.

As such, Brazil is not isolated in this planetary system, nor is it particularly corrupt. But the whole set-up created is indeed heavily corrupted. Data for Brazil, 519.5 billion dollars in terms of offshore capital, are impressive, ranking fourth in the world. These resources should pay the taxes that would allow expanding public investment, and should be applied in fostering the economy where they were generated.

A second particularly interesting study is from the Global Financial Integrity, coordinated by Dev Kar, *Brazil: capital flight, illicit flows and macro-economic crises, 1960-2012*. This is a drain of resources by evasion estimated at 100 billion Brazilian reais per year between 2010 and 2012 (over 2 per cent of GDP). These are resources which in turn feed a good part of the 519.5 billion dollars in tax havens seen above. According to the report, 'the [Brazilian] government should engage more efforts to combat both under-invoicing of exports and

Table 13.1 Unrecorded capital flows, offshore assets, and offshore earnings, 1997–2010: Latin America and Caribbean (LAC) region foreign debt adjusted for currency changes, rescheduling, and arrears (nominal and real $2000 billions – 40 countries in region – 33 with data)

Period	Country	Original outflows Σ Real#B Σ Nom $B. ($2000) $		Offshore earnings (Σ$B.) ($2000) $	CF/GN1 sources period medians %	CF/ period medians %	Fligth stock ($B2010) (nominal) $	External debt ($B2010) nominal $	CF stock/ Ext. Debt. %	Offshore earning % Outflows %
1970–2010	Brazil	345.0	362.6	247.3	1.7	43	519.5	324.5	160	68
1970–2010	Argentina	213.9	259.3	272.8	3.4	68	399.1	129.6	308	105
1970–2010	Mexico	223.7	263.5	299.1	1.8	36	417.4	186.4	224	113
1970–2010	Venezuela	269.1	278.2	202.0	5.7	82	405.8	55.7	728	73
1970–2010	All Others (29)	205.1	211.9	169.1	1.7	41.5	316	317	100	87
	LAC Total	1,254.8	1,375.5	1,190.3	2.5	51	2,058.3	1,013.4	203	87

Source: World Bank IMF/UN/Central Bank/CIA (data); JSH analysis

Adjusted for Currency Composition of Debt; 75% Reinvestment Rate; Ave Yield = $US 6 mos CD rate (http://www.taxjustice.net/cms/upload/pdf/Appendix%203%20-%202012%20Price%20of%20Offshore%20pt%201%20-%20pp%201–59.pdf)

overpricing of imports, actively adopting additional deterrent measures rather than retroactive punishment.' Here, multinational companies prevail. Kofi Annan believes that this mechanism drains about 38 billion dollars a year from the African economies. The mechanism is known as mispricing, or trade misinvoicing, and has been intensely debated in the January 2015 African Union meeting, with the yearly evasion estimated at 53 billion dollars (GFI 2014 and 2015).

First of all, it is important to understand the limits of government's action. At international level, while the American and European elites will indeed continue to tolerate tax havens, including in the USA itself as in the case of the State of Delaware, and in Europe as in the case of Luxembourg and Switzerland, real control will be almost impossible. Tax evasion became too simple, and the ability to locate the illegal capital is remote. The order may, however, be significantly improved by controlling the outputs, transfer pricing and the like. The GFI report mentioned above, points out these possibilities and acknowledges strong advances by Brazil in recent years. On an international level, the BEPS (*Base Erosion and Profit Shifting*) endorsed by 40 countries representing 90 per cent of world GDP, gives us some hope as to the onset of a reduction of the planetary system of tax evasion by transnational corporations (OECD 2016).

The numbers have to be put in perspective. The Rio+20 summit hoped to raise 30 billion dollars to fund sustainability worldwide, while Brazil alone has over 500 billion dollars in tax havens, and the overall tax haven money is more than 20 trillion dollars. The Paris COP-21 summit on climate change hoped to drum up 100 billion dollars until 2020. Our hopes are to depend on Bill Gates or Mark Zuckerberg? However generous or simply conscious some people can be, we need a social and political compact to build sustainable development, and the figures that are being discussed are ridiculous. To quote Stiglitz again, 'new geopolitical realities demand new forms of global governance'.

Sustainable development in Brazil

Brazil has come a long way. Ending the corporation-supported military dictatorship in 1985 brought us back to civilization, the 1988 Constitution generated basic civil rules of the game, the breakdown of hyperinflation in 1994 (it reached 80 per cent a month!) brought us back to normal accounting and planning practices. And in 2003 the Lula government organized the first countrywide and long-term integral programme to fight poverty and social exclusion. Bolsa Família is widely known, but the Ministry for Social Development coordinated 149 such programs, involving 'Luz para Todos' which brought electricity to whole deprived regions, a more decent minimum wage law, large-scale stimulation of formal jobs, social security for people in informal and rural activities and so forth.

The results can be seen in the large-scale study by IBGE (the Brazilian Institute of Geography and Statistics), IPEA (the National Economic Research Institute),

FJP (Fundação João Pinheiro – an academic research foundation) and UNDP (United Nations Development Program). The results are impressive. In 1991, 85 per cent of the 5,570 municipalities in Brazil were listed in the lowest Human Development Indicators, the 'very low' class, under 0.500 HDI. In 2010, only 0.6 per cent of the municipalities remained in this class. Life expectancy rose from 65 years in 1991 to 75 in 2014. Brazilians earned ten more years to complain. Child mortality fell from 30 to 15 per thousand. Overall 36 million people were pulled out of poverty, the Gini inequality indicator fell from 0.58 to 0.49, curiously approaching the Gini indicator in the US which reached 0.45 and climbing. In 1991 only 13 per cent of 18–20 youths had completed secondary schooling, in 2010 the proportion reached 41 per cent. In two decades, coming from very low, Brazil has been transformed. And yet, so much has yet to be built.

From an environmental point of view, the picture is not so rosy. The basic comforting figure is that the destruction of the Amazon forest fell from 28,000 square kilometers in 2002 to roughly 4,000 – still a tragedy, but a huge victory, presently being pushed back by the agro-industrial complex. And Brazil covers around half its energy consumption with clean sources, well above the world average.

According to La Rovere (2012) 'the participation of renewable sources in domestic Brazilian energy supply (47%) is well above the world average, of 12.9% in 2006, and higher than the average of OECD (Organization for

Table 13.2 Recent evolution of gross domestic energy supply and current situation in Brazil: 1990 to 2009

Energy source	1990		2000		2005		2009	
	Mtoe	%	Mtoe	%	Mtoe	%	Mtoe	%
Petroleum and by-products	57.7	40.7	86.7	45.5	84.5	38.6	92.4	37.9
Natural gas	4.3	3.1	10.3	5.4	20.5	9.4	21.1	8.7
Mineral coal and by-products	9.6	6.8	13.6	7.1	13.7	6.3	11.6	4.7
Nuclear energy	0.6	0.4	1.8	0.9	2.5	1.1	3.4	1.4
Subtotal non-renewable	69.7	50.9	112.4	59.0	121.3	55.5	128.5	52.7
Hydroelectric and hydraulic	20.0	14.1	30.0	15.7	32.4	14.8	37.1	15.2
Firewood and vegetable coal	28.5	20.1	23.0	12.1	28.5	13.0	24.6	10.1
Sugarcane by-products	18.9	13.4	20.8	10.9	30.1	13.8	44.4	18.2
Other renewable	2.1	1.5	4.4	2.3	6.3	2.9	9.2	3.8
Subtotals renewable	69.7	49.1	78.2	41.0	97.3	44.5	115.4	47.3
Total	142.0	100	190.6	100	218.7	100	243.9	100

Source: MME/EPE, "National Energy Balance", 2006 and 2010.

Economic Co-operation and Development) countries, of 6.7%, due to renewable biomass and hydroelectricity.'

On the other hand, it is hugely difficult to face the necessary change in the consumption part of the energy matrix, where car corporations have been steadily pushing for more cars in the place of electric mass transportation systems, and government has paid only lip service to other energetic alternatives. Brazil has only recently invested in railroads and ship building, and transporting commodities over large distances by truck is obviously not only very expensive but deadly in terms of CO_2 emissions. The rapid urbanization – Brazil presently has 85 per cent of its population living in cities – has raised the overall energy intensity of development.

Other key issues are also daunting, such the overuse of chemicals in agriculture, with toxic substances reaching an absurd level on the consumer's table, and the impressive lack of basic sanitation in more than a third of households and so forth.

The general picture is that the recent process shows very strong and sustained progress in social, economic, and less so environmental issues, for two decades, and this has deeply changed the quality of life, particularly for the poorer half of the population. The basic mechanism was linked to the convergence of strong public social services and conditioned transfers to the poor and to the less developed regions of Brazil, while the traditional oligarchy found its economic interest in the expansion of the internal market. High international prices for commodities helped. But the marriage would not last. Understanding what went wrong, and how a very solid and long-term virtuous cycle was interrupted, is important not only for Brazil, but for different initiatives and countries which have invested in sustainability and inclusive development.

Thus the important initiative to promote inclusion, jobs and unrequited transfers to the poor during the Lula and Dilma administrations has produced excellent results. But the financial system has caught up with the initiatives and is stalling the Brazilian economy through huge interest rates on consumers, entrepreneurs and the public debt. The impressive financial profits are invested not in the real economy, but in other financial papers internally and through tax havens. The mechanism is certainly not very sophisticated but it drains the resources we need for sustainable development. In John Kenneth Galbraith's words, 'the way banks earn money is so simple that it is repugnant.' In 2015 we defined the Sustainable Development Goals in New York and the climate change goals in Paris. We did not define much in the Addis Ababa summit on how to fund these goals. Regaining control and ensuring our resources are used for these goals is essential, however unpleasant and down to earth the money business can be.

The gears of the financial system

The different parts of the system are well known, what we have done here is to put them together so as to show how the gears work together and the paralyzing impact on the Brazilian economy. We will look at credit in commercial chains, credit cards, banks (both for individuals and legal entities), the public debt, taxes and financial outflows. Much research is still to be done with this outlook, but the orders of magnitude of how the real economy is being drained by financial intermediaries becomes quite clear. Consider this as the Brazilian dimension of the global financial mess. Banks created the 1929 world crisis, the 2008 crisis we are still scrambling out of, they broke so many countries, generated the austerity mess to compensate their losses with our taxes, why should they be shy in Brazil? Not only CO_2 goes up in the air.

Let us start with a quick overview so the systemic aspect is clear. The numbers are quite clear. According to the Brazilian Central Bank, outstanding credit in the banking system in July 2015 reached 3,111 billion Brazilian reais (roughly 840 billion dollars), 54.5 per cent of GDP. The average interest rate on these operations was 28.4 per cent (the European equivalent would be around 3–5 per cent). This means the volume of interest money banks reap on this credit represents around 880 billion Brazilian reais, 15.4 per cent of GDP. Such a mass of financial resources captured by intermediaries deserves a closer look (BCB ECOIMPOM 2015). Therefore, it is important to understand the origin and destination of these resources. We will call this the integral financial flow.

Consider the installment plans in a retail chain, such as Casas Bahia. When you buy household ware on credit you will be paying a little over 100 per cent interest. Since so many, particularly the poor recently incorporated into the economy, have to buy on credit, their final purchasing capacity will be cut by half. Add to this, the fact that intermediaries charge an average 400 per cent on credit card 'revolving credit' and more than 230 per cent on overdraft, and we see that well over half the purchasing power of consumers is drained here to financial intermediaries, thereby sterilizing much of the economy's stimulation on the side of demand.

The result is that the population becomes heavily indebted while purchasing very little. In March 2005, 19.3 per cent of the average family income went to debt service. Ten years later, almost half this income, 46.5 per cent, is drained into the financial system. The installment plan retailers present to the consumer 'fits in your pocket' according to the TV commercials, but it overloads that pocket for a long time. Thus the demand stimulus on the economy is jammed.

Similar results are found on the investment side of the process, because if in the reproductive cycle most of the profit goes to financial intermediaries, the producer's capacity to expand production is thwarted, with the double restriction of reduced demand and restricted self-financing: they get paid very little for their product, when facing the huge commercial chains which have become basically financial intermediaries more than providers of commercial services. This hits particularly the small and medium enterprise.

In banks, personal credit average interest rate is 103 per cent according to ANEFAC (National Association of Executives in Finance, Administration and Accounting), which is staggering. Interest for legal entities is also prohibitive, in the order of 40 to 50 per cent, and to start a business under these conditions is not feasible. There are official credit lines in public banks which operate with more reasonable interest rates, but they only partially compensate for the appropriation of results by the private financial intermediaries.

The third item in the gear is the 'Selic' rate, the official and reference central bank interest rate, paid to owners of the public debt. With a GDP of 5.5 trillion Brazilian reais, one percent of GDP is 55 billion. If the debt service is set at 5 per cent of the GDP, for example, this means over 250 billion Brazilian reais of taxes are transferred essentially to financial groups, each year. For 2015 this figure is expected to reach 400 billion Brazilian reais, most of it obviously accruing to the existing debt, carving a deeper hole. Thus, a very significant part of the government's capacity to finance more infrastructure and social policies, another key mechanism to stimulate the economy, is sterilized.

Furthermore, the high 'Selic' discourages productive investment in companies as it is easier – zero risk, total liquidity – to profit from the huge public interest rates, 14.25% for an inflation of 8 per cent in July 2015. And for banks and other intermediaries, it is easier to profit from public debt than to promote the economy by funding productive initiatives, where you have to identify opportunities and do your project analysis homework. The large profits in financial intermediation end up contaminating a whole set of economic agents, all sold in the name of protecting the population from the inflation monster, the endlessly repeated argument in the media, whether popular or specialized.

It is thus understandable that we have this strange situation of a 12 month financial profit growth of the giant Itaú Bank of 22 per cent while GDP growth remains stalled or retreating. The economy is being drained by installment plans, credit card costs, bank interest rates for personal credit, interest rates for legal persons, and the high 'Selic' (official) interest rate. This is the Brazilian dimension of the global financialization. In the global financial speculation system, someone has to bring real value in, and this is how it works in Brazil. And through Santander, City, HSBC and other international banks heavily involved in Brazil, but also through the international outreach of Brazilian banks such as Bradesco and Itaú, the country joins the world casino.

To close the circle, we have tax evasion. With the global crisis we have little more than some hand-slapping as regards the financial regulation system, but at least we have more information: as we have seen, Brazil has roughly 520 billion dollars, about a third of its GDP, according to the Tax Justice Network research. Which means that resources which should be reinvested in the development of the economy are not only diverted to financial games internally, but also migrate to tax havens where they do not pay taxes. For example, we now have some data on Itaú and Bradesco in Luxembourg, while the Global Financial Integrity studies show over 35 billion dollars (roughly 2.3 per cent of GDP) illegally drained from Brazil every year through misinvoicing and mispricing. In another line of investigation, ICIJ (International Consortium of Investigative Journalists) has identified around 8,600 Brazilian fortunes in the Geneva HSBC asset management unit.

Legal or illegal transfers to tax havens represent only the external part of the drain, since the Brazilian tax system is heavily skewed, with the poor paying 32 per cent of their income in taxes, while the rich pay an average 20 per cent. Thus the regressive Brazilian tax system (there are no taxes on fortune and inheritance taxes are ridiculous) represents a formally legalized internal tax avoidance system, while the tax havens solution managed by the banks themselves represents the external mechanism. Join these various pieces together, and the dimension of the systemic deformation becomes quite obvious.

Brazil has made a huge effort to include the poorer one third of the population into the economy, around 60 million people. The financial intermediaries managed to paralyze the economy and to thwart the growth impact that this enormous effort had generated. The financial intermediaries adapted quickly to siphon away the new economic capacity at the bottom of the social pyramid, and the new technologies allow financial corporations to reach out cheaply to the small money which traditionally did not interest the banking system.

The numbers we present here unfortunately match. The data are known, all we have done is to put together different lines of research that usually do not communicate, with help coming from different institutions. Interestingly, organizing the data to evaluate the integral financial flow system had not been tried in Brazil. Most people comfortably sleep with the idea that financial mechanisms are beyond comprehension. But it is not so complex to present how the gears work together, and to understand how the financial system manages to push a big economy down the drain. In fact, we need many more people to understand how the economy is being deformed. No GDP can progress with such an amount of resources drained away from the productive cycle, out of the real economy. And no sustainable development will be achieved if we do not control our resources. This is obviously not a particularly Brazilian issue, but understanding how it works in a concrete country may help understanding also more global issues.

To wrap up the logic of this mechanism, consider that change towards a less unequal for society and less destructive for the planet development process involves simultaneously economic and political change. If the economy is growing, economic change is made politically more manageable. If the economy is stalled, and the rich earn less money, this is called a crisis. Ten per cent of world population going hungry, and roughly 5 million children dying from ridiculous causes and particularly from lack of access to food or to clean water, is not called a crisis, it is a challenge to be discussed. French economist Delavoye reminds us that it is easier to take essential resources from the poor than superfluous resources from the rich.

Curiously, we must make the present system work so that we can change it, however dangerous this may be. In this line, consider that economic growth in the present system relies on four engines: mass family consumption to stimulate the demand side; private investment to produce more goods and services; public investment in social policies and infrastructure; and the external sector, the so-called export driven growth. In Brazil, the first three engines of growth have been stalled by financial costs, while the last one, with the collapsing world commodity prices, only deepens our woes. Here we will concentrate on the first three. Let it be sufficient to say that focusing on exports is no alternative in the present chaotic world commodity instability.

By the end of 2015, President Dilma Rousseff's government was experimenting with a 'fiscal adjustment' centered on the reduction of public expenditure, which bankers love, but a broader fiscal and financial adjustment is unavoidable to put the economy on its feet. The consumers, the real economy entrepreneurs, and the public administration in its capacity as provider of infrastructure and social policies, could be winners in the straightening up of a deeply skewed system. But by trying to change the financial system, she started a political war.

In an example we registered in a shop in Joinville, a TV set was being sold at 674 reais in one payment, or 1670 reais in installments. The 'pocket-sized' monthly payment is around 69 reais. Interest rate over 120 per cent! You pay more than double. Consumption capacity is curtailed for two years. The producer receives little, and will invest little. Further, the consumer can buy little due to the burden of the interest rate. It is the so-called toll economy that jams the productive system, on the producer's side as well as on the consumer's, to benefit the intermediary. You can buy today, but the consumption capacity is curtailed for two years. A similar commercial network in Europe, MediaMarkt, charges 13.4 per cent a year, not 122 per cent! A 600 euros purchase in 18 installments will show a final total cost of 699 euros. And they have good profits.

Installment plans in commercial chains

Let us begin with the interest rates to final borrower, an individual, as practiced in trade, the so-called installment plans. The ANEFAC (National Association of Finance, Administration and Accounting Executives) shows the data for June 2014.

First of all, a methodological observation: in Brazil, the interest is almost always presented as 'monthly rate,' as shown in the first column in Table 13.3. It is technically right, but commercially and ethically wrong. It is a way of confusing the borrowers, because no one is able to mentally calculate the compound interest. Worldwide, the annual interest rates are used. The Itaú Bank, for example, presents on its website interest rates only in their monthly format, since per year they would show to be as extortionate as they truly are. In the photo above we see the TV set offered at an interest rate of 6.87 per cent per month and only in small print do they show the real interest rate of 122 per cent per year, which is an affront.

Table 13.3 Behaviour of installment plan interest rates by sector

Sectors	May/14		Jun/14		Variation %	Percentage points by month
	Monthly rate %	Yearly rate %	Monthly rate %	Yearly rate %		
Large chains	2.49	34.33	2.51	34.65	0.80	0.02
Medium chains	4.80	75.52	4.83	76.13	0.63	0.03
Small chains	5.41	88.18	5.45	89.04	0.74	0.04
Tourism companies	3.74	55.37	3.75	55.55	0.27	0.01
Household items	6.14	104.43	6.16	104.89	0.33	0.02
Eletro-eletronics	4.60	71.55	4.64	72.33	0.87	0.04
Imported items	5.23	84.36	5.24	84.57	0.19	0.01
Vehicles	1.80	23.87	1.78	23.58	−1.11	−0.02
Gym items	6.49	112.67	6.52	113.39	0.46	0.03
Inf. technology	4.35	66.69	4.38	67.27	0.69	0.03
Mobile phones	4.04	60.84	4.08	61.59	0.99	0.04
Int. decoration	6.30	108.16	6.33	108.87	0.48	0.03
Overall average	4.62	71.94	4.64	72.33	0.43	0.02

Source: ANEFAC (2014). www.anefac.com.br/uploads/arquivos/2014715153114381.pdf

The average interest charged on installment plans, of 72.33 per cent, simply means that this type of trade, instead of supplying decent commercial services, essentially became a banking service. It takes advantage of the fact that people do not understand the financial calculation, and have little cash availability which opens the way for extortion. Here, the retail seller of 'household items,' by charging interest of 104.89 per cent jams the demand, as it will be curtailed for 12 or 24 months while installments are being paid, and hampers the producer, who receives very little for the product. It is what we have described as the toll economy. Ironically, the stores announce that they 'facilitate'. In this whole procedure, consumer purchasing power is divided by two, and the reinvestment capacity of the producer is at a standstill.

Interest rates for personal credit

Consumers do not restrict themselves to buy by installments, whose average rate of 72.33 per cent is reproduced in the first line in Table 13.4. They also use credit cards and other financial mechanisms little understood by the vast majority of consumers.

Based upon the June 2014 data, these ANEFAC figures found that financial intermediaries also charge 238.67 per cent on the credit card, 159.76 per cent on overdraft, and 23.58 per cent in car purchases. Personal loans cost on average 50.23 per cent at the banks and 134.22 per cent at the financial institutions. We are leaving out the street usury that exceeds 300 per cent. Inflation in Brazil at the time was about 6 per cent.

Table 13.4 Personal credit interest rates

Credit line	May/2014		Jun/2014		Variation %	Variation of percentage points
	Monthly rate %	Yearly rate %	Monthly rate %	Yearly rate %		
Com. rates	4.62	71.94	4.64	72.33	0.43	0.02
Credit card	10.52	232.12	10.70	238.67	1.71	0.18
Overdraft	8.22	158.04	8.28	159.76	0.73	0.06
Dcc. banks – financing cars	1.80	23.87	1.78	23.58	−1.11	−0.02
Personal loan banks	3.41	49.54	3.45	50.23	1.17	0.04
Personal loan financial institutions	7.29	132.65	7.35	134.22	0.82	0.06
Overall average	5.98	100.76	6.03	101.90	0.84	0.05

Source: ANEFAC. (2014) (www.anefac.com.br/uploads/arquivos/2014715153114381.pdf)

It is noteworthy that the ABECS (Brazilian Association of Credit Cards and Services Companies) considers that the average interest on credit cards is 280 per cent, thus well above ANEFAC assessment. The ABECS considers that 50.1 per cent of consumer credit is taken on cards, 23.5 per cent in payroll loans, 13.1 per cent in installment sales of vehicles, and 13.3 per cent 'other.' In the case of cards, it concerns around 170 billion Brazilian reais. It is important to keep in mind that even without entering into card's credit, typically, a store has to pay about 5 per cent of the value of purchases to the bank, in addition to rental of the machine. This 5 per cent may be less for large stores having bargaining power with the financial system. Nevertheless it is a giant private tax on half the consumer credit, drastically reducing consumer purchasing power.

The ABECS considers that this portfolio 'is responsible for fostering consumer credit in the country.' This is a positive way of presenting the issue, however it encourages credit, not consumption. In the case of frequent access to revolving credit, people pay three or four times the value of the product. Miguel de Oliveira, Director of ANEFAC, summarizes the situation well: 'The person who cannot pay the bill and needs to pay step by step, or enter into the revolving credit, is actually funding the credit card debt with another type of credit. The problem is that this debt is endless. People end up not realizing the interest charged' (DCI 2014).

Obviously, with these interest rates, people, when buying on credit, spend more on interest than on the actual value of the product. We do have figures on how deep in debt families are (as seen above, roughly 46.5 per cent of family income is deviated to debt service), but this is not sufficient information. Because in this instance, families not only become heavily indebted, they do so buying very little. The numbers are clear: in practice, they pay almost double, sometimes even more. In other words, they buy half of what their money could buy, if it were in cash. And the cash purchases already include the profits of commercial intermediation.

The villain is not the taxes we pay, which is always what the media would like us to believe (and government bashing is such an easy sport). Albeit the overwhelming burden of indirect taxes only worsens the situation, it is the shift from the purchasing power to the payment of interest that really dampens the economy. Families were spending much more, as a result of the high level of employment and increased purchasing power of the grassroots. However, the interest rates sterilized the economy's dynamic capacity that this mass consumption could represent. Thus one of the main vectors mobilizing the economy is jammed. An economy of financial intermediaries was generated. Families that need goods and services are harmed, and indirectly, the effectively producing companies see their inventory grow. Much of the impact of economic strengthening by redistributive policies the government has stimulated is lost. Payroll loans help, but reach only 23.5 per cent of the consumer credit (DCI, 2014), and are also found within the range of 25 to 30 per cent interest, which seems to be reasonable for Brazilians only because of the exorbitant levels affecting other forms of credit.

Interest rates for legal persons

Interest rates for legal entities do not lag far behind. The study of ANEFAC presented an average rate of 50.06 per cent practiced per year, 24.16 per cent for working capital, 34.80 per cent for discount of trade notes, and 100.76 per cent for secured account. No one in their right mind can develop productive activities, start a business, face the time-to-market and balance accounts paying such interest. Here, the private investment is directly affected.

Banking activity may be quite useful to finance economic initiatives that will be profitable. But this implies that the bank will use the money from deposits (besides of course the leverage) to promote business initiatives, whose outcome will bring about legitimate profit to the investor, further permitting repayment

Table 13.5 Interest rates for legal persons

Credit line	May/2014		Jun/2014		Variations %	Variation of percentage points by month
	Monthly rate %	Yearly rate %	Monthly rate %	Yearly rate %		
Working capital	1.84	24.46	1.82	24.16	−1.09	−0.02
Discount of trade notes	2.48	34.17	2.52	34.80	1.61	0.04
Secured account	5.92	99.40	5.98	100.76	1.01	0.06
Overall average	3.41	49.54	3.44	50.06	0.88	0.03

Source: ANEFAC, (2014)

of the loan. The basic activity of a bank would be to add up depositors' savings to convert them into financing for economic activities, but this is no longer within the scope of activity of these banks. The economy, blocked on the side of demand by the type of consumer credit seen above, in the banks as well as in installment plans, therefore is equally blocked on the side of financing to the producer. Thus, demand as well as investment, the two main engines of the economy, are jeopardized.

Here, the rules of the game become extremely distorted. Large transnational corporations now have unbelievable comparative advantages as they may be financed from abroad with interest rates typically five or six times smaller than their domestic competitors. Many Brazilian companies may find funding at rates that could be considered normal, for example by BNDES, and other government banks, but without the capillarity which would allow irrigation of the huge mass of small- and medium-sized companies scattered throughout the country. It is noteworthy that in Germany 60 per cent of savings are administered by small local savings banks (*sparrkassen*) that generously irrigate small economic initiatives. Poland, which according to *The Economist* best faced the crisis in Europe, has 470 cooperative banks that finance activities of the real economy. One of the country's leading economists, J. Balcerek, comments wryly that 'our outdated banking system saved us from the crisis.'

Interest on the public debt

A third deformation is brought forth by the immense drain on public resources through the public debt. In 2015, roughly 7 per cent of GDP will be paid from our taxes to financial intermediaries, which invest in the public debt. This money could instead be used to finance public investments, infrastructure and social policies. This is very convenient for banks because, instead of having to identify good business projects, to promote investments and to evaluate and follow up productive initiatives, in short, to do their homework, they invest in government bonds of high yield, total liquidity, with no risk, ready cash, and a very attractive profit. Thus the public tax system is used to transfer our taxes to banks, with no productive impact, on the contrary, by draining public investment capacity.

Here, the effect is doubly detrimental: on the one hand, because with the profitability achieved with simple investment in public debt, banks no longer seek to foster the economy, but make investments in government bonds instead of irrigating the economic activities with loans. On the other hand, many productive companies, instead of investing more also apply their surplus in government bonds. The economic machine thus becomes hostage to a system that is profitable for those who invest in financial papers, but not for those investing in the real economy. For the government, it is even comfortable as it is easier to borrow than face the so badly needed tax reform. In 2015, the expected toll on the government is expected to reach 400 billion Brazilian reais, 7 per cent of GDP, bringing to a standstill the third engine for an economy to thrive, which is government capacity to finance infrastructures and social policies.

Table 13.6 Taxable income of the public sector (% of GDP)

Year	Primary outcome	Interest	Nominal income	"Selic"
2002	3.2	−7.7	−4.5	19.2
2003	3.3	−8.5	−5.2	23.5
2004	3.7	−6.6	−2.9	16.4
2005	3.8	−7.4	−3.6	19.1
2006	3.2	−6.8	−3.6	15.3
2007	3.3	−6.1	−2.8	12.0
2008	3.4	−5.5	−2.0	12.5
2009	2.0	−5.3	−3.3	10.1
2010	2.7	−5.2	−2.5	9.9
2011	3.1	−5.7	−2.6	11.8
2012	2.4	−4.9	−2.5	8.6
2013	1.9	−5.1	−3.3	8.3
2014	1.5	−6.0	−4.5	11.0

Source: Central Bank, (2014) (forecast 2014 Amir Khair)

A systemic deformation

This has been going on for quite a few years, gradually squeezing out productive activities. The table below shows that the real rate of interest to individuals (adjusted for inflation) charged by HSBC in Brazil is 63.42 per cent, while it is 6.60 per cent by the same bank for the same line of credit in the UK. For Santander, the corresponding figures are 55.74 per cent and 10.81 per cent. For Citibank they are 55.74 per cent and 7.28 per cent. Itaú charges solid 63.5 per cent. For legal entities, a vital area because it would involve promoting

Table 13.7 Overall real total yearly interest rate★ on personal loans in banking institutions in selected countries, during the first week of April 2009

Institution	Country	Real interest rate (%)
HSBC	United Kingdom	6.60
	Brazil	63.42
Santander	Spain	10.81
	Brazil	55.74
Citibank	U.S.A.	7.28
	Brazil	60.84
Banco do Brasil	Brazil	25.05
Itaú	Brazil	63.25

Source: data supplied by the banking institution for the interest in OCDE and BCB for inflation in selected countries and Brazil.

★ Interest accrued by administrative services, risks of default, profit margin and taxing.

the productive activities, the situation is equally absurd. For a legal entity, HSBC, for example, charges 40.36 per cent in Brazil, and 7.86 in the UK (IPEA 2009).

An IPEA study comments: 'For loans to individuals, the differential may be almost 10 times higher for the Brazilian in relation to the credit equivalent abroad. For legal entities, the differentials are also worthy of attention, since they are damaging to Brazil. For loans to companies, the cost difference is smaller, but still it is more than four times higher for the Brazilian.'

We therefore face a structural deformation of the financial intermediation system. There is no great mystery in the process: the global financialization, with its various forms of organization depending on the country and the laws, acquired specific methods to bleed the economy in Brazil, a national dimension of a nowadays planetary deformation.

The Brazilian Constitution, in Article 170, defines as principles of economic and financial order, among others, the social function of property (III) and free competition (IV). Article 173, paragraph 4, states that 'the law shall repress the abuse of economic power aimed at domination of markets, elimination of competition and the arbitrary increase of profits.' Paragraph 5 is even more explicit: 'The law, without prejudice to the individual responsibility of the legal entity leaders, will set forth the responsibility of the latter, liable to punishments compatible with its nature, for acts performed against the economic and financial order and against the people's economy.' Cartels are illegal. Exorbitant profit without corresponding productive contribution will be 'repressed by law' with 'compatible punishments.'

The practical result is a systemic deformation of the entire economy, which jams the demand on the consumption side, weakens investment, and reduces the government's ability to finance infrastructure and social policies. If we add the deformation of tax system based mainly on indirect taxes (embedded in prices), with a fragile burden on income and assets we have, here, the complete framework of an economy damaged in its foundations. After an impressive progress during the Lula and Dilma administrations, it is presently paralyzed by a less and less sustainable dead weight.

Balancing the system

Internally, the measures cannot be straightforward. ANEFAC clearly states the shortcomings of a system that is formally ruled by private law: 'We emphasize that the interest rates are free and they are stipulated by the financial institution and there is consequently no price control or ceilings for charges made. The only commitment for the financial institutions is to inform the client which fees will be charged should any type of credit be required.' Of course, as it is a cartel, the credit borrower has no choice. The ANEFAC recommendations are very simple: 'If possible delay purchases in order to have money to buy the same thing with cash payment avoiding interest.' In other words, do not use credit. This, recommended by the 'Association of Finance Administration and Accounting Executives,' is impressive.

However the government has powerful weapons. The first is to resume the gradual reduction of the 'Selic' official interest rate, which would force banks to seek alternative investments, bridging the gap between the financial system and entrepreneur initiatives, and reducing the outflow of public funds to the banks. The second is to reduce interest rates to end borrowers in the network of public banks, as was tried in 2013, but this time persisting in the dynamics. It is the best way to introduce market mechanisms in the financial intermediation system, contributing to weaken the cartel and forcing it to reduce the stratospheric interest rates: the final borrower would again have some options.

The third is to rebuild the fossil tax system: the objective is not to raise taxes, but to rationalize their incidence. The INESC research shows that 'in Brazil, the tax on assets is almost irrelevant, since it is equivalent to 1.31 per cent of GDP, accounting for only 3.7 per cent of the tax revenue of 2011. In some of the central capitalist countries, taxes on assets represent more than 10 per cent of tax revenue, for example, Canada (10%), Japan (10.3%), Korea (11.8%), Great Britain (11.9%) and the USA (12.15%)' (INESC 2014, 21). If we add the low incidence of income tax, and the fact that indirect taxes represent 56 per cent of the tax collection, we have a situation that claims for change.

'It should be stressed that the tax burden is very regressive in Brazil as it is concentrated on indirect and cumulative taxes that mostly burden the workers as well as intermediation system, the population needs to be properly informed. One of the most awesome aspects of this vital area for the country's development is the ominous silence not only of the media but also of the academia and research institutes, on the scandalous process of deformation of the economy by the financial system.'

Whichever way we look at it, the fact is the Brazilian economy is being drained by intermediaries who produce little or nothing. If we add up the interest rates of family consumption, the cost of installment plans, interest rates on business credit, the drain through the public debt and the tax evasion by means of tax havens and illicit transfers, we have a structural deformation of the production processes. Efforts to boost the economy while dragging this speculative waste load bound to our feet are not sufficient. There are more ills

Table 13.8 Tax incidence in Brazil: 2011

Tax incidence	% of collection	% of GDP
Consumption	55.7	19.7
Income	30.5	10.8
Assets	3.7	1.3
Others	10.1	3.6
Total	100	35.4

Source: INESC – implications of the Brazilian tax system, September 2014.

in the economy, but here we are addressing a huge mass of resources, which are necessary for the real economy. It is time that the business world itself – that actually produces wealth – wakes up to the imbalances, and assigns responsibilities where they should be. Ensuring the productive use of financial resources has become a key issue for Brazil.

Sustainable economics

Here we concentrated on the money side of the issue, based on the Brazilian example. We can easily see that the different gears of the financial system, both private and public, not only do not contribute to sustainable development, but managed to paralyze the very positive first steps. Basically, not only is the financial corporate world not interested in stimulating sustainable development, but it actually diverts money from the existing production process into financial gains, and manages not to pay taxes by means of tax havens. The political and moral legitimacy of this process is nil. As for banks, Polychroniu sums it up very simply: 'Banks should return to doing what they were created to do in the first place: offer a safe environment for people's savings and provide capital to business for development purposes.'

In 1997 Congress approved a law which allows corporations to fund elections. In Brazil we now have organized groups of congressmen linked to the car industry, to banks, to big constructors, to the four media giants, to the national and international agro-industrial complex. There remains very little citizen representation. This large scale corruption of the political process has been declared illegal by the Supreme Court in 2015, for the next elections. But the present Congress has been approving laws that support corporate interests, while taking down democratic progress we had managed to build during the last decades. Policies supporting the poor and promoting sustainability are in the front line of attack. What little progress we managed to make has led to an organized rightwing attack which generated, at of the end of 2015, economic ruptures and political chaos, culminating with the impeachment of the legitimately elected president Dilma Rousseff in 2016.

Going back to the beginning of this paper, our challenges are unfortunately easy to define: we are facing the necessary paradigmatic change in the way we deal with the planet, and this means we have to use our resources to fund another type of development; and we must make sure that this development is for everyone. We have the technologies, the money, and people who know how to go about it. But we do not have the corresponding political power and decision process. During 2015 we have seen in Addis Ababa, in New York and in Paris that people are very much aware of what should be done. And much is being done. But the time window we have both in the environmental and the social areas is short, and change is desperately slow. The October 2015 UNEP report states this in simple words: 'To achieve the sustainable development we want will require a realignment of the financial system with the goals of sustainable development.'

References

ANEFAC – Associação Nacional de Executivos em Finanças, Administração e Contabilidade, *Report on interest rates* (2014). Tables on pages 2, 3 and 5 www.anefac.com.br/uploads/arquivos/2014715153114381.pdf Accessed 20 October 2015.

Banco Central (BCB) – *History of interest rates* (2014) – Selic www.bcb.gov.br/?COPOMJUROS. Accessed 15 September 2015.

Banco Central (BCB) – *Política monetária e operações de crédito do SFN* (23/09/2015) – www.bcb.gov.br/?ECOIMPOM. Accessed 25 September 2015.

BIS Quarterly Review (2013) – www.bis.org/publ/qtrpdf/r_qt1306.pdf Accessed 20 March 2014.

DCI – A metade do consumo é financiada por cartões (Half the consumption is financed by credit cards) – August 20, 2014, pg. B1.

Dowbor, L. *Os estranhos caminhos do nosso dinheiro* (2014) Editora Fundação Perseu Abramo, São Paulo, 2014, http://dowbor.org/blog/wp-content/uploads/2012/06/13-Descaminhos-do-dinheiro-p%C3%BAblico-16-julho.doc   Accessed   25 September 2015.

Dowbor, L. *Intelligent alternatives for energy use in Brazil*, http://dowbor.org/blog/wp-content/uploads/2012/06/Intelligent-alternatives-of-energy-use-LDowbor.docx Accessed 20 October 2015.

Furtado, C. *Para onde caminhamos* (Which way are we going?) – Paper published in JB November 14, 2004. www.centrocelsofurtado.org.br/arquivos/image/201411191728100.Dossier%20CF%2020%20nov%202014%20ArtigoJBNovembro2004.pdf   Accessed 20 February 2013.

GFI (2014) – Global Financial Integrity – Brazil: flight of capital, September – www.gfintegrity.org/wp-content/uploads/2014/09/Brasil-Fuga-de-Capitais-os-Fluxos-Il%C3%ADcitos-e-as-Crises-Macroecon%C3%B4micas-1960-2012.pdf Accessed 20 September 2014.

GFI (2015) – Global Financial Integrity – *Illicit Flows from Africa*, January www.gfintegrity.org/press-release/au-un-high-level-panel-report-prioritizes-curbing-trade-related-illicit-flows-calls-sdgs-follow-suit/ Accessed 17 March 2016.

Henry, J. *The Price of off-shore revisited* (2013) – www.taxjustice.net/cms/front_content.php?idcat=148; *The Economist*, Special report, *The missing $20 trillion, 16/02/2013*.

ICIJ – International Consortium of Investigative Journalists (2013) www.icij.org/offshore/how-icijs-project-team-analyzed-offshore-files

ICIJ – International Consortium of Investigative Journalists (2014), Luxemburg Tax Files – November, 2014 – www.theguardian.com/business/2014/nov/05/-sp-luxembourg-tax-files-tax-avoidance-industrial-scale

Illicit Flows from Africa, January 2015, https://mail.google.com/mail/u/0/?ui=2&ik=4dd873709b&view=pt&search=inbox&th=14b438c80396501d&siml=14b438c80396501d Accessed 30 January 2015.

INESC – As implicações do sistema tributário brasileiro na desigualdade de renda (2014) – September, 2014, www.INESC.org.br/biblioteca/textos/as-implicacoes-do-sistema-tributario-nas-desigualdades-de-renda/publicacao/

IPEA – *Transformações na indústria bancária brasileira e o cenário de crise* (2009). Transformations in the Brazilian banking industry and the crisis scenario – Official Report by the Presidency, April, 2009, p. 15 www.ipea.gov.br/sites/000/2/pdf/09_04_07_ComunicaPresi_20_Bancos.pdf

Khair, A. (2015) – *A borda da cachoeira* – OESP, 01/02/2015 – http://economia.estadao.com.br/noticias/geral,a-borda-da-cachoeira-imp-,1627819

La Rovere, E. (2012) *Alternative Energy in Brazil: a favourable heritage* – http://dowbor.org/blog/wp-content/uploads/2012/06/Alternative-Energy-in-Brazil-a-favourable-heritage-Emilio-La-Rovere.docx

MME/EPE Ministério de Minas e Energia/Empresa de Pesquisa Energética, Balanço Energético Nacional (2006) – Ano base 2005 – Relatório final – Rio de Janeiro: EPE, 2006. https://ben.epe.gov.br/downloads/BEN2006_Versao_Completa.pdf Accessed 2 March 2016.

MME/EPE Ministério de Minas e Energia/Empresa de Pesquisa Energética, Balanço Energético Nacional (2010) – Ano base 2009 – Relatório final – Rio de Janeiro: EPE, 2010. https://ben.epe.gov.br/downloads/Relatorio_Final_BEN_2010.pdf Accessed 2 March 2016.

OECD – ICIJ – BEPS: *Base Erosion and Profit Shifting* (2016) – http://publicintegrity.us4.list-manage1.com/track/click?u=8dc6eceed67f7f012462d0b12&id=f388dc1436&e=d256201ac5 Accessed 2 August 2016.

Oxfam – *Having it all and wanting more* – 2015 – http://policy-practice.oxfam.org.uk/publications/wealth-having-it-all-and-wanting-more-338125.

Polychroniu, J.C. (2014) *Reconceiving change in the age of parasitic capitalism*, Truthout, www.truth-out.org/opinion/item/25974-reconceiving-change-in-the-age-of-parasitic-capitalism

Rodrigues, F. (2014) Folha de São Paulo November 5, 2014 www1.folha.uol.com.br/mercado/2014/11/1543572-itau-e-bradesco-economizam-r-200-mi-em-impostos-com-operacoes-em-luxemburgo.shtml

Schneyer, J. (2013) – Commodity Traders: the Trillion Dollars Club –http://dowbor.org/2013/09/joshua-schneyer-corrected-commodity-traders-the-trillion-dollar-club-setembro-201319p.html/ or www.reuters.com/assets/print?aid=USTRE79R4S320111028

Stiglitz, J. (2015) www.theguardian.com/business/2015/aug/06/joseph-stiglitz-america-wrong-side-of-history

Tax Justice Network – James Henry, *The Price of off-shore revisited* – www.taxjustice.net/cms/front_content.php?idcat=148; Data on Brazil are in Appendix III, (1) pg. 23 www.taxjustice.net/cms/upload/pdf/Appendix%203%20-%202012%20Price%20of%20Offshore%20pt%201%20-%20pp%201-59.pdf

Tax Justice Network (2014) www.taxjustice.net/wp-content/uploads/2014/06/The-Price-of-Offshore-Revisited-notes-2014.pdf

Tax Justice Network (2011) *The Cost of Tax Abuse: the Cost of Tax Evasion Worldwide*, www.taxjustice.net/2014/04/01/cost-tax-abuse-2011/

The Economist, December 7, 2013 *The rise of Black Rock*, www.economist.com/news/leaders/21591174-25-years-blackrock-has-become-worlds-biggest-investor-its-dominance-problem

The Economist, February 16, 2013 *The missing $20 trillion*, Special Report on Offshore Finance.

UNEP (2015) Aligning the financial system with sustainable development – http://web.unep.org/inquiry file:///C:/Users/Ladislau%20Dowbor/Downloads/Inquiry%20-%20Pathways%20to%20Scale%20FINAL%2020150119.pdf

Vitali, S., Glattfelder, J.B. and Battiston, S. (2012) ETH, *The Network of Global Corporate Control*. http://arxiv.org/pdf/1107.5728.pdf

14 Climate change and the integration of public policies

Marcel Bursztyn and Maria Augusta Bursztyn

The debate on climate change requires consideration of more general issues, such as regulation, public policy and planning. Despite persistent (though declining) research, discourse and decision-making, which challenge the veracity of manmade climate change, successive reports of the Intergovernmental Panel on Climate Change (IPCC), and the publication of increasingly solid studies have expanded the degree of certainty regarding the issue. A testimony to this convergence is the mobilization of the international community around agreements and protocols that help address vulnerabilities, create coping mechanisms and promote actions to mitigate climate change impacts (in the sense of reducing the causes of the problem in the long run).

As a result, the structures of government (and governance) also reflect the trend toward institutionalization of regulatory actions that are aimed at tackling climate change. This process is similar to what happened during the 1970s and 1980s, when the international debate, and national experiences coping with the challenges of environmental management, inspired the development of institutional bodies and the definition of environmental policies (McNeil, Verburg and Bursztyn 2012). From a theoretical point of view, the idea of policy integration requires consideration of the objectives of a given policy in the development and implementation of other policies (Toledo-Filho 2014). At least three criteria must be met at all stages of the public policy cycle, from design to evaluation, according to Underdal (1980): comprehensiveness, consistency and aggregation.

The purpose of this chapter is to present existing conditions and discuss the challenges inherent in the adoption of climate issues as a national public policy priority. The chapter also deals with climate as an opportunity to promote sector policy integration in order to improve planning. In this sense, the incorporation of climate issues in public policy may be a strategy to increase the effectiveness of the regulatory function of the state and to reduce the effects of contradictory public intervention strategies. This chapter focuses on the Brazilian experience.

The climate issue becomes a political priority

In Brazil, the climate issue has been present in public policy since the introduction of the National Environmental Policy Act (Law no. 6938, 1981). But this initial presence was indirect:

- By controlling vehicle emissions in order to reduce air pollution, there is also a decrease in emissions of greenhouse gases (GHGs).
- By reducing the use of CFCs (chlorofluorocarbons) by industry, as laid down by the Montreal Protocol, this also contributes to the reduction of negative climatic effects.
- By fighting deforestation and creating protected areas, there is improved balance between emissions and carbon sequestration.
- By requiring control measures on water eutrophication processes in reservoirs, there is avoidance of methane generation.

It is worth noting that long before the environmental legislation of 1981, since the beginning of the twentieth century in fact, Brazil has implemented public works to counteract the effects of droughts in the semiarid Northeast.

More explicit consideration of climate policy in Brazil began to emerge at the end of the first decade of the present century. Some important developments included (Bursztyn and Bursztyn 2013, 449–451):

- In 2000, the federal government created the Brazilian Forum on Climate Change (FBMC is the acronym in Portuguese), chaired by the President. The Forum members include State Ministers, heads of regulatory agencies, state secretaries of the environment, representatives of the business sector, civil society, academia and non-governmental organizations.[1]
- In 2004, Brazil submitted its Initial National Communication to the Climate Convention, containing its first inventory, which presented data on GHG emissions from 1990 to 1994.
- In 2007, the Inter-Ministerial Committee on Climate Change (CIM is the acronym in Portuguese) was established, consisting of 17 federal agencies and coordinated by the Office of the Chief of Staff of the Presidency, and its Executive Group, made up of eight ministries and the FBMC, coordinated by the Ministry of the Environment. These agencies had the main functions of developing and implementing the National Policy on Climate Change (PNMC is the acronym in Portuguese) and the National Climate Change Plan.
- In 2008, the National Climate Change Plan was launched, which aims to encourage the development and improvement of mitigation actions in Brazil, as well as the creation of internal conditions to address the negative impacts of global warming. The plan is designed to go through revisions and seasonal evaluations of results. It is divided into four areas: mitigation opportunities; impacts, vulnerability and adaptation; Research and

Development; and education, training and communication. Among the highlights were targets to reduce by 80 percent the rate of deforestation in the Amazon by 2020 and doubling planted forest areas from 5.5 million ha to 11 million ha (of which 2 million ha with the use of native species). It should be noted that these targets are conditioned on the existence of new national and international resources for the supervision and economic reorientation of forested regions.

• In 2009, Law 12,187 was enacted, which regulates the National Policy on Climate Change (PNMC). The law sought to reduce anthropogenic emissions and remove GHGs through the use of sinks. It also established that sectors of the economy should have GHG emission reduction targets. Art. 12 called for the country to reduce GHG emissions between 36.1 and 38.9 percent, by the year 2020. These figures were presented as voluntary targets at COP-15 in Copenhagen in 2009.

• In 2010, Decree No. 7390 formally established the National Climate Change Plan.[2] In Art. 1, the Plan states that "The principles, objectives, guidelines and instruments of public policy and governmental programs should, whenever applicable, be compatible with the principles, objectives, guidelines and instruments of the National Policy on Climate Change." This legal determination points to the need to integrate different strands of public policy, as shown by Art. 3°:

    I -   Action Plan for the Prevention and Control of Deforestation in the Legal Amazon – PPCDAm;

    II - Action Plan for the Prevention and Control of Deforestation and Fires in the Cerrado – PPCerrado;

    III - Ten Year Plan for Energy Expansion – PDE;

    IV - Plan for the Consolidation of a Low Carbon Agricultural Economy; and

    V - Steel Emissions Reduction Plan.

• In 2015 Brazil launched the National Plan for Adaptation to Climate Change,[3] whose preparation involved the participation of civil society, the private sector and state governments, and had the involvement of various ministries. The scientific community played an important role in drawing up the plan, through the FBMC, the Climate Network and the National Center for Monitoring and Early Warning of Natural Disasters (CEMADEN is the acronym in Portuguese). Eleven sectors were considered in the National Plan for Adaptation to Climate Change: agriculture, water resources, food security and nutrition, biodiversity, cities, disaster risk management, industry and mining, infrastructure, vulnerable communities and populations, health and coastal zones.

The climate issue gains importance among national institutions

The development of the National Plan on Climate Change, 2008, was of a political nature. Commitments under the UN Framework Convention on

Climate Change (UNFCCC) did not require Brazil to reduce GHG emissions. As in other developing countries, Brazilian actions aimed at climate tend to focus on adaptation measures. However, internal pressure from environmentalists, combined with a desire to show the international community the country's engagement in mitigation actions, led the government to establish instruments aimed at climate goals (May and Vinha 2012). Such actions are considered voluntary and fall into the Nationally Appropriate Mitigation Actions (NAMA) category.

The Plan makes clear that coping with climate change promotes public policy integration through the participation of various sectoral plans. In some cases, such as in the areas of health, water resource management and urban development, the involvement should occur through adaptation actions. The industrial, infrastructure and forest sectors, among others, should work on mitigation activities. Agriculture should act to both mitigate activities that cause climate change and adapt to potential impacts.

The ABC Plan (Low Carbon Agriculture) was launched in 2010[4] to cope with climate change in the agricultural sector. The ABC Plan included seven programs:

- Recovery of Degraded Pastures;
- Integration of Crop-Livestock-Forest and agroforestry systems;
- No-tillage System;
- Biological Nitrogen Fixation;
- Planted Forests;
- Processing of Animal Waste; and
- Adaptation to Climate Change.

Other sectors also included climate change in their plans. In 2007, the Ministry of Science and Technology (currently Ministry of Science, Technology and Innovation, MCTI is the acronym in Portuguese) created the Climate Network (Brazilian Research Network on Global Climate Change) whose mission is to generate and disseminate knowledge so that Brazil can respond to the challenges represented by the causes and effects of global climate change. Starting with a set of 10 thematic sub-networks, in 2015 the Climate Network increased the number of sub-networks to 15 and distributed them among a wide range of universities and research institutes in Brazil.[5] The main issues were:

- Agriculture
- Biodiversity and Ecosystems
- Cities and Urbanization
- Natural disasters
- Regional development
- Scientific disclosure
- Economy
- Renewable energy

- Climate modeling
- Oceans
- Water resources
- Health
- Ecosystem services
- Land use and coastal zones.

The 15 thematic areas of the Climate Network mobilized a broad universe of researchers and have served as a lever to increase the amount of Brazilian scientific production on climate, in addition to supporting public decisions.

In its first phase, the Climate Network operated in a decentralized manner, with each sub-network concentrating on its own subject area. From 2015, the sub-networks started to act in an integrated manner, according to three major integrative projects:

- Climate modeling
- Food security, water and energy
- Human dimensions of climate change.

Following the government's determination to integrate climate change into sectoral policies, in 2013 the Ministry of Industrial Development and Foreign Trade (MDIC) launched the Sector Plan for Mitigation and Adaptation to Climate Change for the consolidation of a Low Carbon Economy in the secondary industry.[6]

In 2014, the Secretariat of Strategic Affairs (SAE is the acronym in Portuguese) of the Presidency, whose mission was to provide subsidies to help governing bodies define their long-term actions, created a Strategic Thinking Group on Climate Change.[7] The SAE was abolished in 2015 after political wrangling and institutional redesign. This decision was, to some extent, a reflection of the second-tier status of long-term strategies (the amount of time beyond a given government's mandate) as compared to initiatives aimed at the short term.

Institutional volatility in the Brazilian state apparatus is noteworthy because executive bodies often experience leadership changes, sometimes are terminated, or are incorporated into other institutions of less prestige and power. With executive administrative changes, research activities can provide continuity, as well as qualified staff to government.

The above examples illustrate the incorporation of the climate change issue into the policies and practices of Brazilian government agencies. The process is similar to what occurred after the adoption of the National Policy for the Environment in 1981: vertical development, through the creation of agencies and the establishment of policies in federated states and municipalities (decentralization); horizontal development, through the integration of environmental themes into agencies and sectoral policies at the federal level (de-concentration). If environmental policy continues to evolve, it is expected that states and municipalities will follow a similar horizontal expansion.

The greatest challenge of climate policy is not just the *convergence* of sectoral actions in relation to tackling climate change, but the *integration* of policies. Convergence already represents an advance by incorporating the climate change issue into different axes of sectoral policy. By integrating the issue, it can be addressed in an interdisciplinary manner, take advantage of positive synergies and avoid zero-sum games. Negative synergies occur when a sector policy axis causes loss of effectiveness or neutralizes the actions of another axis.

Contradictions between climate policies and other axes of public policy

Viewed from the perspective of evaluating the effectiveness of public policies, adding another element that guides sectoral policy development from the top-down, similar to the provisions of the PNMC, is not a guarantee of success. In practice, the tangle of missions, policies and instruments that deal with the many sectors of public regulation brings a risk that the new priority (of supra-sectorial character) will be confined to rhetoric. Several examples illustrate this assertion.

While the Ministry of Agriculture plays a leading role in tackling climate change, with its adoption of the ABC Plan, its responsibilities in the promotion of other national priorities, such as increasing production of commodities by agribusiness, have also been strengthened.

The national economy is increasingly dependent on the export of agricultural products, particularly soybeans, corn, cotton and beef. These products benefit, directly and indirectly, from public policies. Examples of these public policies include: credit for production, infrastructure construction and adaptation of environmental policy instruments (such as the Forest Code) allowing agribusiness to avoid complying with previously established rules.

In just over a decade manufacturing exports lost substantial space to the primary commodity sector (including mining). In 2000 the export of manufactured goods and commodities were of the same order of magnitude. However, in 2011 the former group represented less than half of the latter, in terms of value (Verissimo and Xavier 2014). This reflects a relative de-industrialization process that, in a way, extends the power of political lobbies associated with the primary sector of the economy, which has hindered progress implementing environmental legislation to control deforestation.

In the energy sector, although the National Plan on Climate Change set a goal to expand the domestic consumption of ethanol by 11 percent over ten years (which is very conservative considering the positive Brazilian experience in ethanol production),[8] little else has been done to integrate climate, energy, and agricultural policies.

In 2004 the National Production and Use of Biodiesel Program (PNPB is the acronym in Portuguese) established a mandatory percentage of biodiesel blended with mineral diesel, that considered technical measures and domestic biodiesel production capacity (Garcez and Vianna 2009; La Rovere, Pereira and Simões 2011). The idea was to unite two axes of public policy: renewable energy production and support for family farming. The government established a Social Fuel Seal, which gives plants that produce biodiesel a subsidy to acquire vegetable oil from small producers. Recent data show that there was an increase in the number of small farmers producing vegetable oil in the first five years of the program. But after 2011[9] the expansion of biodiesel production occurred mainly through the purchase of vegetable oil from owners of large crop areas, which are intensive consumers of chemical inputs and whose expansion of cultivated areas has a direct or indirect effect on deforestation.

Brazil decided to rely on the potential of newly discovered oil reserves and natural gas sources in deep ocean waters, which would allow it to more than double its production of hydrocarbons in a relatively short time. High oil prices on the international market served to stimulate an energy policy, which ultimately neglected renewable sources, thus leaving Brazil behind most countries that stand out in the world economy. The decline of oil prices (from about $115 a barrel in May 2014 to less than $30 in early 2016) exposed the fragility of Brazil's energy strategy. Nonetheless, the subsequent economic recession has not helped reverse the delay in exploring alternatives sources of energy.

Despite initiatives such as PNPB and the development of flex-fuel engines (adapted to gasoline or ethanol in any proportion), the transport sector energy model remains based on the use of petroleum products. The use of biofuels has not proven sufficiently competitive. Mendes and Rodrigues Filho (2012) show that between 1990 and 2008 (thus before the oil exploration in deep ocean waters by Brazil), the growth in absolute emissions from the oil and gas sector was 115 percent, which represents an annual increase of 8.2 million tCO_2e.

In terms of electricity generation, Brazil has increased GHG emissions per kilowatt produced. In 1990, 93 percent of all electricity produced in Brazil came from hydroelectric plants. In 2011, that figure had fallen to 81 percent.[10] This shows the effect of strict environmental policy and pressure from environmental groups, which have shown many of the negative impacts of the construction of large dams. But this decline also reveals a paradox, which is the increase in GHG emissions.

Other goals of the National Plan on Climate Change will likely not be satisfactorily met, although they still have not been verified through monitoring and evaluation. Some of these goals are:

•   double the area of planted forests, to 11 million hectares in 2020, with 2 million ha of native species; and
•   increase recycling of municipal solid waste by 20 percent by 2015.

Land policy also contradicts climate policy. There is a determination to reduce deforestation in the Amazon by 80 percent. However, there has also been a proliferation of settlements, which were promoted by the federal government agency responsible for increasing the colonization of unoccupied areas for agrarian reform and land regulation (INCRA), as well as an expansion in agribusiness. These contradictions constrain the possibility of success. Yanai et al. (2015) estimate that 41 percent of the forest area in agrarian reform settlements was cleared by 2013. Le Tourneau and Bursztyn (2010) also addressed the taboo topic (aspects of social policy that are in opposition to environmental policy) concerning the contradictions between land and environmental policies in the Amazon. The authors show that the strategy adopted since the early 1960s, and which continues to the present day, is transferring the "land issue" issue (i.e. the demand for land by small farmers when there is limited land availability) from other regions to the Amazon. This process inevitably leads to pressure on the forest, and therefore, on climate dynamics.

Mesquita (2015) focuses on another contradiction between public policies. Studying food security strategies in the semi-arid Northeast, the author points out the relationship between the persistent vulnerability of farmers to climate effects and their limited participation in the Food Acquisition Program (PAA is the acronym in Portuguese), whose mission is to support family farmer members, who are the most vulnerable to climate effects. Frequent and prolonged droughts reduce the possibility of these farmers selling to the government procurement program, which supplies food to schools and public hospitals. Accordingly, government demand ends up being satisfied by larger scale producers that possess better technical knowledge, in other regions.

Toledo-Filho (2014) provides an overview of UK and German experiences compared with those in Brazil. The former two countries excel in integrating climate and energy policies. The study shows that in Brazil, chronic bottlenecks caused by difficulty reconciling missions and sectoral instruments with supra-sectoral commitments, limit success.

It is important to note that Brazil's history reveals many examples of the contradiction between public policy axes, which shows that this is not a specific problem of climate policy. Examples include:

- The Amazon occupation strategy since 1970, which represented a clear conflict with environmental policies, which were also established in that period.
- The promotion of export agriculture, which generated tensions with environmental policies. Considering the characteristics of international trade, this promotion has limited the effectiveness of industrial policy that adds value to primary resources, thus generating increasing demands for new land.
- Conflicts between multiple uses of water resources, where demand for hydro power production competes with other uses, such as agriculture and even urban water supply.
- Urban housing policy (or lack of policy) – there is a financing program for construction, but not a demographic or economic development strategy in chosen locations – that conflicts with urban environmental policy and its regulatory instruments (such as zoning and building codes); the effective

absence of a national plan for mitigation and adaptation of cities to climate change allows new neighborhoods to be built in inappropriate areas that are vulnerable to problems such as flooding, erosion and landslides.

- Policies "against drought," which result in the creation of modern irrigation areas (Oasis type), contrasting with the massive coexistence initiatives (such as the construction of cisterns for rainwater storage).
- Policies (or strategies) that cause demographic pressure on vulnerable natural environments and on poor infrastructure, such as occurred (and still is in some places) during the occupation of the Amazon frontier.

Factors that limit policy integration

The challenge of addressing climate change exposes a persistent vulnerability of public policies: how to move from sectoral action to *convergence* around supra-sectoral questions and, especially, how to ensure that *convergence* leads to *integration*. Although this is a universal problem, it is critical in Brazil, where a set of circumstances increases the fragility of institutions, generating risks that public actions may result in zero or even negative-sum games. Among the factors that inhibit integration policies are:

- The corporate bureaucratic culture of sectoral bodies, which leads to self-centered practices and even institutional conflicts with other public entities, whose missions are different. The environmental licensing requirement, and the respective conditionalities related to certain projects, as laid down by law, are examples of conflicts between institutions.
- The political culture of government sectors split among government allies, leading to a dissociation between institutional mission and interests.
- The "capture" of public intervention mechanisms by certain stakeholders at the local level. As an example, for more than a century, drought coping policies in the semi-arid Northeast have been controlled largely by the local elite, with poor results.
- The co-optation of public officials through intimidation or corruption, leading to non-compliance with institutional missions.
- The existence of a broad set of priorities, many of them contradicting each other. Good planning practice teaches that when there are many priorities there are really no priorities. Defining a hierarchy of priorities at the right scale is the responsibility of government.
- Emphasizing short-term priorities, which is unhealthy for democracy. Leaders who are unable to envision a future beyond the duration of their term, are bound to limit their actions to just conjuncture measures. They build schools, but do not improve education; they act on the consequences of serious problems, but do not care for dealing with structural causes.

Short-term measures before economic crises (and they are frequent) are an enemy to structural changes. Whenever GDP growth rates are low, or negative, the government approach is commonly to favor sectors that have momentum, even if the sector's actions and plans do not align with what should be long-term strategies. One example is the recent increase in the share of the primary sector in the Brazilian economy, which has caused de-industrialization and puts Brazil on a different economic trajectory in relation to the most successful emerging countries.

Between rhetoric and priority

The specialized literature on the subject indicates that one challenge of the planning process is priority setting. It is clear that the more complex a society, the greater the number of issues that require government regulation and long term strategies. However, although some of these issues are priorities for particular agencies, not every issue has the same degree of relevance when actors are faced with other problems. For example, how should health and energy priorities be compared? When making resource allocation decisions, what should be the criteria? Each subject is relevant to its respective ministry, but a higher-level decision-making body must determine an overall hierarchy of importance.

Recent participatory planning and governance practices show that an expansion of the universe of actors involved in decision-making, by including different sectoral spheres and levels of government, as well as representatives of civil society, tends to lead to a higher degree of accuracy, legitimacy and effectiveness. But there are also risks, including the action of lobbies, lack of contextual knowledge (Fonseca, Bursztyn and Moura 2012), clientelist practices and dilution of responsibility (accountability).

The same kind of conflict between different sectoral bodies occurs internally in these institutions. For example, primary and technical education are priorities in education, but which one is more relevant politically? A posteriori, the effective allocation of funds is a good thermometer. But the lack of a priori definition is a problem.

Priority setting among different levels of government is even more complex. What is of interest to the country as a whole may not meet municipal needs or expectations. A classic example of this is the creation of protected areas, which are of national interest, but "immobilize" local productive spaces.

Considering the above context, climate policy tends to reproduce the problems and limitations that have happened in other priority axes, such as sustainable development: there is progress through convergence (horizontal and vertical), but impasses in integration remain. In this sense, rhetoric tends to nullify effectiveness. When an issue is called a priority, but is not actually a priority, there is inaction. This phenomenon was studied by Fonseca and Bursztyn (2009) in terms of sustainability, a concept that tends to be assimilated by all (or almost all) policy areas, without necessarily changing practices or attitudes. When this happens, it opens up space for actors (government or otherwise) that incorporate politically correct speech, as a way to legitimize practices that contradict the content of the speech itself (Fonseca, Bursztyn and Moura 2012).

Rethinking planning

The complexity of dealing with climate change in the context of public policy brings the debate back to the foundations of planning theory. In terms of its interdisciplinary nature, its institutional character, its international coverage and even its intergenerational timeframe, climate change policy is a typical case of what Rittel and Webber (1973) called a "wicked problem". Unlike the "domesticated" or "benign" problems, wicked problems are not easily resolved. Further, there is no mathematical formula to assess whether or not the problem was resolved completely.

The construction of a bridge, for example, requires a set of procedures, and if 99 percent of the work is carried out, missing only the access ramps, the project is not successful. Reducing greenhouse gas emissions requires a complex web of decisions and procedures. But if the results are only 50 percent of planned targets, one cannot conclude that there was total failure. On the other hand, even if the bridge is 100 percent complete (and the degree of success of the work is 100 percent), that does not mean the objectives are achieved automatically. If the completed bridge does not lead anywhere, then it was a poorly conceived venture and it would have been better had the bridge not been built.

The solution of a mathematical problem can be true or false. On the other hand, the solution to a complex problem is relative, and solutions can be good, bad or good enough. Faced with the study of governance forms adopted for the treatment of social issues, Grindle (2004 and 2007) realized that a set of indicators, increasingly numerous and sophisticated, was used as a parameter to measure if the processes were good or bad (good governance or bad governance). Her conclusion was that the list of attributes required to consider such processes "good" was large and growing, thus had little utility in practice. The fact that these problems have no absolute solution led the author to suggest the concept of "good enough" governance.

The bridge is a "domesticated" problem in the view of Rittel and Webber (1973). The reduction of GHG emission levels is a "wicked" problem. Generally domesticated problems can be placed within the responsibility of sectoral agencies. The problems of greater complexity, in turn, tend to require the coordinated action of various institutions. So they are dependent on some kind of supra-sectorial structure, as well as policy decisions, which determine their priority. The challenge for planning is to match the advantages of sectoral specialization with the imperatives of coordination and integration of the sectors.

Final considerations

Climate change explicitly became a Brazilian priority more than a decade ago through a series of strategic actions involving various government sectors. This has led to the inclusion of the topic in sectoral policies, featuring a convergence process. However, we cannot, at the moment, say that there has been effective integration between policies. There are obvious signs that in the overall universe of public policies, there are other priorities, which are antagonistic to efforts to reduce emissions of GHGs.

For now, actions to mitigate climate change are just one of many national priorities. The involvement of sectoral bodies helps, but is not enough to solve the complex challenges of reducing emissions. Immediate issues such as job creation, increased exports, fighting inflation, reducing the infrastructure deficit or power generation are also priorities that have garnered attention and gained legitimacy. Addressing such issues by traditional means can neutralize mitigation initiatives, confining climate policy to adaptation practices.

The gap between climate policy and other aspects of public regulation provides an opportunity to update a debate that is as old as the very practice of planning: coordination between various State missions and responsibilities.

To plan is to prioritize. This does not mean simply that less pressing issues should be ignored. But it means that a hierarchy must be developed and effects must be harmonized (direct and indirect, positive and negative, short- or long-term). For one to avoid negative or zero sum games, different strands of public policy must be integrated.

In terms of climate policy integration with other axes of sectoral public policies, Brazil shows advancement towards the three criteria mentioned by Underdal (1980): comprehensiveness, consistency and aggregation. However, in terms of consistency, there is still a long way to overcome the actual weaknesses. Sectoral policies, while incorporating climate-related policies, still maintain characteristics and objectives that conflict with climate change coping strategies.

Notes

1 www.forumclima.org.br/pt/home. Accessed 23 April 2015
2 www2.camara.leg.br/legin/fed/decret/2010/decreto-7390-9-dezembro-2010-609643-norma-pe.html Accessed 23 April 2015.
3 www.mma.gov.br/clima/adaptacao/plano-nacional-de-adaptacao Accessed 13 November 2015.
4 www.agricultura.gov.br/desenvolvimento-sustentavel/plano-abc Accessed 23 April 2015.
5 http://redeclima.ccst.inpe.br Accessed 23 April 2015.
6 www.desenvolvimento.gov.br/arquivos/dwnl_1371044607.pdf Accessed 23 April 2015.
7 http://pesquisa.in.gov.br/imprensa/jsp/visualiza/index.jsp?jornal=1&pagina=6&data=25/07/2014 Accessed 23 April 2015.
8 Shortly after the 1973–1974 oil crisis, Brazil imposed a gasoline replacement policy for ethanol produced from sugarcane (Pro-Alcohol).
9 www.mda.gov.br/sitemda/secretaria/saf-biodiesel/legisla%C3%A7%C3%A3o. Accessed 30 October 2015.
10 www.oxfordenergy.org/2014/08/sustainable-energy-in-brazil-reversing-past-achievements-or-realizing-future-potential/ Accessed 15 October 2015.

References

Bursztyn, M. A. and Bursztyn, M. (2013) *Fundamentos de Política e Gestão Ambiental: caminhos para a sustentabilidade*. Rio de Janeiro, Garamond, 606 p.

Fonseca, I. F., Bursztyn, M. and Allen, B. S. (2012) "Trivializing sustainability: Environmental governance and rhetorical free-riders in the Brazilian Amazon," *Natural Resources Forum*, 36 28–37.

Fonseca, I. F. and Bursztyn, M. (2009) "A banalização da sustentabilidade: reflexões sobre governança ambiental em escala local," *Sociedade e Estado*, 24, 17–46.

Fonseca, I. F., Bursztyn, M. and Moura, A. M. (2012) "Conhecimentos técnicos, políticas públicas e participação: o caso do Conselho Nacional do Meio Ambiente (Conama)," *Revista de Sociologia e Política*, 20, 183–198.

Garcez, C. A. G. and Vianna, J. N. S. (2009) "Brazilian Biodiesel Policy: Social and environmental considerations of sustainability," *Energy*, 34, 645–654.

Grindle, M. (2004) "Good enough governance: poverty reduction and reform in developing countries," *Governance: An International Journal of Policy, Administration, and Institutions*, 17(4), 525–548.

Grindle, M. (2007) "Good enough governance revisited," *Development Policy Review*, 25(5), 553–574.

La Rovere, E. L., Pereira, A. S. and Simões, A. F. (2011) "Biofuels and Sustainable Energy Development in Brazil," *World Development*, 39(6), 1026–1036.

Le Tourneau, F-M. and Bursztyn, M. (2010) "Assentamentos rurais na Amazônia: contradições entre a política agrária e a política ambiental," *Ambiente e Sociedade*, 13(1), 111–130.

McNeil, D., Verburg, R. and Bursztyn, M. (2012) "Institutional context for sustainable development," in McNeil, D., Nesheim, I., Brouwer, F. eds, *Land Use Policies for Sustainable Development: exploring integrated assessment approaches*, Edward Elgar, Cheltenham, 24–44.

May, P. H. and Vinha, V. (2012) "Adaptação às mudanças climáticas no Brasil: o papel do investimento privado," *Estudos Avançados*, 26(74), 229–246.

Mendes, T. A. and Rodrigues-Filho, S. (2012) "Antes do pré-sal: emissões de gases de efeito estufa do setor de petróleo e gás no Brasil," *Estudos Avançados*, 26(74), 201–218.

Mesquita, P. S. (2015) Segurança alimentar, mudanças climáticas e proteção social no semiárido brasileiro (Cariri, Ceará) Unpublished PhD thesis Centro de Desenvolvimento Sustentável, Universidade de Brasília.

Rittel, H. W. J. and Webber, M. M. (1973) "Dilemmas in a general theory of planning," *Policy Sciences*, 4, 155–169.

Toledo-Filho, D. F. (2014) Integração da política climática: segurança energética e proteção climática, lições das experiências da Alemanha e Reino Unido Unpublished PhD thesis Centro de Desenvolvimento Sustentável, Universidade de Brasília.

Underdal, A. (1980) "Integrated marine policy: what? why? how?" *Marine Policy*, 4, 159–169.

Verissimo, M. P. and Xavier, C. L. (2014) "Tipos de commodities, taxa de câmbio e crescimento econômico: evidências da maldição dos recursos naturais para o Brasil," *Revista de Economia Contemporânea*, 18(2), 267–295.

Yanai, A. M., Nogueira, E. M., Fearnside, P. M. and Graça, P. M. L. A. (2015) "Desmatamento e perda de carbono até 2013 em assentamentos rurais na Amazônia Legal," in Anais XVII Simpósio Brasileiro de Sensoriamento Remoto. www.dsr. inpe.br/sbsr2015/files/p0978.pdf Accessed 23 April 2015.

15 Environmental policy and governance in Brazil

Challenges and prospects

Adriana Maria Magalhães de Moura

Brazil is a country endowed with abundant natural wealth: it holds nearly 13 per cent of all surface water in the world, is the second leading country in terms of forest[1] area and is first in tropical forest area. Moreover, it is extremely rich in biodiversity. The country possesses at least 13 per cent of the entire world's species, many of which exist exclusively in the country (Lewinsohn and Prado 2006; FAO 2006). However, those numbers, which indicate enormous natural capital, hide numerous environmental problems that affect the population.

Water, though abundant, is unevenly distributed in the territory: about 80 per cent of available water is concentrated in the Amazon region, where only 4 per cent of the population resides, while the Northeast, particularly in the semiarid region, suffers from frequent droughts (ANA 2011). Also in the Southeast, in metropolitan areas such as São Paulo, there was a recent collapse of water supply due to low rainfall.[2]

About 34 million Brazilians have no access to a piped water supply. Furthermore, 37 per cent of distributed water is wasted because of poor distribution systems. In many places, especially near metropolitan regions (São Paulo, Rio de Janeiro, Belo Horizonte, Salvador and Porto Alegre) water quality is poor, mainly due to the dumping of sewage. In 2013, less than half of the population was served by sewage collection and the treatment rate was even lower: only 39 per cent.[3] Lack of sanitation is the cause of numerous deaths and hospitalizations due to diseases such as gastrointestinal infections. It is estimated that if 100 per cent of the population had access to sewage collection there would be a reduction, in absolute terms, of 74,600 annual hospitalizations (*Trata Brasil*, 2014).

Another urban environmental problem that affects the people's health is poor municipal solid waste management, a major challenge to be faced by the country in coming years and that has been already solved by developed countries. From 2010 to 2014 the production of municipal solid waste grew by 29 per cent, as a result of population growth, changes in the habits of the population and an increase in consumption. Over 40 per cent of waste produced (around 81,000 tons per day) is still destined for garbage dumps, which contaminate soil and groundwater (Abrelpe 2014).

With regard to forests, Brazil has adopted a development model based on the growth of commodities from agriculture and extractive industries, sectors that are natural resources intensive (IPEA 2012). Between 1999 and 2010, the agricultural sector accounted for 42.5 per cent of total Brazilian exports, which highlights the importance of the sector in Brazilian trade (Conceição and Conceição 2014). In the same period, there were high rates of deforestation in the Amazon[4] and the Cerrado forests, which may have resulted in biodiversity loss and reduction in the supply capacity of ecosystem services of these biomes.

The significant devaluation of the Brazilian currency (real) will intensify pressures to expand this model based on commodity export, with consequent impacts on vegetation cover, soil and water resources. Interestingly, the Southeast and South regions are irrigated seasonally by evapotranspiration of the Amazon rainforest, the so-called 'flying rivers'. Therefore, deforestation of the Amazon, which is still high, contributes to water scarcity in these regions. Thus, the proper management of abundant Brazilian forests and associated biodiversity are critical for maintaining the water system, the energy supply and the climate balance in the country.

Considering the economic crisis facing the country, additional challenges are posed to Brazilian environmental policies in addressing critical issues related to degradation and scarcity of natural resources. Although the so-called 'environmental conscience' of society has increased in recent decades, promoting favourable conditions for the implementation of environmental policies, due to the worsening of the current socio-economic problems, the issue has been overshadowed, as seen at Rio + 20, an occasion during which the country showed little commitment to adoption of measures that could lead to concrete changes.

The aim of this chapter is to provide an overview of the trajectory of environmental policy in Brazil at the federal level and contribute to reflection on its current challenges with environmental governance. The concept of governance, which considers a healthy environment as a public good, is a common responsibility of both government and society and its institutions. The Brazilian Federal Constitution also recognizes that environmental conservation is a public issue that depends not only on State action, but on the entire community (Environment Chapter, art. 225).

This chapter is divided into four parts, including this introduction. The second part discusses the trajectory of federal environmental policy in Brazil, especially the institutional structure, cooperation of central and local government, legal aspects, policy instruments and environmental expenditures. The third part discusses the application of environmental governance principles in Brazil. The final section identifies some of the current challenges that must be overcome for achieving better environmental governance in Brazil.

Trajectory of Brazil's federal environmental policy

Brazilian environmental policy began in the 1930s, when the first steps towards the development of legislation related to the management of natural resources were taken, such as the Water Code and the Forest Code, both established in 1934. At that time, some forms of environmental public regulation were already practiced, even before the creation of exclusive agencies for this purpose. Since then, the country has advanced gradually, both in the establishment of important legal frameworks on the subject, and the institutionalization process of environmental policy.

The role of institutions in environmental policy making

The public service institutionalization related to the environment began in 1973, with the creation of the Special Secretariat for the Environment – SEMA. The creation of a specific institutional *locus* to deal with environmental issues, from the absorption of some functions of other existing institutions, gave more focus to the issue, providing an 'institutional signature' and a place to deal with environmental policies (Bursztyn and Bursztyn, 2013).

Since then, the institutional framework for environmental policy has been structured gradually. The National Environmental Policy, NEP (Law 6.938/1981), has been the main instrument to structure institutions at the three levels of government, and was organized in the form of a National Environmental System, Sisnama (Table 15.1).

The Sisnama is a network of existing environmental institutions at all levels of public administration, but does not exist by itself: its primary function is facilitating communication or interaction between the parts (Milaré, 2009).

Although it was established more than three decades ago, the Sisnama is still not effectively structured and articulated as a national system. Some examples of failures in the Sisnama are: centralization of tasks at the federal level, in MMA and IBAMA; the overlap of work done by federal and state institutions, lack of definition of the role of municipal environmental institutions and conflicts between municipal and state institutions (Araújo, 2013).

The restructuring of institutional organization at the federal level took place through the closure and merging of institutions (such as the creation of

Table 15.1 Structure of the National Environmental System – Sisnama

1. Governing Council
2. National Environmental Council (CONAMA)
3. Ministry of Environment (MMA)
4. Brazilian Institute of Environment and Renewable Natural Resources – IBAMA
5. State environmental agencies
6. Municipal institutions of environment

Source: Law 6.938/1981.

IBAMA), and sometimes as new institutions emerged from older institutions (as in the case of ICMBio, which emerged from a division of IBAMA). New institutions were also created to meet unmet gaps left by existing institutions, such as the National Water Agency, ANA, created to implement the instruments of the National Water Resources Policy.

This institutional restructuring sometimes conflicted, and required changes to accommodate newly created institutional roles and missions, as well as for institutional development (hiring and training, physical structure, equipment, logistics, etc.), which continue to the present day.

In general, institutional development is still a work in progress in the environmental field. One of the main difficulties is hiring qualified technical staff. While created in 1992, only in 2004 did the MMA hire its first public servants. Before that, the Ministry worked through contracts operated by international organizations (such as UNDP). This policy damaged the performance of the agency, and resulted in a lack of team stability and loss of technical personnel, which persists to the present day. Table 15.2 provides a summary of the current institutional framework for environmental management at the federal level.

In addition to those environmental institutions, some environmental programmes involve the participation of other ministries and institutions of the federal government. The Ministry of Planning selected the following environmental programmes for the PPA[5] 2016/2019: Water Resources, Conservation and Sustainable Use of Biodiversity, Climate Change and Environmental Quality. The climate change programme has the participation of the Ministry of Science, Technology and Innovation – MCTI. Actions on solid waste are jointly developed by the MMA and the Ministry of Cities.

Coordination and cooperation of central and local governments

To understand the current institutional framework, which implements environmental policies in Brazil, one must consider its federal structure, which requires the coordination of the three levels of government. In this scenario, state and municipal governments have autonomy to set policies with their own priorities, within their borders and areas of competence.

Responsiveness to environmental problems is seated on a complex institutional structure, with several federal entities that implement environmental policies, as degradation and pollution of natural resources transcend political boundaries. In Brazil, the dimensions of the territory, inter-regional inequalities and the structure of government impact the implementation of federal environmental policy.

The success of federal environmental policies depends on entities at the state and municipal level, particularly when policies are formulated, because states and municipalities are the ones that implement many policies. To be effective, federal policies require good relations and cooperation between different levels of government. The state and local governments are obliged to comply with

Table 15.2 Institutional environmental structure at the federal level

Institution/law	Purpose/mission
Governing Council (Law 6.938/1981 – NEP)	Highest authority in Sisnama. Consists of all the Ministers of State and the Union Attorney General. Its function is advising the President in the formulation of national policy and governmental guidelines for the environment. However, it is practically inactive until today.
National Environmental Council – CONAMA (Law 6.938/1981 – NEP)	Collegiate body with deliberative and consultative function. It is responsible for establishing criteria and environmental standards; and possesses the political and strategic role of articulating environmental policies and promotes the objectives of the NEP.
Ministry of the Environment – MMA (Law 8.490/1992)	Its function is to plan, coordinate, supervise and control the actions related to the environment and to formulate and implement the national environmental policy, with the aim of preservation, conservation and rational use of renewable natural resources.
Brazilian Institute of Environment and Renewable Natural Resources – IBAMA (Law 7732/1989)	Performs the national environmental policy, promotes the preservation, conservation, rational use, supervision and controls the use of natural resources. Ibama focuses on control measures, monitoring, enforcement and environmental licensing for potentially polluting activities.
Brazilian Forest Service (SFB) (Law 11.284/2006)	Its mission is to promote economic and sustainable use of forests and operates exclusively in the management of public forests. Maintains the National Forest Information System and manages the National Register of Public Forests.
Chico Mendes Institute for Biodiversity Conservation – ICMBio (Law 11.516/2007)	Federal agency under the MMA that performs the actions of the national policy of federal conservation units. Also has the function to carry out research programmes in biodiversity conservation and environmental education.
National Water Agency – ANA (Law 9.984/2000)	The agency is part of the National System of Water Resources Management, responsible for the implementation of the instruments of the National Water Resources Policy (PNRH), with the aim of promoting sustainable use of water.

Source: Author's elaboration.

national legislation, but only take part in programmes proposed by the federal government on a voluntary basis.

Under the Sisnama, federal environmental policy only becomes effective if federal entities establish a joint agreement to cooperate with each other. This interdependence demands shared rules that may counteract the typical autonomy (self-regulation) in federal regimes. Such interdependence requires a shared management structure and coordinated effort (also called an 'environmental federative pact'), which is focused on protecting the environment (Silva 2013; Neves 2012).

The Federal Constitution provides, in Article 23, that actions related to the environment are a common responsibility of the Union, States, the Federal District and Municipalities. Law 140/2011 establishes rules for cooperation between all federal entities, in order to harmonize and standardize the work between them, avoid duplication and make environmental policies more effective. Before this law, many jurisdictional conflicts occurred between federal entities, causing Sisnama to perform inefficiently.

The law does not solve the issue of intergovernmental cooperation, but is a guide forward. Advances depend on the political will of managers. Formally, the Brazilian federation is cooperative. But in practice the connections between, and actions of, the levels of government are sometimes diffuse and contradictory, marked by tensions between centralizing and decentralizing tendencies (Almeida 2011).

The country still needs to harmonize interests, through negotiated 'co-responsibility' or interdependence, to coordinate the responsibilities of states and municipalities with the global coordination of the federal government, since intergovernmental cooperation is *sine qua non* for viable environmental policy.

Obstacles to cooperative action must be addressed to enhance high-level coordinating mechanisms within the structure of Sisnama. The creation of new mechanisms is not needed, only the revitalization of the Governing Council and the Environmental Policy Committee which is part of Conama. These entities will work to unite strategies and encourage joint action at various levels of government to face environmental issues.

Legal framework

During the last three decades, Brazil has advanced its legal framework related to environmental management, moving to a more comprehensive and integrated approach to environmental management from a limited approach, restricted to few natural resources.

The Brazilian Constitution is advanced in environmental matters. There are also significant federal laws, which treat general topics, such as the structuring of environmental policy (Law 6.938/1981) and the establishment of penalties for environmental crimes (Law 9.605/1995). It also treats specific resources and topics, such as forests, water, genetic resources, protected areas, environmental licensing and pesticides, among others.

Table 15.3 highlights some of the major federal environmental laws during certain periods. There are also laws on related topics such as urban planning and natural disaster management, with strong interface with land use and land management. In addition, Conama has edited over 400 resolutions that regulate a wide range of environmental issues.

Although Brazil has a legal framework that covers nearly all relevant environmental issues, some legislators see difficulties implementing these laws. According to them, legislation is severely weakened when articles conflict with each other or are ambiguous, vague or too specific. This could lead to 'regulatory pollution' (Milaré 2009). Therefore, the creation of an Environmental Code is intended to consolidate existing legislation and avoid conflicts of interpretation that generate delays in enforcement.

Milaré (2009) points out that, even though Brazilian environmental legislation has been celebrated, in practice, environmental standards have not reached their goals of restraining activities that degrade the environment, e.g. they are unable to reconcile economic growth with environmental protection.

The inefficiency in enforcement is not exclusive to the environmental area, as the judiciary still suffers from excessive bureaucracy, delays and inaccessibility. The exacerbating factor in the environmental sphere is a lack of expertise on the topic among those working in the legal sphere. Difficulties begin in the identification of irregularities, due to inspection complications encountered by the responsible environmental protection agencies. Other problems in the enforcement of environmental laws are due to overlapping responsibilities between federal entities and cultural factors, since society still does not consider the environment a priority and is poorly organized to confront environmental causes, which are mostly of collective interest (IPEA 2010a).

According to Neves (2007), Brazilian environmental legislation, although vast, in many cases takes only a symbolic character and may act as an alibi. Legislators may develop standards to meet the expectations of society, without the conditions necessary to enforce compliance. This is a critical consideration. For example, the flexibility introduced in some laws, such as the new Forest Code (Law 12,651/2012), which became less restrictive, has, in practice, granted amnesty to several entities responsible for illegally deforesting certain areas. We are quite aware that the flexibility of the law resulted from the need to accommodate antagonistic social and economic interests.

Thus, in general, Brazilian environmental law is relatively advanced, although there is a lack of systematization and regulation, which causes difficulties in interpretation and implementation (compliance).

Brazilian environmental law requires incentives in order to prevent environmental irregularities. Incentives can bring beneficial results for the environment with less public spending, since they do not require the entire judicial apparatus to impose punishments. Moreover, a penalty does not promote environmental benefits after a crime has been committed, unless it is accompanied by a remedy for the environmental damage. Incentives may be in the form of economic instruments that benefit those who do not cause damage

Table 15.3 Evolution of Brazilian environmental legislation: 1930–2012

Federal legislation	Issue
1930–1960	
Decree n° 24.643/1934	Water Code.
Decree–Law n° 25/1937	Organizes the protection of historical and artistic heritage sites. It includes as national heritage natural monuments, sites and landscapes of outstanding value.
Law n° 4.771/1965[1]	Institutes the new Forest Code.
Law n° 5.197/1967	Protection of fauna.
1970–1979	
Law 6.225/1975	Protection of soil and combats erosion.
Decree–Law 1.413/1975	Control of pollution of the environment caused by industrial activities.
Legislative Decree 56/1975	Approves the Antarctic Treaty.
Law 6.453/1977	Civil and criminal liability for nuclear damage.
1980–1989	
Law 6.803/1980	Industrial zoning in critical polluted areas.
Law 6.938/1981	National Environmental Policy (NEP) – outlines the purposes and mechanisms of formulation and application.
Law 6.902/1981	Creation of Ecological Stations and Environmental Protection Areas.
Law 7.661/1988	Institutes the National Coastal Management Plan.
Law 7.347/1985	Regulates civil action liability for damage caused to the environment.
Law 7.805/1989	Regulates prospecting of mineral activities, by requiring prior environmental license.
Law 7.797/1989	National Fund for the Environment.
Law 7.802/1989	Regulates the disposal of waste and pesticide containers.
1990–1999	
Law 8.171/1991	Agricultural policy – includes environmental protection among its objectives.
Law 8.723/1993	Pollutant emission reduction by motor vehicles.

Law 9.433/1997	National Water Resources Policy and National Water Resources Management System.
Law 9.605/1998	Environmental crimes law.
Law 9.795/1999	National Environmental Education Policy.

2000–2012

Law 9.985/2000	National Nature Conservation Units System – SNUC.
Law 9.966/2000	Prevention, control and monitoring of pollution caused by the discharge of oil.
PM 2.186–16/2001	Regulates access to genetic resources and their conservation and use.
Law 10.650/2003	Regulates public access to environmental information on Sisnama member entities.
Law 11.105/2005	Establishes safety standards and activities of oversight mechanisms involving genetically modified organisms – GMOs.
Law 11.284/2006	Regulates the management of public forests for sustainable production.
Law 11.428/2006	Regulates the use and protection of native vegetation of the Atlantic Forest biome.
Law 11.460/2007	Regulates the planting of genetically modified organisms in protected areas.
Law 11.445/2007	National guidelines for basic sanitation.
Law 11.794/2008	Procedures for the scientific use of animals.
Law 11.828/2008	Tax measures applicable to donations intended to prevent, monitor and combat deforestation.
Law 12.114/2009	Creates the National Fund on Climate Change.
Law 12.187/2009	National Policy on Climate Change.
Law 11.959/2009	National Policy for Sustainable Development of Aquaculture and Fisheries.
Law 12.305/2010	National Policy on Solid Waste.
Law 140/2011	Rules for cooperation between the Union, the states, the Federal District and municipalities in administrative proceedings in the exercise of common responsibility on environment protection.
Law 12.512/2011	Support Program for Conservation and Development Program for Rural Productive Activities.
Law 12.651/2012	Protection of native vegetation (new Forest Code).

Source: Author's elaboration.

1 The Law No. 4.771/1965 was expressly repealed and replaced by Law No. 12.651/2012.

or who develop practices with positive impacts on the environment (IPEA 2010b).

Brazilian environmental policy instruments

Brazil has a significant number of instruments to implement environmental policy, most notably the 13 instruments of NEP (Law No. 6.938/81). Besides these, there are various instruments distributed in the main Brazilian environmental legislation, such as the National Policy on Solid Waste (PNRS) – Law 12.305/2010, the National Policy on Climate Change (NPCC) – Law 12.187/2009 and the National Policy on Water Resources (NPWR) – Law 9.433/1997, which are aimed at achieving specific objectives set out in these laws.[6] They are instruments of a regulatory nature (command and control), but also information,[7] economic, or cooperation (voluntary) instruments.[8]

Some instruments already show significant positive impact, while others lack sufficient regulations to be operationalized. However, there are few studies that clearly demonstrate the results achieved, and the costs and benefits of these instruments.

The use of regulatory environmental instruments still predominates (command and control). These instruments include environmental licensing of potentially polluting activities, the establishment of environmental quality standards and environmental zoning, which establishes or defines a more suitable distribution of economic activities and areas that should be preserved in the territory.

Such instruments are still needed. They have advantages such as simplicity and the possibility of immediate application; furthermore, they transmit a political message of strong or rigorous action on the part of the government in protecting the environment, which can dissuade or prevent undesirable behaviours. However, regulatory instruments also have disadvantages, such as lack of flexibility, lack of incentives to go beyond the minimum established in a given regulation and the necessity of a complex and costly institutional apparatus to enforce, monitor and develop a sound legal structure (Strauch 2008).

There is no doubt that Brazil must advance in the use of less reactive instruments and move towards more preventative, proactive and flexible mechanisms, such as economic instruments.[9] These are more oriented to adapt to the market, providing, in many cases, a 'win-win' approach for the public and the private sector.

The NEP mentions forest concession, environmental servitude and environmental insurance, as notable economic instruments. But others can be created, which can gradually become part of the national environment policy. Table 15.4 presents some of the economic instruments currently being applied in Brazil.

As emphasized by Ganem (2013), economic instruments are the new frontier in the expansion of legal regulations for the protection of natural resources.

Table 15.4 Environmental economic instruments applied in Brazil

Forest concession – The government provides public forests for sustainable exploitation by corporations. It can significantly contribute to the public management of forests, reducing illegal exploitation and illicit trade in wood products (Law No. 11.284/2006).

Environmental servitude – Authorizes the self-limitation of the owner of the land use for environmental preservation. The area can be used by another landowner to compensate for the lack of legal reserve (area to be protected, according to the Forest Code), ensuring mutual benefits to both economic agents (new Forest Code – Law No. 12.651/2012).

Environmental insurance – Aims to transfer the risk of environmental recovery in the event of an accident that generates environmental degradation, to insurance companies. Reduces the risk of unanticipated costs and ensures the necessary payment to recovery or environmental compensation. In general, environmental insurance in Brazil is mostly used by companies that use fuel or polluting substances. One of the factors that hinder the growth of this type of insurance is the lack of inspection and punishment for companies that damage the environment.

Charging for water use – Tool introduced by the National Water Resources Policy (NWRP), in order to encourage the rational use of water and generate financial resources to invest in the preservation of watersheds.

Green Tax on Circulation of Goods and Services (ICMS) – implemented in some states which apply environmental criteria to allocate part of ICMS paid to municipalities. It has been used mainly to encourage the creation and maintenance of protected areas.

Support for Conservation Program (Law 12.512/2011), which is based on the principle of payment for environmental services.

Sustainable Public Procurement (SPP) – incorporates sustainability criteria in public procurement and enables the public sector to achieve environmental and social goals without the need to allocate additional resources in its budget. In Brazil there are already a number of resolutions that support the practice of SPP at the federal level and in states like São Paulo and Minas Gerais (Moura 2011).

Source: Author's elaboration.

Several propositions of economic instruments are being debated in the National Congress, such as those relating to Payment for Environmental Services (PES) and Reducing Emissions from Deforestation and Forest Degradation (REDD).

Environmental expenditures in the public budget

The allocation of financial resources by the central government budget toward environmental policies indicates the level of priority given to the issue and directly impacts the ability of federal programmes to meet proposed goals. However, IPEA research on public environmental expenditures in Brazil[10] indicates that despite worsening environmental problems, resources have been proportionally decreasing in the federal budget in recent years.

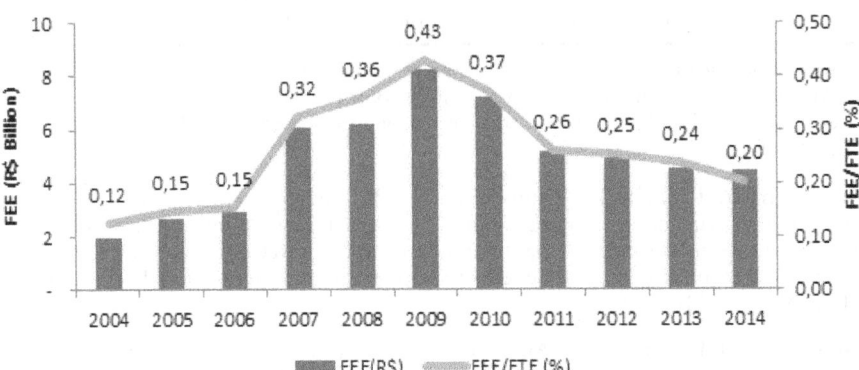

Figure 15.1 Federal Environmental Expenditures (FEE)★ and Federal Environmental Expenditures in relation to the Federal Total Expenditures (FTE): 2004–2014

Source: Author's elaboration based on data from IPEA 2016.

★ Amounts adjusted for inflation to 2014 prices, based on the General Price Index (IGP-DI), calculated by the Getúlio Vargas Foundation – FGV.

The research analyzed the expenditures of the federal government on environmental activities from 2004 to 2014. It used a methodology adopted by the United Nations System of Environmental-Economic Accounting that classified environmental activities (European Commission 2012). The research used a structured database with budget information from the 'Portal Siga Brasil',[11] in which all federal government programmes (across ministries) were classified as environmental or non-environmental activities (the state-owned enterprise expenses are not included).

It was found that the Federal Environmental Expenditures (FEE) gradually increased in absolute terms in the period 2004–2009, from 1.96 billion reais in 2004 to 8.27 billion in 2009. In proportion to Federal Total Expenditures (FTE), the FEE also increased by almost four times, from 0.12 to 0.43 per cent.

However, from 2010 until 2014 the FEE began to decline year by year, both in absolute terms and in relation to the FTE. Expenditures declined to 4.47 billion reais in 2014 and represent only 0.20 per cent of federal budget expenditures, a reduction of about 50 per cent from 2009 figures, which was the year of the greatest spending on environmental activities between 2004 and 2014.

Thus, in spite of the rhetorical concern with sustainability, there was decreasing environmental expenditure in the public budget (Figure 15.1).

Environmental governance principles

Environmental governance has been selected as one of 21 critical environmental issues for the twenty-first century (UNEP 2012). This priority points to the apparent discrepancy that now exists between the nature of environmental challenges and the capabilities of the governance system, which includes, in addition to governments, a multiplicity of social actors.

Environmental governance can be defined as the processes and institutions through which societies make decisions that affect the environment (Loë et al. 2009). Thanks to the cross-scale nature of environmental problems and the dispute over values and interests related to the use of natural resources, governance strategies of various involved actors are particularly important to the success of environmental policies.

There are several principles for good governance, which are complementary to each other and include: accountability, legality, participatory decision-making and the triad of efficiency, effectiveness and efficacy. Successful governance must also fulfil some fundamental functions: setting objectives and goals (direction), coordination of goals, implementing actions necessary to achieve goals and evaluation. The identification of these functions allows us to understand the origins of possible flaws in the process and guides the mechanisms for improving the quality of governance performance (Peters 2013). Below is a discussion of progress in some of the principles and governance functions applied to environmental issues in Brazil.

Planning, setting objectives and targets – guidance

Planning is still fragile in Brazilian environmental policy. Greater investment is necessary to establish medium- and long-term plans with goals and objectives based on the identification of strategic areas of focus, which must consider not only environmental issues, but also the social impact of environmental problems.

CONAMA is responsible for formulating the National Agenda for the Environment, which is a planning instrument to be proposed as recommendations to the institutions of Sisnama. The agenda was made only once (2007–2008), and has been criticized for being too extensive and not very objective. Thus, the instrument has not been considered an effective planning instrument for the CONAMA or federal environmental agencies (Fonseca and Moura 2011).

The MMA recently prepared its first strategic plan, for the years 2014 to 2022, which establishes its mission, vision, values and key strategic goals. The initiative is important for the planning process of the ministry (Brasil 2014). However, as an example of how other governments have long developed such plans, the US Environmental Protection Agency (United States Environmental Protection Agency – EPA) has produced, since 1995, a regularly updated four-year strategic plan, which sets the agency's priorities for the period (previously submitted to the Congress). These plans are divided into annual action plans, which are then evaluated through progress reports.

Accountability

Accountability involves evaluation, accounting and transparency in decision making and in the use of public resources, as well as a clear definition of roles and responsibilities in relation to the achieved performance. For the effectiveness of this principle, some tools include: the definition of both checks and balances on the part of state institutions and efficient monitoring and evaluation systems.

The environmental area is no exception to rules of accountability applied to the entire federal government. However, environmental accountability has been more focused on the legality of the procedures and the financial implementation of environmental programmes (efficiency) and is still limited in terms of effectiveness.

Improved accountability in the environmental area requires information and data monitoring of natural resources in the country. Unlike the social and economic fields, environmental issues have only recently been the focus of researchers and there is not a long tradition of producing statistics, making it difficult to build indicators in this field (IBGE 2015).

Thus, there are two main difficulties to be overcome in assessing the responsiveness of the State to environmental issues: the incipient nature of primary databases on the environment and also methodological and conceptual difficulties. Such problems in measuring progress or retrogression make the performance of environmental policy unclear, both for the manager and for the general public and can lead to unthoughtful environmental policy planning in the country (Veiga 2007).

Studies on evaluation capacity of environmental policies in Brazil have shown that there is not a mature culture of evaluation of these government policies. Federal environmental agencies responsible for environmental policies (highlighted in Table 15.2) must maintain their own evaluation structures. However, they are still poorly structured to carry out systematic and comprehensive assessments on their policies. The National Environmental Council (CONAMA) has among its responsibilities evaluating the implementation and enforcement of environmental policy. However, the Council has failed to fulfil this mission, since it has not done, to date, reviews of federal environmental policies (Moura 2013; Fonseca and Moura 2011).

Lacking a structured evaluation system, with reliable indicators and perennial methodologies that allow for the development of historical series, broader analysis of the effectiveness of federal environmental policies does not have the necessary foundation. Only time-specific analysis on specific topics is possible. Without these tools, the effectiveness of Brazilian environmental policy (the quality of the environment in Brazil indicates that we are on the path to sustainability?) appears to be inaccurate or biased. Moreover, without these accountability mechanisms, decisions cannot be taken effectively and decision makers cannot identify flaws that must be corrected in implementing policies. This is not a new concern, but perceived advances are still limited.

Participatory decision-making and social control

Social pressure on the State plays an important role in the implementation of policies and in the provision of public goods. In addition, environmental protection is a shared obligation of the government and the community. However, Brazil still has a limited tradition of popular participation in decision making, a reality that has changed slowly in recent decades.

Participation in the development and implementation of environmental policies has some fundamental issues that make the process more complex. Olson (1999) states that in relation to collective goods, such as environmental ones, it is important to consider that while the cost of involvement in the process is only individual, the benefits are diffuse, e.g., available to the whole population. This can cause 'free ride' behaviour with little incentive for the individual to participate in the development and implementation of environmental policies.

There are doubts, also, that individual participation can effectively impact complex problems whose dimensions tend to transcend the local level, both in terms of causes and solutions. Speech presented to the public, ranges from alarmist warnings, which often do not materialize, to rhetorical and superficial information or highly technical information, which may be difficult for a lay audience to understand. This information asymmetry can generate doubts or apathetic behaviour from the public (Siqueira 2008).

Social participation, or the exercise of citizenship in relation to the environment, depends on a number of key requirements: the public's awareness and value of the environment, access to environmental data and information and also access to instruments to act on environmental issues.

Brazil adopted the National Environmental Education Policy (Law 9.796/1999), in order to encourage the training of human resources, study and research on the subject, as well as the production of educational material on environment issues. The law established that environmental education should not be implemented as a specific discipline in the school curriculum, but as content to be addressed in various disciplines. However, it is unknown to what extent this strategy has been effective in supporting knowledge on specific issues while allowing an integrated view of environmental issues (Viana 2013).

Environmental agencies are obligated to allow the public access to documents and data that deal with the matter (Law 10.650/2003). The law also provides that Sisnama agencies should publish annual reports on the quality of the environment, a provision that has not been fulfilled regularly. Although the law advances environmental education, it presupposes an active pursuit of information, which is not always simple and does not replace systematic dissemination and public access for all citizens. Thus, in general, the impact of this legislation is still limited.

The effective social control over environmental issues depends on the availability of means to this control. Society evaluates government performance especially regarding the quality of the services provided or the results obtained.

Thus, the lack of qualitative information is an obstacle to societal control of environmental policy since there is no way for the public to participate in something they only partially understand.

Among the formal instruments available to participate in environmental policies, we highlight the environmental collegiate, watershed committees, management boards of protected areas and environmental funds, public audiences and the holding of national conferences. Table 15.5 shows the most important environmental collegiate bodies, the main one being the National Environmental Council (CONAMA).

Public hearings in the environmental licensing process can have a great impact on the environment. They constitute an important instrument of social participation. However, in general, hearings occur in the late stages of the decision-making process, with complex and difficult information to be absorbed for a constructive debate with stakeholders. In addition, suggestions for the improvement of environmental policies are often not considered, leading to only formal participation in the legal licensing process, with low effectiveness (Siqueira 2008).

Another recent instrument of environmental participation is the National Conferences for the Environment (NECs), established in 2003. They must take place every two years and are designed to create an opportunity for social convergence for the formulation of a national environmental agenda. Since 2003, there have been four NECs with specific topics in order to focus debate. These conferences have involved a significant number of people on local stages (about 70,000) as well as on the national stage (on average 1,500 people).

In each NEC, participants present many proposals. However, these proposals are rarely considered as part of an agenda for Brazilian environmental policy. The discussions and interactions among participants are not enough, since the results of these deliberations, when not seriously considered, or outright ignored, can lead to disbelief in the efficacy of this participatory process as an instrument that contributes to the formulation of the National Environmental Agenda, under the responsibility of CONAMA.

Public participation also takes place through non-governmental organizations (NGOs). These organizations must be registered in the National Register of Environmental Organizations to be recognized by the government to be entitled to send representatives to CONAMA. According to the records of the

Table 15.5 Collegiate related to environmental issues at the federal level

National Council of Environment (CONAMA)
Amazon National Council (CONAMAZ)
National Water Resources Council (CNRH)
Deliberative Council of the National Environment Fund
Council to the Genetic Heritage Management
Committee on Public Forest Management
National Forestry Commission (Conaflor)

Source: Author's elaboration.

Council there are a total of 654 non-governmental organizations that work on environmental issues.[12]

These entities act as broad channels of participation, each with diverse aims and objectives: some focus on research; others monitor government programmes. Sometimes they work as representatives of some segments of the population or marginalized groups in decision-making centres of political power.

However, this representation lacks electoral legitimacy, and effective surveillance for identifying possible private interests in the action of some NGOs. Although they bring positive contributions, and are seen as important channels for participation and engagement of the population, these organizations need closer monitoring, and greater transparency, since many receive not only international resources and/or voluntary donations, but also government funds (Siqueira 2008; Lopez and Abreu 2014).

Conclusion – current challenges for a better environmental governance in Brazil

The trajectory of Brazilian federal environmental policy shows that advances have occurred, both in terms of institutional structure, as in the establishment of important legal frameworks. But there are still great challenges to be faced in relation to the improvement of environmental governance and the application of its principles, which, taken together, will contribute to the increase in responsiveness of the state to environmental problems.

Beyond what has been addressed, the following two steps are important for the improvement of Brazilian environmental policy: (1) the coordination and coherence in the governance structure of environmental institutions; and (2) the articulation of domestic priorities in conjunction with the global agenda.

Coordination and coherence in the governance structure of environmental institutions

In Brazil, government responsiveness to environmental issues is based on a complex institutional structure, with several levels of government that must coordinate in order to implement environmental policies.

Further progress is needed in the development of federal environmental institutions. These institutions were established relatively recently, and they have not yet reached the desired efficacy and maturity. These institutions (highlighted in Table 15.2) need to improve their performance, invest more in training their workforce (with structured and attractive careers) and increase enforcement capacity, thus ensuring compliance with the law and the decisions of environmental agencies.

The federal government must also support improved structuring and training of environmental agencies at the regional (states) and local levels (municipalities). Deficiencies at these levels constitute major obstacles to the successful implementation of policies emanating from the federal level. Bureaucratic inefficiencies must also be addressed, as many institutions of Sisnama are focused on processes (such as the issuance of environmental licenses) rather than results related to improving environmental quality.

However, the inherent complexity of environmental issues requires not only the development of environmental institutions in the strict sense, i.e. the institutions that compose the Sisnama. Reasonable articulation and communication is essential among institutions whose policies affect the environment. As in other countries, in Brazil the agencies responsible for environmental policies (ministries) often function as bureaucratic islands, isolated from other institutions that affect the environment. In this way, other ministries tend to assume that concern for environmental issues and policy is not their responsibility.

Environmental agencies are rarely powerful enough to influence most economic decisions that have significant impacts on the environment. The environmental area suffers particularly during conflicts between public and private interests. Often, the 'rivalry' occurs inside the public institutions themselves, or in different branches of government. In arenas like CONAMA where representatives from government institutions, the productive sector and civil society all participate, it is clear that interests are often not only conflicting, but sometimes opposing and irreconcilable.

Thus, more efficient institutions in the environmental field are not necessarily synonymous with larger institutions with more resources, more staff, or more structure. Due to the specificity of environmental issues, it is desirable that environmental functions are also absorbed and internalized in other federal agencies that make daily decisions affecting the environment and have much higher financial support, such as the energy, transport and agriculture sectors.

Institutional efficiency in the environmental field requires a strong capacity for strategic coordination of environmental institutions with other government agencies and the ability to coordinate crosscutting actions, without the risk of getting lost in the management of numerous councils and inter-ministerial committees. That is, unifying action and speech, to give coherence to Sisnama and guide the various ministerial portfolios toward an integrated approach, in relation to the Brazilian environmental policy.

Domestic priorities in conjunction with the global agenda

International financial institutions like the World Bank have shaped Brazil's environmental policies for a long time. The creation of many environmental agencies in Brazil was often a response to external pressures. Under this external pressure, international interests have often been prioritized.

In recent decades, there has been a proliferation of international environmental agreements. The environment is the second most common area for international agreements in Brazil, second only to foreign trade, although some environmental agreements tend to be more declaratory (intentions and principles) than rules to be followed. Brazilian environmental foreign policy has worked well in these agreements, since the country has assumed the role of a key player in multilateral environmental negotiations (Barros 2011).

However, it is important to note that the extensive agenda created by the various international environmental agreements is built with the participation of countries with asymmetric capabilities. This raises questions about the legitimacy of this agenda, which can overshadow national priorities. Leadership in the development of these agreements still belongs to developed countries, which tend to define the global agenda in terms of their domestic interests. Brazil has sought to meet the goals laid out in these agreements, which, despite their importance, can obfuscate the serious internal environmental problems that remain unsolved in Brazil. These issues may not be on the agenda of more developed countries, since they may be further advanced in their solutions or may have resolved them already.

In the late 1980s a Latin America and Caribbean Development and Environment Committee was created to think about environmental problems from a regional view, which resulted in the document 'Our Own Agenda' (UNDP 1991). Similarly, Brazil still needs to think about its own strategies to address its priority environmental problems.

As emphasized previously, there is a serious delay in the implementation of access to basic sanitation (including sewage treatment and solid waste management), which is the most frequent cause of environmental degradation in Brazilian cities. This results in enormous adverse impacts on the health of the population (IBGE 2015). When Brazil organizes its internal environmental agenda only in terms of international agendas, significant issues, such as water and sanitation fail to receive the necessary attention, as these are less prominent on the global agenda, because they are considered domestic problems. Meanwhile, pollution problems in water bodies worsen, as loss of springs and advancing desertification lead to water scarcity and diseases in some areas, resulting in great economic and social damage.

Brazil has a prominent role in the international arena, as a large holder of natural resources (probably the biggest natural wealth of the planet) and an emerging power. However, the country continues to lose its natural resources by exporting non-patented biodiversity, timber and agricultural commodities (Strauch 2008). The country needs to develop a strong domestic agenda, in order to guide its negotiations with other countries and must complement its natural wealth with adequate levels of sanitation and technological development to become less dependent on the use, and extraction of, raw materials. This is an important step for Brazil to avoid accelerated losses of natural resources, some of which may be irreversible, and, sooner or later, will have a great impact on the domestic economy as well as in international trade.

Notes

1 Brazil has about 516 million hectares of forest, equivalent to 60.7 per cent of the country, second only to Russia, where temperate forests dominate.
2 The lack of rainfall has also reduced the level of the hydroelectric reservoirs, responsible for generating about 72 per cent of electricity in the country. Thus, the use of thermoelectric, more expensive and pollutant energy, increased by 18 per cent in 2015 (MME 2015).
3 Latest international comparative data place Brazil in the 112th position in a ranking of sanitation in 200 countries (Trata Brasil 2014).
4 Despite the decline observed in recent years, in 2014 the rate of deforestation in the Amazon was 5,012 km^2 (Prodes/Inpe 2015).
5 The Multi-Year Plan (PPA) is the instrument that brings together all the public policies (programmes) of the Federal Government for a period of four years. It includes objectives, targets, and indicators for the planned programmes.
6 The PNRS, for example, provides 17 instruments, while the PNMC covers a total of 18 instruments. The PNRH brings six instruments, including charging for water use.
7 Such as the Environmental Quality Report – RQMA, to be published annually, the National Information System Environment – Sinima and the provision to any interested citizen to information concerning the environment.
8 Among which are: voluntary environmental audit, environmental self-regulation in the companies through environmental management systems and public programmes of voluntary participation.
9 Economic instruments are intended to induce the economic agents toward favourable environmental changes and may take the form of financial transfers of economic agents to the government or subsidies to activities identified as beneficial to the environment.
10 Preliminary results of the IPEA's research 'Dimensioning and Monitoring of the Federal Environmental Expenditure', coordinated by the author.
11 The 'Siga Brasil' is an information system on government budget maintained by the Federal Senate.
12 www.mma.gov.br/port/conama/cnea/cneaenti1.cfm Accessed on 19 October 2015.

References

Abrelpe – Associação Brasileira de Empresas de Limpeza Pública e Resíduos Especiais (2014) *Panorama dos Resíduos Sólidos no Brasil 2014* Abrelpe.
Almeida, M. H. T. (2011) 'Federalismo, democracia e governo no Brasil: ideias, hipóteses e evidências', *Revista Brasileira de Informação Bibliográfica em Ciências Sociais*, São Paulo, n.51, 13–34.
ANA – Agência Nacional de Águas (2011) 'Relatório de conjuntura dos recursos hídricos no Brasil – Informe 2011' Brasília: ANA/SPR, 2011. 112 p.
Araújo, S. M. V. G. (2013) 'Os fundamentos legais da Política Nacional do Meio ambiente', in Ganen, R. S. eds, *Legislação brasileira sobre meio ambiente: fundamentos constitucionais e legais*. Brasília: Edições Câmara, v. 1.
Barros, A. F. G. (2011) 'O Brasil na governança das grandes questões ambientais contemporâneas, país emergente?' *Texto para Discussão* n. 40 Brasília: CEPAL and IPEA.
Brasil, Ministério do Meio Ambiente (2014) *PNIA 2012 – Painel Nacional de Indicadores Ambientais: referencial teórico, composição e síntese dos indicadores da versão-piloto*, Brasília.

Bursztyn, M. A. and Bursztyn, M. (2013) *Fundamentos de política e gestão ambiental: caminhos para a sustentabilidade* Garamond, Rio de Janeiro.

Conceição, J. C. P. R. and Conceição, P. H. Z. (2014) 'Agricultura: evolução e importância para a balança comercial brasileira', *Texto para Discussão* n. 1944 IPEA Brasília.

European Commission (2012) *System of Environmental-Economic Accounting Central Framework*. Co-edited: European Commission; Food and Agriculture Organization; International Monetary Fund; Organisation for Economic Cooperation and Development; United Nations; World Bank.

Fonseca, I. and Moura, A. M. M. (2011) 'Processo político e decisório no âmbito do Conselho Nacional do Meio Ambiente – Conama' Relatório de pesquisa, IPEA, Brasília.

FAO (2006) *Global Forest Resources Assessment 2005*. Rome, Italy: Food and Agriculture Organization of the United Nations, v. 147.

Ganen, R. S. (2013) 'Legislação brasileira sobre meio ambiente'. *V.1. Fundamentos constitucionais e legais*. Câmara dos Deputados, Edições Câmara, Brasília.

IBGE – Instituto Brasileiro de Geografia e Estatística (2015) *Indicadores de desenvolvimento sustentável: Brasil 2015*, IBGE, Rio de Janeiro.

IPEA – Instituto de Pesquisa Econômica Aplicada (2010a) 'Direito ambiental brasileiro: princípios e limites de implantação', in Alvares, A. and Mota, J. A. eds, *Sustentabilidade ambiental no Brasil: biodiversidade, economia e bem-estar humano* Instituto de Pesquisa Econômica Aplicada – IPEA, Brasília 209–25.

———(2010b) 'A lei de crimes ambientais frente à evolução do direito ambiental: sanções versos incentivos' in Alvares, A. and Mota, J. A. eds, *Sustentabilidade ambiental no Brasil: biodiversidade, economia e bem-estar humano* Instituto de Pesquisa Econômica Aplicada – IPEA, Brasília 227–45.

———(2012) 'Produtividade do Brasil nos anos 2000–2009: análise das Contas Nacionais' *Comunicados IPEA*, v. 133, pp. 1–18.

———(2016) *Dimensionamento e acompanhamento do gasto ambiental federal*. IPEA, Brasília (to be published).

Lewinsohn, T. M. and Prado, P. I. eds. (2006) 'Síntese do conhecimento atual da biodiversidade brasileira', in Lewinsohn, T. M. ed. *Avaliação do estado do conhecimento da biodiversidade brasileira*. Brasília, DF: Ministério do Meio Ambiente, v. 1, 21–109.

Loë, R., Armitage, D., Plummer, R., Davidson, S. and Moraru, L. (2009) 'From government to governance: a state-of-the-art review of environmental governance,' Final Report. *Guelph: Rob de Loë Consulting Services*.

Moura, A. M. M. (2011) 'Financiando o Desenvolvimento Sustentável: o papel das compras públicas', in *Brasil em desenvolvimento 2011: Estado, planejamento e políticas públicas*. Brasília: IPEA, v. 2, cap.17.

———(2013) 'Governança das políticas ambientais no Brasil: desafios à construção de um sistema integrado de avaliação', *Texto para Discussão, n. 1904*, Brasília: IPEA.

Lopez, F. G. and Abreu, R. (2014) 'A participação das ONGs nas políticas públicas: o ponto de vista de gestores federais', *Texto para Discussão, n. 1949*, Brasília: IPEA.

Milaré, E. (2009) *Direito do ambiente: a gestão ambiental em foco*. 6th edn. São Paulo: Revista dos Tribunais.

MME (2015) 'Boletim de monitoramento do sistema elétrico brasileiro – Março de 2015.' Ministério de Minas e Energia, MME.

Neves, M. S. C. (2007) *A constitucionalização simbólica* 2. ed. São Paulo Martins Fontes.

——(2012) 'Política ambiental, municípios e cooperação intergovernamental no Brasil', *Estudos Avançados*, n. 26, 137–150.

Olson, M. (1999) *A lógica da ação coletiva: os benefícios públicos e uma teoria dos grupos sociais*, Edusp, São Paulo.

Peters, B. G. (2013) 'O que é governança?' *Revista do Tribunal de Contas da União*, Brasília, ano 45, n. 127.

PNUD – Programa das Nações Unidas para o Desenvolvimento. (1991) *Nossa própria agenda* Comissão de Desenvolvimento e Meio Ambiente da América Latina e do Caribe. Nova Iorque: BDI; Pnud.

Silva, C. S. G. (2013) 'Federalismo cooperativo ambiental no Brasil: Notas sobre a Lei Complementar nº 140/2011'. *Jus Navigandi*, Teresina, ano 18, n. 3544.

Siqueira, L. C. (2008) 'Política Ambiental para quem?' *Ambiente e sociedade*, Campinas, v. 11, n. 2, 425–437. <goo.gl/VIylUY> Accessed 30 March 2016.

Strauch, M. (2008) 'Instrumentos da política ambiental', in Strauch, M. and Albuquerque, P. P. eds, *Resíduos: como lidar com recursos naturais*. Oikos, São Leopoldo, 191–212.

Trata Brasil (2014) 'Benefícios econômicos da expansão do saneamento', *Instituto Trata Brasil*.

UNDP (1991) *Our Own Agenda*. Latin American and Caribbean Commission on Development and the Environment (LAC CDE). UNDP and IDB in collaboration with ECLAC and UNEP.

UNEP (2012). *21 Issues for the 21st century: result of the UNEP foresight process on emerging environmental issues*. United Nations Environment Programme – UNEP, Nairobi, Kenya.

Veiga, J. E. (2007) 'Indicadores para a governança ambiental', *Anais do Encontro da Sociedade Brasileira de Economia Ecológica ECOECO* Fortaleza.

Viana, M. B. (2013) 'Exercício da cidadania ambiental', in Ganem, R. S. Ed., *Legislação brasileira sobre meio ambiente: instrumentos da Política Nacional do Meio Ambiente*. Brasília: Edições Câmara, v. 2.

16 Collective forest reserves in agrarian reform settlements

Measures to prevent a non-commons tragedy in the Brazilian Amazon

Peter May, Robert Davenport, Pedro Nogueira and Paulo César Nunes

Context and theoretical basis

Since the military government instigated accelerated occupation of the Amazon by settlers from Brazil's dry Northeast in the 1970s, the region has been the object of settlement to offer "land without people for people without land." Despite its misconception given the region's abundant historical occupation by traditional forest peoples, this did little to impede a dramatic series of resettlement programs. In consequence, many migrants came into conflict with these historical residents, whose occupation had minimal impact on forest cover.

Today, over 5 percent of the Amazon biome in Brazil is occupied by agrarian reform settlements (ARS). Over 160,000 families were granted lots in more than 1,900 settlements by the federal government from 1995–2011 alone (Schneider and Peres 2015). Settlers were initially granted lots on the condition that they clear half of the area and put it into productive use, for which they received a transitional stake for basic needs, inputs and housing. Revisions to the national Forest Code in 1997 responding to global alarm with the pace of deforestation in the Amazon restricted clearing to 20 percent of properties, but these restrictions were only weakly enforced. (In 2012, a rollback in the Code weakened such regulation even further, particularly on small properties such as those found in ARS.) Recent research has shown that deforestation within ARS accounts for over 55 percent of original forest cover, over double that found on neighboring properties, and has become one of the principal sources of continuing deforestation in the Amazon (ibid.).

One measure adopted in an effort to counter further forest loss and degradation, and its ongoing effects on environmental services important both to settlers and society at large, is the establishment of "collective legal reserves." Under the Forest Code, properties are required to maintain a fixed proportion of land under natural vegetation varying by biome, known as a "Legal Reserve" (LR). Collective LRs are formed by demarcating all or part of the area that would formerly have been designated on each lot, to a common area, to be subject to collective management and protection. Individual lots still must protect so-called "permanent protected areas" along streams and on steep slopes, but are at least partially exempted from LR requirements. Yet, the simple creation of areas purported to be under collective protection does not predetermine that management rules will arise of their own accord.

Based on cross-cultural studies, Elinor Ostrom and colleagues (e.g. Ostrom 1990, 2005, 2010) identified numerous cases in which shared rules, norms, and strategies govern collective action and foster trust among resource users. To appraise the obstacles and opportunities for management of collective LRs, an understanding of the socio-economic and institutional basis for establishment and perpetuation of these norms helps us to explain settlers' behavior in our field sites. Such collectively adopted institutions enable settlers to regulate the internal allocation and use of land and resources within a settlement, as well as create mechanisms to mediate conflict (Cronkleton and Larson 2015).

In these studies, however, Ostrom focused on how local or autochthonous rules systems arose for common pool resource management (CPRM), through institutions that operated independently from the State. Her observation of functioning CPRM cases thereby suggested that social norms or trust among resource users, in combination with social capital and stable leadership, was one prerequisite for the viability of collective action. In contrast, in this case study we consider social capital or trust as interdependent with State policy and institutional interventions.

A more nuanced treatment of social capital is necessary in the Brazilian Amazon, because recent colonists and settlers within ARS lack a common social or cultural background. They have migrated to the Amazon from distinct regions within Brazil, some even from urban areas. Very few have any knowledge of the ecological characteristics of Amazonian soils or forests upon arrival. Settler communities within ARS are often fractured, subject to an individualistic frontier culture that has little appreciation for the merit of commonly held assets. As such, they have not had the opportunity or history to enable social trust to evolve in support of collective action. For example, in a previous study of one of the ARS cases under consideration, 78.5 percent of interviewed settlers participated in local associations (Eiró and Tricaud 2010), but such social connectivity, sometimes taken as a proxy for social capital in literature on livelihoods, did not translate into collective action to manage common forest areas.

We suggest that in the instance of ARS in the Brazilian Amazon, social capital or trust cannot be assumed to be a prerequisite "stock" that can be used to achieve CPRM, but rather may emerge out of human experiences on the land and in communication with the State and intervening organizations. Such social norms are the result of socio-ecological systems (SES) that are multilayered in character (Brondizio et al. 2009). Social capital cannot, however, be assumed synonymous with collective action. Criticizing the lack of clarity of the connections between social capital, collective action, and natural resource governance, Ishihara and Pascual (2009) suggest emphasis on "common knowledge": the set of understandings embedded in a given social structure, which enables putting resources to particular use. Given the disorder and conflict present in the Brazilian Amazon, a common property forest, we suggest, can only be sustained among ARS settlers if it were to benefit a critical mass of users (cf. Chhatre and Agrawal 2009), thus leading to common knowledge and to collective action.

This chapter is structured as follows. In the second section, we provide background information on development and occupation in Northwest Mato Grosso (NW MT), describing the two case study settlements and their history of occupation and land use change. In the third section, we outline the land use change, socioeconomic and institutional characteristics that differentiate the two sites, despite their physical proximity and similar logistical conditions. In doing so, we employ Ostrom's framework for analysis of socio-ecological systems (SES) (Ostrom 2007, 2009), in which social capital is only one among over 40 variables. In the fourth section, we discuss the observed experience with collective LR areas, and consider our findings in light of the SES analytical framework. In the final section, we advance and offer caveats for more extensive implementation of collective reserves in ARS as a contribution to environmental governance in the Brazilian Amazon.

The study area and policy context

NW MT is a humid tropical forest (Submontane Open and Dense Ombrophilous Forest, IBGE 2014) region about the size of Guatemala (107,622 km²). The region consists of seven municipalities: Aripuanã, Castanheira, Colniza, Cotriguaçú, Juína, Juruena, and Rondolândia. Its population of about 120,000 is now 62 percent urban.

NW MT was colonized in the 1970s by private land corporations which received public concessions in return for infrastructure development and technical support, enabling them to speculate on land sales to farmers from southern Brazil who hoped to exchange small parcels in the South for significantly larger holdings in the Amazon. Most colonists turned to extensive cattle ranching and unauthorized logging, while others became gold and diamond miners. These enterprises became NW MT's primary land use activities (Tito et al. 2011).

By 2012, although deforestation had proceeded apace, over 80 percent of NW MT's original 104,000 km² of primary forest cover still remained. Forest remnants were located within 11 indigenous areas (37 percent of remaining forest cover) and in nine Conservation Units (5.5 percent of remaining forest cover) (Figure 16.1). The remaining 57.5 percent of forest cover was located on private properties ranging from large ranches to small farms, and in agrarian reform settlements (May et al. 2012).

Beef cattle ranching is the predominant agricultural enterprise in NW MT. Mato Grosso is host to the largest livestock population in Brazil (28.4 million head of cattle); only one municipality in NW MT – site of a major slaughterhouse

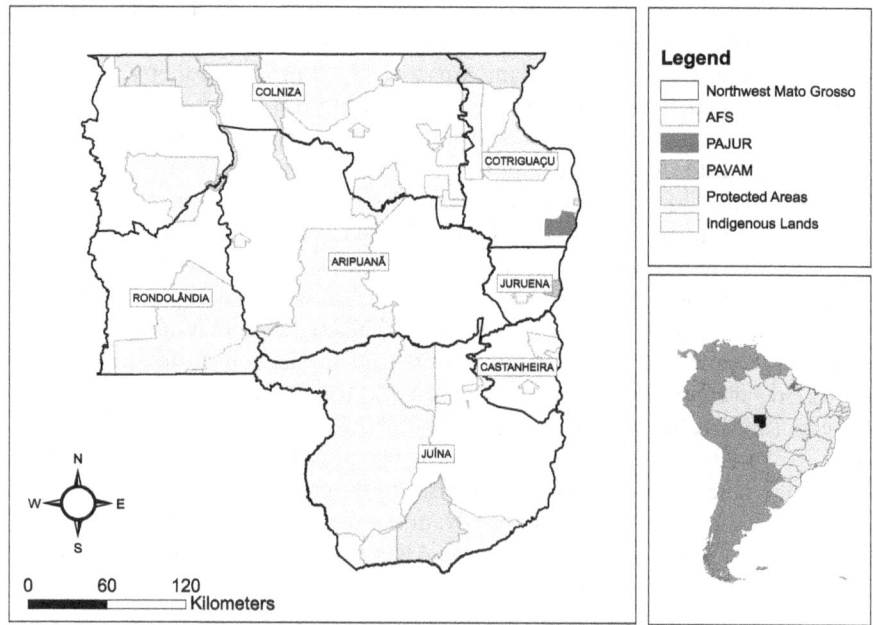

Figure 16.1 Location of ARS in NW MT, identifying two case study areas

financed by the Brazilian National Development Bank (BNDES) – has over 642,500 head (IBGE 2014). Cattle in Brazil are generally managed at low stocking rates (at an average of 1.08 AU[1]/ha in the Amazon) with little attempt at pasture renewal or intensification, implying that growth in the herd alone implies an increase in deforestation. The BNDES has in the past deliberately encouraged this expansion by financing slaughterhouse expansion without heed for the proximity of remaining intact forests, while rural credit to family farmers has fostered herd expansion and fencing with similar results (Smeraldi and May 2009). Despite fervent criticism (Greenpeace 2009) and promises by agribusiness to alter behavior (Campos 2013), public policy and land use practices have continued to fuel deforestation.

Recognizing the threat to remaining forests in NW MT, international donors in collaboration with state and federal agencies pledged support to a sequential series of projects beginning with the "Pilot Program for the Protection of Brazilian Rainforests" (PPG-7) in the 1990s, and a regional Global Environment Facility (GEF) sponsored project implemented jointly by the UNDP and the state government from 2001 to 2010, both of which sought to enhance land use management capacity by local governments and producer associations in NW MT. The PPG-7 and GEF projects both constituted multi-faceted attempts to build conditions for sustainable development in NW MT. They were comprised of support for decentralization of environmental administration; consolidation and increase in the area and number of protected areas; consultation with regional stakeholders in support of mandated social, economic and ecological zoning plans (ZSEE); and agroenvironmental measures. The latter sought to improve land use practices in ARS, many of which lie in the buffer zones of protected areas and indigenous lands.

Agroenvironmental measures included in programs addressed to ARS in NW MT included technical assistance for soil and water conservation, restoration of degraded land and the establishment of silviculture and agroforestry systems. They also provided technical assistance toward sustainable use of non-timber forest products (NTFPs); and for low impact management of forests for timber extraction. Programs for subsidized credit, product certification and incentives for added-value processing of NTFPs and agroforestry products were also included. These projects sought to intervene in the institutional context by putting into practice a set of instruments and aligned state policies, supporting local efforts at social organization, and providing technical assistance on various fronts.

In a previous study (Davenport, et al. submitted), we evaluated the relative success of such measures in stemming deforestation and in improving settler welfare across three ARS in NW MT. In the current chapter, we hypothesize that interaction with these measures contributed substantially to the relative success of collective LR in stemming continued forest encroachment and degradation. The degree to which settlers were engaged in local adaptive application of policy instruments influenced their individual land use decisions, as well as affecting change in land use and institutional arrangements at the scale of entire ARS. When these measures attained scale, they also diffused these practices and institutional innovations to neighboring properties and traditional populations in their area of influence. These differences in stakeholder engagement are reflected in the institutions that have evolved for collective management of remaining forests.

Besides relating the emergence of collective LR to institutional and technical intervention, we sought to understand how cooperative, rules-based systems may emerge, in spite of contradictory policy and socioeconomic trends at a regional scale. Key to explaining land use decisions and resultant forest conditions we are concerned with "how a particular governance arrangement fits the local ecology, how specific rules are developed and adapted over time, and whether users consider the system to be legitimate and equitable" (Ostrom 2010).

Objectives and methods

Our principal objective was to analyze whether and how collective LR was associated with deforestation dynamics and policy interventions in similar contexts. To do so, in addition to historical land use data, we carried out semi-structured interviews with key informants representing practitioners and stakeholders in the field in NW MT aiming to: i) comprehend the process of creation and implementation of the collective LR at Vale do

Amanhecer Settlement Project (PAVAM) and Juruena Settlement Project (PAJUR); and ii) identify socio-economic and institutional factors that positively or negatively affected the implementation and maintenance of the collective LR areas within these settlements, informed by the SES analytical framework (Ostrom 2007, 2009).

Our socio-economic analysis relied, in the case of PAVAM, on data provided in Davenport et al. (submitted), and, for conditions in PAJUR, on research reported by Eiró and Tricaud (2010). The former was concerned with characterizing production systems, labor requirements and net returns associated with adoption of AFS and forest product extractivism, while the latter was concerned with income and land use and the difficulty of administering forest LR, both for individual lots and in collective areas.

An institutional analysis carried out in PAVAM in 2012 and 2013 applied questionnaires and semi-structured interviews with individual farmers, designed to record the experience of farmers' interaction with various government agencies and project entities. Cross-referencing the latter, researchers also conducted a group workshop with PAVAM farmers, in order to appraise the historical development of land use decision-making processes and institutional arrangements. Settlers identified community events, deforestation trajectories and factors involved in land use decisions in each year from 1998 – year of the founding of the settlement – up to 2012. In PAJUR, field research undertaken in 2009 by collaborating researchers had engaged settlers in discussion of their common experience through rapid rural appraisal (Eiró and Tricaud 2010).

To appraise the effectiveness of policy interventions during the study period, we accessed settlement and producer data obtained through monitoring and impact assessment during the PPG-7 and GEF projects (Gonçalves 2009; Tito et al. 2011). We complemented these historical data with further analysis on the correlation of land use policies with deforestation rates within a total of 12 ARS throughout NW MT including the two case study areas (Nogueira 2014).

Forest cover data analysis included baseline and year-by-year changes in forest stocks within each ARS from 1997 to 2012, mapped using PRODES[2] and SEMA/MT[3] annual deforestation data, as compared with regional deforestation rates (Nogueira 2014). For this purpose, we calculated the total area of forest remnants at the outset of settlement, and then evaluated how deforestation progressed, measured in terms of the percentage of original forest cover remaining in each settlement at the time annual monitoring took place (using as a reference the coverage status measured on August 1 of each year).

Results

We here describe settlement and deforestation dynamics for the two selected agrarian reform projects (see Figure 16.2), both of which are under the administration of Brazil's National Institute of Colonization and Agrarian Reform (INCRA):

1 Vale do Amanhecer Settlement Project (PAVAM), municipality of Juruena: 14,400 ha, 243 households, inaugurated in 1998, settled in 1998.
2 Juruena Settlement Project (PAJUR), municipality of Cotriguaçu: 30,792 ha, 467 households, inaugurated in 1997, settled formally in 2002 (though some occupants had moved into the settlement prior to inauguration).

Table 16.1 indicates the remnant forest cover in the two settlements starting from the year of their initiation, while Table 16.2 compares the annual rate of deforestation between these settlements and the overall NW MT region. In 2014, PAVAM had 54 percent of original forest area remaining, 16 years after settlers began to occupy their lots in 1998. In contrast, PAJUR had only 37 percent of original forested area in 2014, despite the fact that its effective occupation had begun four years later than PAVAM (though some residents had occupied areas prior to formal inauguration). In Table 16.2 it is clear that both PAVAM and PAJUR exhibit deforestation rates greater than that of the NW region as a whole, following the trend in the Amazon region. However, the trend in PAVAM is toward a decline in annual deforestation close to the regional mean, while PAJUR continues to deforest relatively rapidly, at a velocity greater than that of the region as a whole.

We also considered the historical development of socio-economic conditions within the two settlements in our case study. In PAVAM in 2005, for 70 percent of its households, over half of their incomes came from working outside of the settlement, often in timber extraction or processing, or in urban areas. Only 12 percent had farm-based incomes at or above the Brazilian

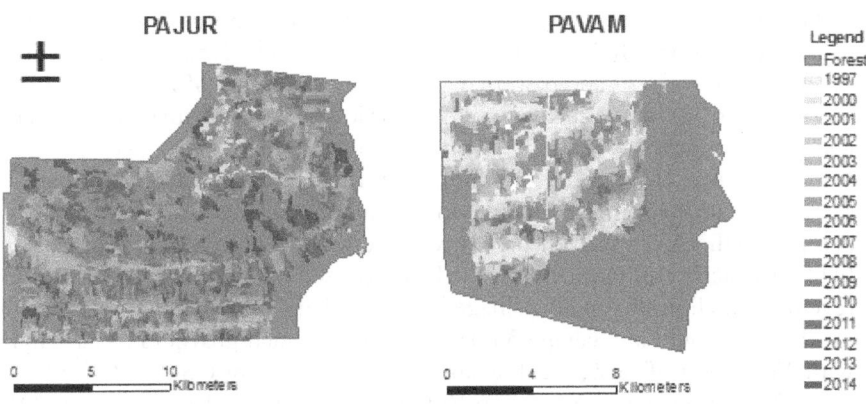

Figure 16.2 Accumulated deforestation (1997–2014) and remaining forest cover (2014) at PAJUR and PAVAM

Source: INPE/PRODES (2016). Own analysis. Note: in PAJUR, the collective LR is separated amongst three non-contiguous areas, one located in the center and two along the western and eastern (Juruena River) fringes of the settlement, while in PAVAM the LR is a contiguous forest area surrounding the settlement along the southern and eastern (Juruena River) border.

Table16.1 Accumulated deforestation (1997–2014) and remaining forest cover (2014) in PAVAM and PAJUR

Settlement	Area (ha)	Forest Remnants in 1997 (ha)	Deforestation 1997–2014 (ha)	Remaining Forest Cover 2014 (ha)	Remaining Forest Cover 2014 (%)
PAVAM	14,400	13,108	6,672	7,727	54
PAJUR	30,792	27,466	19,293	11,499	37

Source: INPE/PRODES (2016). Own Analysis. Note: Year lots began to be distributed by INCRA: PAVAM – 1998 and PAJUR – 2002.

Table 16.2 Rate of deforestation (% of original forest cover degraded per year) in NW MT, PAVAM and PAJUR

	1998	1999	2000	2001	2002	2003	2004	2005	2006	2007	2008	2009	2010	2011	2012	2013	2014
NW	0.9	0.9	0.9	0.6	1.1	1.6	1.6	1.5	0.5	0.6	0.4	0.2	0.2	0.2	0.2	0.3	0.3
PAVAM	3.1	3.1	3.1	5.4	3.0	9.5	2.2	4.5	0.4	2.5	2.6	1.2	0.4	0.4	0.6	0.6	0.1
PAJUR	0.4	0.4	0.4	0.4	1.8	7.5	7.1	9.9	2.6	7.2	4.9	3.4	3.2	1.9	5.2	2.9	3.6

Source: INPE/PRODES (2016). Davenport et al. analyzed economic performance by identifying as a baseline the prevalent operations for cattle and mixed cattle (beef and dairy) systems, comparing this baseline with alternative production patterns including AFS and NTFP extraction and processing, the latter focused on Brazil nuts. The regional revenue baseline for well managed mixed dairy and beef enterprises by smallholders, with an average 33 ha of pasture, yielded gross US$ 212/ha.yr[-1], with 45% of costs, yielding annual net revenues of US $4,547/yr[-1].

minimum wage (US$ 150 in 2005[4]). The majority was below the poverty line (Vargas 2006). Settlers in PAJUR, similarly, derived an average of 66 percent of their incomes in 2009 from off-farm activities in the timber industry or in urban areas. Although livestock was by no means a lucrative proposition due to the prevalence of pests and pasture degradation, 69 percent of settlers in PAJUR also opted to use the land for pasturing cattle as a form of financial insurance (Eiró and Tricaud 2010).

Davenport et al. analyzed economic performance by identifying as a baseline the prevalent operations for cattle and mixed cattle (beef and dairy) systems, comparing this baseline with alternative production patterns including AFS and NTFP extraction and processing, the latter focused on Brazil nuts. The regional revenue baseline for well-managed mixed dairy and beef enterprises by smallholders, with an average 33 ha of pasture, yielded gross US$ 212/ha.yr[-1], with 45 percent of costs, yielding annual net revenues of US $4,547/yr[-1].

By 2012, however the scenario had changed, at least in PAVAM. In 2013, two cooperatives AMCA[5] and COOPAVAM,[6] were operating there, with support from the GEF project and interventions from complementary measures from 2010 onward. These cooperatives were processing Brazil nuts procured from the settlement's collective forest reserve, as well as from surrounding lands. Economic value for Brazil nut flour and oil was US$ 71/ha/yr with 10 percent costs, yielding an estimated US$ 461,880 annual net revenue, based on

Brazil nuts collected solely from PAVAM's 7,200-hectare collective forest reserve. With approximately 70 direct participants in the cooperatives, and assuming 10 percent costs, this would yield US\$ 6,573 user/yr^{-1} net revenue: a 45 percent gain in annual net revenue over the regional baseline for well-managed mixed beef and dairy cattle. As such, livelihoods shifted to be based around Brazil nut trees as commonly held assets within the settlement itself.

In addition, cooperatives based in PAVAM also sourced raw Brazil nuts from six surrounding indigenous territories in 2012, as well as from a regional Extractive Reserve and adjoining forestlands, generating significant additional revenues. In this value chain, indigenous communities supplied raw nuts to the settlement under favorable price contracts: in 2013 the cooperatives' price for raw Brazil nuts had risen to BRL 3.00/kg, compared to BRL 0.60/kg in 2002.

Based on combined remote sensing, socio-economic and institutional analysis, the socio-ecological systemic (SES) characteristics of PAJUR's collective forest areas were compared with PAVAM's collective forest area, applying variables from the SES analytical framework developed by Elinor Ostrom (Ostrom 2007, 2009). We identified a set of 21 "second tier" SES variables that best encapsulated institutional differences between the two settlements' collective forest reserves, detailed in Table 16.3.[7]

Many SES variables illustrated clear discrepancies between PAVAM's and PAJUR's respective socio-ecological systems. The most striking difference is that in PAJUR the collective LR areas do not function as a common pool resource at all, but as *terra nullius* (non-property) or indeed as a "non-commons" (May 1985) subject to appropriation and conversion to *res privae* (private property). Deforestation to secure private land tenure in a risk-laden context is the dynamic characteristic of the Brazilian Amazon frontier (Araujo et. al. 2009); construction of roads prior to the installation of PAJUR's basic infrastructure stimulated unauthorized settlement and consequent land use change. The history of use of the resource (U3 in Table 16.3) was thus key to the evolution of the SES. According to one of our respondents, four families were already living inside the collective areas in PAJUR when the collective LR areas were demarcated under INCRA jurisdiction. Nothing was done to remove the initial settlers, and these and subsequent incursions to the collective areas were made by surrounding loggers, ranchers and smallholders who were aware that INCRA would soon distribute the land to eligible beneficiaries, but who also knew that land use restrictions were not being seriously enforced.

By 2012, approximately 200 families had occupied ~4,000 hectares in PAJUR's collective reserve areas. In terms of ecological outcomes (O1 in Table 16.3; deforestation rates in Table 16.2) we observed that deforestation at the settlement level took a significant uptick in 2012, from 2 percent to 5 percent, whereas continued land tenure insecurity has fostered continued deforestation by occupants with the expectation that formal private property rights will eventually be conceded. Some of the occupants live within the occupied areas, while others only retain farms there, mostly involving beef and dairy cattle ranching, as well as some cocoa and coffee plantations. Those

Table 16.3 Socio-ecological systemic (SES) differences between PAVAM and PAJUR

SES second tier variables	PAVAM	PAJUR
Resource Systems (RS)		
RS3 Size of resource system	7,200 hectares of forest reserve (most of which had been harvested selectively for 'noble' hardwoods prior to settlement, but fundamentally intact and contiguous)	4,000 hectares of pasture and ~11,000 hectares of remaining forest within collective reserve boundaries (assuming 15,000 hectares total for collective LR i.e. roughly half the size of the settlement)
RS4 Human-constructed facilities	Brazil nut processing plant consisting of several buildings and facilities involving several distinct operations (oil, flour, food processing); roads in relatively good condition; avoiding APP outside of the collective reserve	Roads in precarious condition, in some cases crossing permanent protected areas in lots outside of the collective reserve
RS9 Location	One continuous area along southern and eastern sides of settlement	Reserve separated in 3 distinct areas; complex geography
Resource Units (RU)		
RU4 Economic value	Brazil nuts processed for flour and oil: US $71/ha $\star$ 103 hectares per user (7,200 ha/70 user families) = estimated net income US$ 6,573 per year per user (assuming 10% costs)	Mixed beef and dairy: US $212/ha $\star$ 20 hectares per user = Net income – US$ 2,332 per year per user (assuming 45% costs)
Governance Systems (GS)		
GS1 Government organizations	ARS created with strong presence by INCRA; aim to be a model in terms of forest management and conservation in the Amazon region; SEMA-MT and IBAMA (gold mining fiasco in 2004 and its repetition in 2012) also involved; difficult relationships with local government	ARS created with low presence of INCRA, minimal presence of SEMA-MT or IBAMA; good relationships with local government

GS2 Nongovernment organizations	Overlapping sequence of integrated projects provided long term support for activities within settlement and regional partnerships: Instituto Pró-Natura (1992–2005); GEF/UNDP (2001–2010); Poço de Carbono Juruena (2010–2014); Petrobras; Sentinels of the Forest (2014 to present); BNDES/FAM	Short term support from the French Office Nationale de Forêts Internationale (ONFI) for local conservation projects and access to Brazil nuts in neighboring property
GS4 Property-rights systems	Collective forest and Brazil nut trees are common property in usufruct under authority of INCRA	Unauthorized land tenure with informal boundaries between users, enforced by local fiat; individually owned cattle demarcated by branding
GS5 Operational rules	800 Brazil nut trees are geo-referenced; prices for Brazil nut products secured through contracts; risk distributed across value chain; tax relief on cooperatives' reported production	Black market land sales and land accumulation; volatile prices; producer risk; taxable invoices for cattle sales difficult to obtain
GS6 Collective-choice rules	1 association representing common interests within the settlement	6 associations representing fragmented interests within the settlement
GS7 Constitutional rules	Collective Environmental License tied to the collective LR, which provides environmental legality to the entire settlement, except as regards APPs on individual lots;	Settlement-wide environmental registry (CAR), but 150–200 families occupying lands within the collective reserve may be 'invisible', since their occupancy is not tied to demarcated lots; CAR rules are out of sync with the historical occupation of the collective reserve
GS8 Monitoring and sanctioning processes	State agencies provide some limited support in monitoring of reserve; sustainable management certification of NTFP products harvested on the collective LR; outside of the collective reserve, 170 of 250 (68%) of demarcated lots have INCRA recognized beneficiaries (as of 2012)	A settlement wide CAR will likely be unable to attribute deforestation to particular users inside the illegally occupied collective LR areas; no legal mechanism enables government to expel unauthorized users from collective reserve; no products from the collective reserve are certified; outside of the collective reserve, 461 of 524 (88%) demarcated lots have INCRA recognized beneficiaries

Table 16.3 continued

SES second tier variables	PAVAM	PAJUR
Users (U)		
U1 Number of users	Roughly 70 families participate in Brazil nut cooperatives	~200 families occupy collective LR, each occupying approx. 20 hectares, mostly in pasture
U3 History of use	Settlers arrived in ~1999/2000, after lots and collective reserve had been demarcated by INCRA	Settlers arrive in 1997/8 before INCRA demarcated lots and collective LR areas in 2003
U5 Leadership/ entrepreneurship	PPG-7, GEF projects link settlement with development of cooperative enterprise based on extractivism and added-value processing; cooperative enterprises underway with local entrepreneurship involve Brazil nut harvesting both within the collective LR and regional forests	Initial efforts to create a cooperative disintegrated and led to more informal association for Brazil nut harvesting on properties external to the settlement
U6 Norms/Social capital	Social recognition and legitimacy of reserve, linked to state policies; norms function independent of municipal politics and alliances (municipal authorities in cahoots with gold miners in 2004–5); boundaries of the forest reserve are known; settlers have informal system for monitoring the LR	Social norm is the occupation of the collective reserve through productive use to secure tenure, which is linked to municipal politics; no social recognition or legitimacy of collective reserve
U7 Mental models	Recognition of the collective LR as a viable commonly-managed asset	No appreciation of the collective LR areas as a commons

Interactions (I)

I5 Investment activities	US $300,000 supplied by international and state agencies (GEF, SEMA, INCRA) for Brazil nut processing infrastructure, and additional CONAB working capital finance ($1.1 million)	Local municipal government provides resources for road maintenance, including unpaved roads within collective LR
I8 Networking	Settler association and cooperatives are linked with Mato Grosso state and Federal agency programs	Networking with municipal governments for road maintenance and in obtaining taxable sales invoices for cattle and other product sales (even though land tenure is unrecognized)
I9 Monitoring activities	Collective LR monitored against invasion through community-based arrangements	Individual occupants secure rights to use land through fiat and protect against access by usurpers

Outcomes (O)

O1 Social performance measures	Sustainable commons; social and material infrastructure favoring sustained, long term economic returns; system functions regardless of land tenure on individual lots outside of the collective reserve	Concern for individual economic returns to minimal investment; Unsustainable, lack of property rights encourages ongoing 'frontier' dynamic; private land accumulation and socio-economic inequality are increasing
O2 Ecological performance measures	Deforestation rates reduced to a minimum by 2010; 54% forest cover retained for the settlement as a whole (as of 2014)	Significant uptick in deforestation rate from 2% to 5% in 2012; 37% forest cover retained for the settlement as a whole (as of 2014)

Source: Own analysis based on Ostrom (2007, 2009). Details in text, according to SES codes.

settlers that only farm in the collective LR areas tend to reside in formally demarcated lots outside the collective LR areas, or in nearby towns. Interview respondents offered several factors associated with occupation of the collective LR areas and the failure to establish common pool arrangements: i) a lack of discussion with settlers to explain the collective LR and to raise awareness about social responsibility for their protection and sustainable use (U7); ii) little cultural appreciation for commonly held assets (U6 and U7); and iii) lack of enforcement by INCRA (GS8), even after certain settlers' complaints about the invasion of the collective LR areas.

PAJUR saw minimal investments (see RS4 and I5) into the institutional and economic development of collective forest areas. While settlers in PAJUR had entered into an agreement to gather Brazil nuts on a neighboring property managed as a carbon sink by the French overseas forestry service ONFI, participants declined to organize production cooperatives around the collective forest reserve areas inside the settlement itself, whether due to perceived risks of informal tenure, or the fact that most productive Brazil nut trees had already been deforested by occupants. Thus, the formal mandate to conserve collective LR areas inside PAJUR has had little relevance in the context of property-rights rules (GS4) based on the de facto private appropriation of land. As such, the operational rules (GS5) in PAJUR's collective LR areas have included black market land sales and land re-concentration, volatile prices, producers assuming most economic risk individually, and undocumented production. The PAJUR collective LR thus does not appear conducive to a stable extractive forest economy.

Despite this apparent chaos, settlers in PAJUR within collective LR areas established an association that was especially active after the collective LR was formally demarcated in 2003. Although they initially enjoyed little support from local public agencies in attending to basic needs, including upkeep of a precarious road infrastructure, the municipal government currently communicates with and supports the occupants of the collective LR areas. Collective representation has enabled them thereby to address road maintenance (I5) and the issuance of taxation invoices for farm products (I8), even with the illegal character of such land occupation. However, they lack the organizing principles that would permit such representation to lead to more sustainable collective use of the common LR.

Concerning public policy, INCRA has started to register its territories, including PAJUR, within the nationally mandated environmental regulatory system known as the *Cadastro Ambiental Rural* (CAR) or rural land use registry (GS8). As of 2015, INCRA plans to apply the CAR to entire settlements in Mato Grosso, on a collective basis, rather than for individual demarcated lots. Although the collective CAR will regularize the observance of environmental requirements by the settlement as a whole, it will not document individual land tenure. As a result, the settlement-wide CAR in PAJUR will not be able to attribute responsibility for deforestation to specific occupants within the collective LR areas (GS8), but rather would only identify non-observance with environmental requirements by the settlement as a whole. Our respondents reported that INCRA's current view regarding the originally designated collective LR areas in PAJUR is that they should no longer be administered as collective LR areas at all; rather, the environmental deficit incurred by the conversion of these areas into individual lots should be compensated by offsetting deforestation with other forested areas elsewhere in the biome.[8]

The other ARS analyzed in this chapter, PAVAM, demonstrated development of an alternative SES organized around common property arrangements. In this case, INCRA was deeply involved in the design and planning of the settlement (GS1), with support of a decade-long series of conservation and development projects. INCRA sought to create an ARS there that could serve as a model in terms of forest management and conservation in the Amazon region. Lots were demarcated before legal beneficiaries arrived, and roads were built parallel to small streams, in an effort to avoid degradation of permanent protected areas in accordance with environmental regulations. Settlers' perspectives in PAVAM were synergetic with formal government rules, reflected in social esteem (U7) for this settlement's collective forest reserve of 7,200 hectares, mapped and authorized for forest extractive production through a collectively held state environmental license or *Licença Ambiental Única* (LAU), signed by SEMA-MT in May of 2012.

The LAU held by PAVAM is a comprehensive environmental license encompassing long-term environmental planning for the settlement as a whole, while it legally certifies settlement cooperatives to engage in sustainable production of Brazil nut products and other processing and marketing activities. The LAU is designed as a more rigorous successor instrument to the CAR in that it goes beyond registering land use to legalizing and certifying sustainable forest management. To our knowledge no other ARS in Mato Grosso has been issued a settlement-wide LAU. For PAVAM, securing the LAU entailed a six-year process, involving significant investments and commitments by institutional actors in the GEF project, as well as by SEMA-MT and INCRA.

Under the LAU issued for PAVAM, the forest area is managed as a single contiguous area with an inventory of 800 Brazil nut trees whose population is geo-referenced. Certification for NTFP production was tied to the mapping and ecological maintenance of the collective reserve, along with monitoring against invasions by community members (I9). Interestingly, the license is effective regardless of formal land tenure conditions on individual demarcated lots in the settlement. The collective rights and responsibilities regarding the forest reserve are fully documented by the state and are viewed as legitimate by the settlers themselves (U6).

Our results indicated that the outcomes of this system included (O2) the reduction of deforestation rates to a minimum by 2012 (see Table 16.2) and (O1) the sustainability of the common property arrangements. Such stark differences can also be attributed to the operational rules (GS5) involved in production and market contracts. In contrast to PAJUR, economic value (RU4) accruing to the collective forest resources was due to human-constructed facilities (RS4) for added value processing and the development of market contracts. The operational rules (GS3) reflected the distribution of economic risk between producer, intermediary and consumer. To finance the procurement of raw Brazil nuts by the two settlement cooperatives (COOPAVAM and AMCA), the National School Meals Program (PNAE) and the Anticipated Acquisition Program (PAA) – both managed by the Federal Supply Company (CONAB) program within the Ministry of Agriculture – provided US$ 1,160,000 in advance credit for 2013. These contracts with CONAB have allowed the cooperatives to process Brazil nut derived foods, including pasta and cookies, to reach 40,000 persons in seven local municipalities in NW MT, as well as selling oil and other products for cosmetics manufacturers and consumers. In addition, these cooperatives were granted significant tax relief on reported production.

Discussion

In this section, we discuss the observed experience with collective LR areas, and consider our findings in light of the SES analytical framework. Our focused SES comparison serves to draw out specific variables relevant to the success or failure of developing CPRM in the Brazilian Amazon. Key to our discussion is our focus on the social structure and function of the formally designated collective forest areas in the two settlements. This is distinct from focusing on land use on individually demarcated lots, which is subject to somewhat different rules. Remarkably, land tenure security on demarcated individual lots is actually slightly higher in PAJUR compared to PAVAM: respectively 88 percent vs. 68 percent of the occupants on demarcated lots are INCRA recognized beneficiaries (GS8). However, the formal situation of land tenure on individual lots demarcated officially within these settlements, we suggest, has not been a determining variable in the rather stark difference between their respective SES involving the collective forest reserve areas.

We suggest that the sustainability of the common property system in PAVAM, and its apparent failure in PAJUR, has been a function of the degree of alignment between social norms and formal governance design. We observed that the integrated conservation and development projects in PAVAM required at least five years (roughly 2006–2011) to foment collective action alternative to the typical frontier pattern of illegal timber extraction, squatting and extensive cattle ranching. In the case of PAVAM, this took the form of aligning state policy instruments with local socio-institutional interventions. In this case these interventions and policy instruments were applied in the following overlapping sequence: 1) setting priorities by mapping the potential of the remaining forest; 2) training and technical assistance; 3) cooperative social organization; 4) legal certification of sustainable production in the form of a community forest resource management plan; 5) material investments in buildings and equipment; 6) market development, credit financing and the elaboration of contracts with surrounding indigenous communities; 7) contracts with private companies and

CONAB; 8) public and political exposure through national and international recognition of effectiveness; and 9) efforts and support to expand Brazil nut production (Davenport et al. submitted). The temporal sequencing of interventions and instruments, served to overcome threats associated with political insecurity and public agency dysfunction. Bureaucratic obstacles included, among other factors, limited public resources at SEMA-MT for the processing of environmental licenses for sustainable production.

In PAVAM the collective reserve took on social legitimacy, and thereby became a recognized mode for collective action, through the application of state policies to strengthen a community-based SES. Although cooperative processing and marketing of Brazil nut products in PAVAM involves only approximately 70 households, just under a third of the total in the settlement, we observed a broader social legitimacy for common pool arrangements, reflected in a dramatic shift in local production practices by settlers there, to the extent that by 2012 no farmers were observed setting fires to manage pasture or to clear additional forest. This shift is all the more remarkable given the fact that the majority of settlers, coming from southern states of Brazil, had no cultural experience with collective property and little incentive to develop cooperatives.

Further regarding local perception of the functionality of the SES, settlers in PAVAM identified the practical influence of specific technical and institutional interventions. Settlers valued highest courses and training (GS2), provision of credit (RU4), technical assistance (GS2), social organization (U6), the elaboration of a settlement-level environmental license (GS7), the installation and outfitting of the Brazil nut factory (RS4), and market channels (U5). All of these interventions supported and complemented the maintenance and protection of the collective reserve.

In contrast, our respondents reported that settlers inside PAJUR's collective LR have had a greater incentive to deforest in order to secure land tenure, than those located on demarcated lots. The principal areas subject to recent burning and deforestation are in fact located within the collective LR areas. Property-rights (GS4) within the collective LR areas are unauthorized, with informal boundaries between private users, enforced by local fiat, whereas operational rules (GS5) include black market land sales and land accumulation, volatile prices, producer risk, and undocumented product sales. Furthermore, according to our respondents, no legal procedures exist to evict undocumented occupants of the collective LR areas. Paradoxically, the formal rules that designated collective areas for forest conservation have served to produce an opposite situation from what was intended.

The failure to establish CPRM in PAJUR is therefore not due to the lack of social capital, but should be attributed to the absence of synergies between government rules and the local reality of a socio-ecological system. In much of the Brazilian Amazon, among smallholders there is little legitimacy for the state's environmental policies, including those aspects of the Brazilian Forest Code that stipulate the maintenance of 80 percent of land under private management in legal forest reserves. Data from several studies in NW MT (e.g. Eiró and Tricaud 2010; Davenport 2013; Rodrigues 2015) attest to the "inevitability" of land use transformation to cattle ranching in this region as well as the social illegitimacy of command and control rules restricting forest conversion to pasture for livestock. The mismatch between state command and control and the outlooks of the local population, also involves NGO and multilateral project intermediaries (GS2) and other locally based organizations who are challenged in bridging the divide between de facto land use patterns and de jure systems for land and environmental administration.

Conclusion

Regardless of their apparent advantages, structural factors are at loggerheads with spontaneous adoption of CPRM by settlers in the Brazilian Amazon forest. Such structural conditions include the need for cooperative infrastructure, technical assistance and appropriate financing, and the possibility to defer returns on investment for several years in order to permit added value industries to start generating yields. Our findings concur that a focus solely on technology adoption is not as effective as one that includes attention to institutional arrangements (Richards and Overseas Development Institute 1997). In addition, unauthorized land occupation in and around INCRA[9] settlements is a longstanding social practice, encouraged by social and political arrangements favoring illegal logging prior to the inauguration of INCRA settlements (Schneider and Peres 2015). Such de facto occupation begins to escalate ~4 years prior to the official decree of INCRA settlement status (ibid.). Frontier settlement patterns thereby subvert state rules to the extent that such rules legitimate individual systems of resource appropriation.

We suggest that the establishment of sustainable common pool resource management institutions in the Brazilian Amazon, especially by recent colonists, can be a difficult process, in that CPRM typically involves a long-term process of adaptive land use management involving rules of access and use coordinated among actors with a view toward institutional and social development. In this context, collective action around CPRM could emerge as a result of continued intervention, empowering local leadership and adaptively applying compatible policy instruments and market avenues. The mere demarcation of collective forest areas is only one in a sequential series of steps needed to achieve sustainability. But do frontier settlers have the time and patience to wait for such arrangements to be thrashed out? Given the prevailing set of practices and perceived technical options – illegal logging and extensive pasture establishment through clearcutting and burning – settlers see themselves on a trajectory toward capital accumulation with limited responsibility for social benefit and the environment.

Therefore, we argue that it is impractical to posit social capital as a primary prerequisite to CPRM in ARS, when social norms in the region are aligned with the dynamic of the frontier. In PAVAM, systematic accomplishments occurred *in spite of* prevailing cultural norms and outlooks regarding economic development and the state. As such, the principal finding of our case study is that effective conservation of forest cover, and sustainable environmental management in PAVAM has occurred through concerted attention to multiple components of the SES – physical, economic, legal and social – that over time permitted that CPRM serve as the basis to the formation of alternative livelihoods. While PAVAM's collective forest reserve appeared on INCRA's maps, it only became a recognized mode (common knowledge) for collective action, we argue, because local, NGO and state officials and technicians attended to the multiple ecological, technological and administrative aspects of managing a forest for added value goods.

The cooperative orientation of this process should be appreciated in the context of various state, municipal and NGO efforts to implement Forest Code regulations. We caution that prioritizing implementation of the CAR without integrating this policy instrument with other interventions and instruments for sustainable development, fails to appreciate how SES involves both economic and governance rules.

Similarly, market instruments to govern forest use rights, such as tradable rights to forest reserves, known as *Cotas de Reserva Ambiental* (CRA), are currently being promoted in Mato Grosso and other regions of Brazil by policy strategists as a mechanism to exchange forest reserve surpluses and deficits established by the CAR (Andrade et al. 2015; May et al. 2015). The CRA mechanism incorporated to allow even greater flexibility of LR strictures in the recently revised Forest Code, presumes that state bureaucracies or an NGO could set themselves up as efficient brokers over such rights and responsibilities, thereby creating appropriate incentives for compliance. As previously pointed out, INCRA's proposal that would allow PAJUR to offset deforestation in the collective LR in order to revert environmental liabilities could backfire. The CRA, by offsetting forest losses in the collective reserve, could have minimal bearing on the underlying structure of livelihoods. Furthermore, although a CRA system would serve to legalize deforestation already incurred by drawing up new maps with substitute forest reserve areas elsewhere on privately held land, it would not accrue additional economic value to the newly designated forests. Such accretion of added value would require additional instruments, such as labeling, environmental services payments or credit incentives.

Concerning the role of formal state policies, the viability and impact of policy instruments is a function of how they interface with other instruments in the context of socio-ecological systems. This is why applying policy instruments compatible with on the ground socio-institutional interventions can be fundamental as a governance strategy. Truly local interventions from within the landscapes in question, we suggest, are better positioned to adapt to volatile social and political circumstances, concentrate on the relevance of policy instruments to livelihoods, and support cooperative investments that can gradually transform natural resource based economies.

Environmental degradation does not necessarily have to be a handmaiden of economic development at the forest frontier. Appropriate policy combinations and local institutional innovation have the potential to foment collective action for forest conservation and sustainable use in ARS in the Brazilian Amazon, based on the developments observed in PAVAM and experiences in other settlements and tenure systems. Careful initial design and planning of ARS land use, in conjunction with gradual approaches to ensure the social viability of forest commons, represents an unusual opportunity to address such systemic socio-ecological challenges and so combat a non-commons tragedy.

Acknowledgements

The authors would like to acknowledge funding from the POLICYMIX project (http://nina.policymix.no) funded by the European Commission, Directorate General for Research, within the 7th Framework Program of RTD, Theme 2 – Biotechnology, Agriculture & Food (Grant no. 244065), and the National Institute for Science and Technology on Public Policy, Strategies and Development (INCT-PPED). We also would like to express our gratitude to: Lucila Nunes de Vargas at INCRA for her support in carrying out the workshop at PAVAM; Hamilton Castanha, Secretary of Environment at the Cotriguaçu city hall and Solene Tricaud at ICV for their profound knowledge of the history and current situation in PAJUR; and particularly to the smallholder farmers and extractivists who gave generously of their time and knowledge to our research efforts. The authors also express appreciation to UNDP/GEF project BRA/00/G31, and to the Amazon Fund (BNDES) for the opportunity to consolidate a long-term relationship between indigenous and settler extractivists, and in this way to connect Brazil nut production carried out on indigenous lands and in the PAVAM Legal Reserve, so as to guarantee the conservation of 870,000 hectares of Amazon forests (http://www.coopavam.org.br).

Notes

1 AU = unit equivalent to one full-grown head of cattle.
2 PRODES – Legal Amazon Deforestation Monitoring System based on Landsat imagery and developed by the Brazilian Institute of Space Research – INPE.
3 Deforestation Monitoring System based on Landsat imagery and developed by the State Environmental Agency – SEMA/MT.
4 www.debit.com.br/consulta30.php?indice=salario_minimo
5 Associação de Mulheres Cantinho da Amazônia, formed by a group of 120 women from PAVAM.
6 Cooperativa dos Agricultores do Vale do Amanhecer.
7 Ostrom's SES approach outlined in Ostrom (2007) consists of multitier variables characterizing different degrees of decomposition. "Second tier" variables refer to indicators at a fine grain which contribute to the effectiveness of CPR schemes, as more detailed aspects of "first tier" variables of which four are identified: i) Resource System, ii) Resource Units, iii) Users and iv) Governance Systems. The analytical

framework calls for consideration of interactions among these variables as well as between the delimited resource system and broader social, institutional and economic spheres.

8 Brazil's new forest code provides for such compensation through the emission of Environmental Reserve Quotas (CRA). However, who would pay the opportunity cost of such forestland compensation is not clear in this case.

9 INCRA was severely defunded as a government agency beginning in the mid-1970s, when the Brazilian military government policy shifted away from supporting landless migrants in the Amazon to support more capitalized interests. The agency has suffered for the lack of funding ever since, and does not possess the legal techniques to forcibly evict settlers from collective LR areas. As such, the argument offered recently in the literature (Schneider and Peres 2015) that INCRA administration has resulted in ecological calamity rests on attributing causality to the statistical correlation between INCRA territories and rates of deforestation. Such approaches fail to consider more deep seated causes, i.e. precisely the absence of the state and of institutional coordination involving other sectors – e.g. agricultural extension, environmental – in the effective governance of INCRA territories. For instance, the superintendent's office for INCRA for NW MT is located in Cuiabá, over 900km distant from the settlements under consideration, whereas in 2012 there was a single INCRA administrator tasked with land tenure regularization for the entire NW MT region, which contains 14 ARS and over 10,000 settler families.

References

Alston, L.J., Libecap, G. D. and Mueller, B. (1999) *Titles, conflict, and land use: the development of property rights and land reform on the Brazilian Amazon frontier*, Ann Arbor, University of Michigan Press.

Andrade, J., Bernasconi, P., Abad, R. and May, P. H. (2015) "O potencial de CRA em um município no Arco do Desmatamento: Cotriguaçu-MT," *Boletim da Sociedade Brasileira de Economia Ecológica 32/33*, 36–41.

Araujo, C., Bonjean, C. A., Combes, J-L, Motel, P. C. and Reis, E. (2009) "Property rights and deforestation in the Brazilian Amazon," *Ecological Economics 68*, 2461–2468.

Ballet, J., Sirven, N. and Requiers-Desjardins, M. (2007) "Social capital and natural resource management: a critical perspective," *Journal of Environment & Development 16 (4)*, 355–374.

Brondizio, E. S., Ostrom, E. and Young, O. R. (2009) "Connectivity and the governance of multilevel social-ecological systems: the role of social capital," *Annual Review of Environment and Resources 34 (1)*, 253–278.

Campos, A. C. (2013) Pacto pela pecuária sustentável evita que consumidor compre carne de área desmatada. Brasília: Agência Brasil (Empresa Brasileira de Comunicação-EBC). Available at: www.ebc.com.br/noticias/brasil/2013/12/pacto-pela-pecuaria-sustentavel-evita-que-consumidor-compre-carne-de-area

Chhatre, A. and Agrawal, A. (2009) "Trade-offs and synergies between carbon storage and livelihood benefits from forest commons." *Proceedings of the National Academy of Sciences 106(42)*, 176–70.

Cronkleton, P. and Larson, A. (2015) "Formalization and collective appropriation of space on forest frontiers: comparing communal and individual property systems in the Peruvian and Ecuadoran Amazon," *Society & Natural Resources: An International Journal, 28(5)*, 496–512.

Davenport, R. B., Vivan, J. L., May, P. H., Nunes, P. C. and de Vargas, L. N. *Environmental Policy and Governance (accepted for publication)* "Pilot outcomes and outlooks for environmental governance in agrarian reform landscapes in Northwest Mato Grosso, Brazil."

Davenport, R. B. (2013) Feasibility and legitimacy of land use regulatory instruments in three agrarian reform settlements in Northwest Mato Grosso: the influence and role of integrated conservation and development projects. Unpublished MS thesis, Centro Agronómico Tropical de Investigación y Enseñanza, Turrialba, Costa Rica.

Eiró, F. and Tricaud, S. (2010) "Gestão Ambiental de Assentamentos na Amazônia Estudo de caso do Projeto de Assentamento Juruena." Paper presented at V Encontro Nacional da Anppas, 4–7 Oct., Florianópolis, Brasil.

Gonçalves, A. L. (2009) *Sistemas agroflorestais da agricultura familiar, Noroeste de Mato Grosso: Sistematização e análise.* Project report, UNDP, Project No. BRA/00/G31 – Promotion of Sustainable Use and Conservation of Biodiversity in the Frontier Forests of NW Mato Grosso, Brazil.

Greenpeace (2009) *Slaughtering the Amazon.* Part 1. Available for download at: www.greenpeace.org/international/Global/international/planet-2/binaries/2009/7/slaughtering-the-amazon-part1.pdf

INPE/PRODES (2016) Project database on deforestation rates www.obt.inpe.br/prodes/index.php Accessed 2 March 2016.

Instituto Brasileiro de Geografia e Estatística (IBGE) (2014) *Produção Pecuária Municipal,* Rio de Janeiro IBGE.

Ishihara, H. and Pascual, U. (2009) "Social capital in community level environmental governance: a critique," *Ecological Economics 68,* 1549–1562.

Katz, E. (2001) "Social capital and natural capital: A comparative analysis of land tenure and natural resource management in Guatemala," *Journal of Land Economics 76,* 114–132.

May, P. H., Andrade, J., Vivan, J. L.. Kaechele, K., Gebara, M. F. and Abad, R. (2012) *Assessment of the role of economic and regulatory instruments in the conservation policymix for the Brazilian Amazon – a coarse grain analysis.* POLICYMIX project report. Available for download at: www.policymix.nina.no

May, P. H., Bernasconi, P., Wunder, S. and Lubowski, R. (2015) "Environmental Reserve Quotas in Brazil's new forest legislation: an ex ante appraisal," Bogor, CIFOR. *Occasional Paper 131.*

May, P. H. (1985) A modern tragedy of the 'non-commons': agro-industrial change and equity in Brazil's babassu palm zone. Unpublished PhD Dissertation Cornell University, Resource Economics, Ithaca NY.

Nogueira, P. (2014) Dinâmica de desmatamento em projetos de assentamentos na região Noroeste de Mato Grosso: uma contribuição para estratégias de baixo carbono. Unpublished MSc Thesis, Institute of Forestry, Federal Rural University of Rio de Janeiro, Sustainable Development Practice.

Nunes, P. C. and Rügnitz, M. T. (2015) Promovendo alternativas econômicas para a conservação das florestas e a valorização dos serviços ambientais indígenas e dos agricultores familiares. Resultados do Projeto Poço de Carbono Juruena. Juruena, Project report, Associação de Desenvolvimento Rural de Juruena (ADERJUR), Juruena, Mato Grosso.

Ostrom, E. (1990) *Governing the commons: the evolution of institutions for collective action,* Cambridge University Press, Cambridge.

Ostrom, E. (2005) *Understanding institutional diversity*. Princeton, NJ: Princeton University Press.

Ostrom, E. (2007) "A diagnostic approach for going beyond panaceas" *Proceedings of the National Academy of Sciences 104 (39)*, 15181–187.

Ostrom, E. (2009) "A general framework for analyzing sustainability of social ecological systems," *Science 325 (5939)*, 419–422.

Ostrom, E. (2010) "Beyond markets and states: Polycentric governance of complex economic systems." *The American Economic Review 100 (3)*, 641–672.

Portes, A. (2000) "The two meanings of social capital." *Sociological Forum 15 (1)*, 1–12.

Portes, A. and Landolt, P. (2000) "Social capital: promise and pitfalls of its role in development," *Journal of Latin American Studies 32 (02)*, 529–547.

Richards, M. and Overseas Development Institute (ODI, London, England) (1997) "Missing a moving target?: colonist technology development on the Amazon frontier," *ODI Research Studies* London, Overseas Development Institute.

Rodrigues, C. H. (2015) "Roubaram nossa honestidade: reflexões sobre as mudanças de uso do solo no PA Nova Cotriguaçu, Mato Grosso." Poster presented at Tropical Conservation and Development Conference "Envisioning a Sustainable Tropics," March 25–28, 2015, U. Florida, Gainesville.

Schneider, M. and Peres, R. (2015) "Environmental costs of government-sponsored agrarian settlements in Brazilian Amazonia," *PLOSOne 10 (8)*, 1–23.

Smeraldi, R. and May, P. H. (2009) *Time to pay the bill: livestock, Amazonia and conjuncture*, São Paulo, Amigos da Terra Amazônia Brasileira.

Tito, M. R., Nunes, P. C. and Vivan, J. L. (2011) *Desenvolvimento agroflorestal no noroeste de Mato Grosso: dez anos contribuindo para a conservação e uso das florestas* Brasília, UNDP.

Vargas, L. N. (2006) Vale do Amanhecer, retrato de um assentamento: modelo de reforma agrária para a Amazônia mato-grossense? Dissertação (mestrado) – Universidade Federal de Mato Grosso, Faculdade de Agronomia e Medicina, Veterinária, Pós–Graduação em Agricultura Tropical.

Vivan, J. L., Davenport, R., May, P. H., Nunes, P. C. and Prins, C. (2014) "Integrated conservation and development projects in the Amazon: an interdisciplinary evaluation with lessons for frontier forests," *Policy in Focus: development without deforestation*. International Policy Center for Inclusive Growth (IPC-IG), UNDP Brazil, Brasília.

Are the multiple social–ecological initiatives the sign of a political and cultural shift?

Philippe Léna and Liz-Rejane Issberner

So-called 'environmental' public policies are closely related to the economic development model in which they fit. Their only ambition is to reduce its negative impacts. They must therefore be compatible with the continuing expansion of the societal metabolism that characterizes the so-called 'development' which is at the root of the Anthropocene. As a result, such measures should neither hinder growth nor employment; they are even at times expected to support them ('green growth', 'green jobs'). They can only, in the best of cases, slow down the rate of physical withdrawals on resources and on the biosphere, or preserve a few selected areas for a time. If the purpose of these measures was to radically transform the relationship between society and environment, they would clash with the global systemic economical logic and its national specificities. As brilliantly demonstrated by the theoreticians of the theory of value (*wertkritik*), the self-valorization of value is at the heart of the system itself and, unlike the planet, knows no limits. The denunciation of the commodification of the world from a moral point of view or as being a by-product of transhistorical anthropological qualities is therefore of little interest. It is merely a mechanism, and it is this mechanism that determines the modalities of capitalist socialization. That being said, the critique of value theory does not offer any guidance to step away from this state of affairs.

Therefore, miracles cannot be expected from environmental governance (both on national and supranational levels), as neither will ever oppose the valorization process. The continuation of this process based on competition, innovation and a considerable increase in the quantity of goods produced is only rendered possible by a growing consumption of energy and matter. The result is an increasing complexification of society as well as a technical, social and societal acceleration, characteristic of what Rosa (2010) calls 'late modernity' (which more or less overlaps with what is known as the second phase of the Anthropocene). The author claims that this acceleration, beyond a certain point, makes it impossible to preserve social integration (Rosa 2010, 35), a role typically devoted to the state in modernity. It causes a desynchronization between the lifespan of policies and the one of economic and social transformations. These transformations are being forced upon politicians, through the acceleration of trade flows (which require the removal of obstacles), the rapid implementation of new technologies or a search for rapid returns on investment. All of these require a constant acceleration of political decisions, which therefore always seem to be produced 'late', as democratic decisions in technology today have a huge potential impact and require numerous assessments. This can be seen in many legislative proposals in Brazil, where the evaluation reports of environmental impacts for major infrastructure works and the legalization of pesticides or genetically modified organisms, among others, are being accelerated. 'Instead of taking *actions*, politics merely *respond* to the demands of situations as they arise' (Rosa 2010, 326). This is also the reason why states have given up on hindering 'progress'; the fear of seeing their nations drop out of the race for 'modernization'. According to Rosa, we are witnessing a reversal: politics, which has been an accelerator of classic modernity, is now an obstacle in late modernity, and progressivists are now supporting deceleration (political control of the economy, environmental protection, etc.), while the conservatives all favour acceleration strategies.[1]

One could say that the 'decelerators' belong to the field described by Gudynas (2014) as of 'alternative to development'. The schools of thought that can be classified in this approach are varied. Their common ground being a proposal to gradually exit economic liberalism and the economic paradigm and to regain control of the economy and the technological choices made by citizens through democratic deliberations. This implies renouncing consumerism, productivism and growth for growth's sake. In Latin America, this proposal can be found among defenders of the bioeconomy, of post-developmentism and of post-extractivism, as well as among economists of the Ecological Economics School, among others.

Like in every major country, Brazilian society is crossed by several movements critical of the hegemonic socio-economic model. Thus, more and more individuals voluntarily limit their consumption and adopt lifestyles that express their dissent. On a collective level, the Transition Towns movement is experiencing some success, especially in the states of São Paulo and Minas Gerais. The Occupy movement could be observed in some capitals such as São Paulo, Porto Alegre or Recife. There are currently more than 100 local currencies as well as systems of exchange, of sharing of goods, and non profit associations of goods or service production. Certain areas of the official Social and solidarity-based economy are engaged in these processes. The rise of agroecology, which became a national policy with the first 'Brazil Ecological Plan 2013-2015 for agroecology and organic production', should also be noted. Diversified (it goes from the simple reduction of inputs to permaculture and certification) and promoted by a new generation of agronomists, it is meeting a growing urban demand.

It is also worth noting the creation of the RBDS (Rede Brasileira pelo Decrescimento Sustentável/Brazilian Network for a Sustainable Degrowth) in 2012, on the occasion of the Conference Rio + 20. The movement was represented at the International Conferences on Degrowth for Ecological

Sustainability and Social Equity of Venice (2012), Leipzig (2014) and Budapest (2016).

The 2013 protests in Brazil and the disillusion with traditional political parties are at the basis of the citizen movement RAIZ (roots), inspired by the Spanish *Podemos* and which proposes to be an expression of the diversity of the social and environmental demands for a transformation of the production/ consumption model and of the political system. It incorporates the 'buen vivir' values of socialenvironmentalism and ecosocialism (Movimento Ecossocialista brasileiro/Brazilian Ecosocialist Movement[2]) as well as the demands from all minorities.

It is in Brazil that socialenvironmentalism was born, created from a rapprochement between indigenous and traditional peoples' activists and environmental activists. The rapprochement between the two categories of activists was not self-evident. It was initiated by the leader *seringueiro* (rubber tapper) Chico Mendes, assassinated in 1988 for defending, with his companions, the forest in which they lived. Since then, inhabited protected areas multiplied and constitute a national policy, despite being threatened by various interests. This rapprochement constituted the starting point of the creation in 1994 of the ISA (Instituto Socioambiental/Socialenvironnemental Institute), a very active NGO in the Amazon. NGOs created subsequently also adopted socialenvironmentalism in its various forms. The IBASE (Instituto Brasileiro de Análises Sociais e Económicas/Brazilian Institute of Social and Economic Analysis, founded in 1981), which played an important role in the organization of the World Social Forum in Porto Alegre, also promotes this approach. The RBJA (Rede Brasileira para a Justiça Ambiental/Brazilian Network for Environmental Justice, founded in 2001 and linked to the global network for environmental justice EJOLT) is responsible for the identification and analysis of socio-environmental conflicts in Brazil. The non-separation of social and ecological dimensions has become inescapable for both environmentalists and social movements.

In the resistance against the Anthropocene, its capitalist mega-machine, its techno-triumphalism and the myths accompanying it, having populations on one's territory whose systems of representation are drastically different, as is the case for Brazil and more generally for Latin America, is definitely an asset. Not only because the respect of their territorial rights allows for the conservation of natural systems and knowledge, but also because of the mental shift and perspective it allows (Danowski and Viveiros de Castro 2014). This is evidenced by the emergence of socialenvironmentalism and of the dissemination of values such as 'buen vivir', taken up by many social and political movements.

It is not possible to mention every movement, NGO, local association and network involved in this 'resistance' and their collective reflection; a mixture of scientific analysis and practical utopia (Altvater 2010, 338). Overall, they have already helped influence national policies and may play a role in certain areas such as the respect of the INDC by pressuring the government to meet its commitments, the consolidation of territorial rights or even by obtaining greater support from governments for agroecology or solar and wind energy development. All this was made possible as long as it did not slow down the advance of the valorization-modernization border, both physically and among society. But can they withstand extractivism, the current unbridled productivism and its probable strengthening as the competition for resources in a finite world becomes more intense? Can they gain in strength and the capacity to make proposals as evictions and destruction progress?

Notes

1   There are some 'accelerationnists' who claim their political affiliation to the left, including Marxism.
2   Shortly following the publication of a World Ecosocialist Manifesto in 2001, a network began to get established in Brazil. It gained most of its notoriety during the World Social Forum of Belém in 2009.

References

Altvater, E. (2010) *O fim do capitalismo como o conhecemos*, Civilização brasileira, Rio de Janeiro.
Danowski, D. and Viveiros de Castro, E. (2014) "L'arrêt de monde", in Hache, E. ed. *De l'univers clos au monde infini*, Ed Dehors, 221–339.
Gudynas, E. (2014), 'Churcar' Alternatives to Development, Alternautas, 1(1), 48–55. www.alternautas.net/blog/2014/7/28/churcar-alternatives-to-development. Accessed 30 March 2016.
Rosa, H. (2010) *Accélération: Une critique sociale du temps*, La Découverte, Paris.

Index

Page numbers in **bold** refer to figures, page numbers in *italic* refer to tables